The Theory of Transformation Groups

The Theory of Transformation Groups

KATSUO KAWAKUBO

Department of Mathematics
Osaka University

Oxford New York Tokyo
OXFORD UNIVERSITY PRESS
1991

This book has been printed digitally in order to ensure its continuing availability

OXFORD
UNIVERSITY PRESS

Great Clarendon Street, Oxford OX2 6DP

Oxford University Press is a department of the University of Oxford.
It furthers the University's objective of excellence in research, scholarship, and education by publishing worldwide in

Oxford New York

Auckland Bangkok Buenos Aires Cape Town Chennai
Dar es Salaam Delhi Hong Kong Istanbul Karachi Kolkata
Kuala Lumpur Madrid Melbourne Mexico City Mumbai Nairobi
São Paulo Shanghai Singapore Taipei Tokyo Toronto

with an associated company in Berlin

Published in the United States
by Oxford University Press Inc., New York

Reprinted 2002

A catalogue record for this book is available from the British Library

Library of Congress Cataloging in Publication Data
Kawakubo, Katsuo, 1942-
(Henkan gunron. English)
the theory of transformation groups/Katsuo Kawakubo.
Translation of: Henkan gunron.
Includes bibliographical references and index.
1. Transformation groups. I. Title.
QA385.K3813 1991 512'.72—dc20 91-2840

ISBN 0-19-853212-1

Preface

The aim of this book is to introduce the reader to the theory of transformation groups—that is, the study of the symmetries of mathematical structures such as sets, rings, fields, topological spaces, and differentiable manifolds. This is a subject which finds wide application throughout many branches of mathematics.

A map from a set X to itself is called a transformation. We do not, in general work with individual transformations, but consider sets of transformations. When a set of transformations forms a group (where group multiplication is defined by the composition of transformations), we call it a transformation group. It turns out that a transformation group can reflect much of the structure of X and of the individual transformations.

The recognition of the importance of describing a set of transformations as a group originated in Galois theory. Indeed, the modern notion of group may be traced back to Galois theory where the concept was first introduced in its present form.

If X is equipped with some mathematical or physical structure, then we assume that the transformations considered preserve this structure. In Galois theory, for instance, the underlying set X is not only a set, but also a field, and each transformation is an automorphism of the field. As another example, X may be a topological space and each transformation a continuous map. Conversely when we study a group itself, we can often gain much information by making it operate on various kinds of sets, thus regarding it as a transformation group.

In 1872, in his inaugural lectures at the University of Erlangen, Felix Klein presented the view that geometry is the study of properties which are invariant under a given transformation group. Therefore, one could say that the starting point of geometry is the theory of transformation groups.

The present book is an invitation to the theory of transformation groups, and hence the mathematical prerequisites have been kept to a minimum; standard courses in general topology and elementary group theory will give adequate preparation for everything before Chapters 5 and 6.

To make this suitable as a guide book, the subjects dealt with are kept to their fundamentals in Chapters 1 to 4 and more advanced topics are presented only in Chapters 5 and 6. The range of topics is selective, rather than exhaustive, in keeping with our aim to present an introductory text. Many examples, counterexamples, and illustrations are also included at every suitable opportunity for the benefit of the reader and each chapter ends with a set of exercises.

Coming to the contents of the book in detail: Chapter 1 presents the basic notions of transformation groups such as fixed point sets, isotropy groups, isotropy types, orbits, orbit spaces, and induced transformation groups.

Chapter 2 gives a brief introduction to the notion of fibre bundles and Chapter 3 is concerned with differentiable manifolds and Lie groups. The study of manifolds from various points of view is one of the principal subjects of modern mathematics and the theory of bundles is indispensable to the study of global structures of manifolds. In turn the study of fibre bundles is one important application of transformation groups. On the other hand, homogeneous spaces of Lie groups provide us with countless examples of manifolds, these may be readily understood from the point of view of transformation groups. For instance, some deep results on characteristic classes of homogeneous spaces can be proved using transformation groups. In general characteristic classes themselves are defined for the structure groups of fibre bundles and essentially lie in the category of transformation groups. To take an extreme point of view, it is difficult to talk about modern mathematics without considering transformation groups; we will always find the topic of transformation groups whenever we open a geometry or algebra book.

Although there are advanced and specialized books about fibre bundles, manifolds, and Lie groups separately, they are often inaccessible to beginners. Consequently, in Chapters 2 and 3 I have tried to give a concise introduction to these three subjects.

Chapter 4 is devoted to the study of the structure of compact Lie group actions on manifolds. This chapter together with Chapter 5 constitutes a significant part of the book. We discuss, for instance, the slice theorem, structure theorems, a decomposition theorem, a blowing-up theorem, existence and uniqueness theorems of principal orbits, and the dimension of compact Lie groups which act effectively. In other words geometric structures are studied by geometric methods.

In Chapter 5 we adopt a different approach from that used so far and consider the algebraization of geometric structures. That is, we abstract algebra which is a discrete concept from geometry which is a continuous one. We present, for example, transfer theorems, the Lefschetz fixed point theorem, and the celebrated Smith theory which says that the fixed point sets of $\mathbb{Z}_p$-actions on $\mathbb{Z}_p$-homology spheres are also $\mathbb{Z}_p$-homology spheres. The main purpose of this chapter is to give formulae for Euler characteristics and Lefschetz numbers of G-complexes and G-manifolds using the Möbius inversion formula and the decomposition theorem of Chapter 4.

The final chapter deals with recent research by the author which presupposes some further reading. We have here, for example, localization theorems, a topological proof of the Atiyah–Singer G-signature theorem, index formulae, equivariant Riemann–Roch theorems, vanishing theorems of equivariant genera (solving a problem of I. M. Singer), and a partial answer to the Petrie conjecture.

This book is based on lectures I have given to undergraduate and graduate students at Osaka University. In writing this book I have greatly profited from four sources: Bredon's *Introduction to compact transformation groups*, Steenrod's *The topology of fibre bundles*, Milnor and Stasheff's *Characteristic classes*, and Adams's *Lectures on Lie groups*.

The reader may notice some overlap with these texts, as I have repeated some material to ensure ready accessibility. I have, however, occasionally modified my version in the hope that it will be more easily understood. I hope that after reading the present book, the reader will be better equipped for a more thorough study of the theory.

I would like to thank Professor I. Tamura of the University of Tokyo for encouraging me to write this book. Thanks are also due to Mr H. Arai of Iwanami Shoten publishers and Mr F. Ushitaki, graduate student of Osaka University, who kindly read an earlier draft and made a number of invaluable suggestions.

I would like also to thank Professor D. W. Sumners, Mrs S. Yagi, and Mrs K. Barth for partially polishing the translation, along with Miss H. Ikehara who tirelessly typed the manuscript for me.

Finally, thanks go to Oxford University Press for agreeing to publish the English translation of my work.

Osaka K. K.
August 1991

Contents

1 Topological transformation groups

In this chapter, we discuss the basic notions of topological transformation groups, such as topological groups, fixed point sets, orbits, orbit spaces, orbit types, isotropy groups, and induced transformation groups. The discussion takes place from the point of view of general topology. These ideas display their real power in the study of smooth transformation groups on manifolds, which we shall deal with later.

This chapter consists of a number of easy lemmas and theorems which are arranged so that the reader will be able to supply most of the proofs himself. The well-prepared reader may skip Chapter 1 and go on to Chapter 2.

1.1 Definitions and fundamental properties of transformation groups

We begin with a number of examples to illustrate the general idea of transformation groups.

Example 1 A linear isomorphism of the n-dimensional real vector space $\mathbb{R}^n$ is called a *linear transformation* of $\mathbb{R}^n$. The set of all linear transformations forms a group which is called the linear transformation group of $\mathbb{R}^n$.

On the other hand, the set of non-singular $n \times n$ matrices over $\mathbb{R}$ forms a group under multiplication, which is called the *general linear group* $GL(n, \mathbb{R})$.

When we fix a basis of $\mathbb{R}^n$, each element of $GL(n, \mathbb{R})$ defines a linear transformation of $\mathbb{R}^n$, and the multiplication of matrices corresponds to the composition of linear transformations. Thus, the linear transformation group is identified with $GL(n, \mathbb{R})$. We then say that $GL(n, \mathbb{R})$ *acts on* $\mathbb{R}^n$ and $GL(n, \mathbb{R})$ is a *transformation group of* $\mathbb{R}^n$.

Example 2 Any subgroup of $GL(n, \mathbb{R})$ acts on $\mathbb{R}^n$. For instance, the orthogonal group $O(n)$ acts on $\mathbb{R}^n$ preserving the length of each vector; hence $O(n)$ acts on the unit disc D^n and the unit sphere S^{n-1}. It follows from linear algebra that for any point x of S^{n-1}, there exists an element A of $O(n)$ such that $x = A \cdot (1, 0, \ldots, 0)$. Moreover, the set of elements of $O(n)$ leaving the point $(1, 0, \ldots, 0)$ fixed is canonically isomorphic to $O(n-1)$. As a consequence, the residue class space $O(n)/O(n-1)$ is homeomorphic to S^{n-1}.

Similarly, $SO(n)/SO(n-1)$ is homeomorphic to S^{n-1} for $n \geq 2$, where $SO(n)$ denotes the rotation group. (For details, see Examples 1 and 2 in section 3.9.)

Example 3 Constructions similar to those in Examples 1 and 2 can be done for complex vector spaces. For instance, both the unitary group $U(n)$ and the special unitary group $SU(n)$ act on S^{2n-1}, and $U(n)/U(n-1)$ and $SU(n)/SU(n-1)$ $(n \geq 2)$ are homeomorphic to S^{2n-1}. In particular, since $SU(1) = \{e\}$, $SU(2)$ is homeomorphic to S^3.

Examples 1, 2, and 3 show that some sets of matrices are better understood if we view them as transformation groups.

Example 4 It is well known that there are only five regular polyhedra: the tetrahedron, the hexahedron, the octahedron, the dodecahedron, and the icosahedron. The groups of motions of regular polyhedra are called regular polyhedral groups. That is, these groups are defined by transformation groups. From a given regular polyhedron, we can obtain another one by taking as vertices the centres of all the faces of the given regular polyhedron. We say that the given regular polyhedron and the one obtained in this way are *dual* to each other. The hexahedron and octahedron are dual to each other, as are the dodecahedron and icosahedron. The tetrahedron is dual to itself. Accordingly, it follows that the groups of motions of dually corresponding regular polyhedra are isomorphic. Usually, one speaks of the octahedral group and the icosahedral group, instead of the hexahedral group and the dodecahedral group.

Example 5 Let p be a prime number. The cyclic group G of order p acts freely on an odd dimensional sphere S^{2n-1}. (The definition of free action is given at the end of this section.) Does G act freely on an even dimensional sphere? The solution will be given in Section 5.6.

Example 6 The following fact is a consequence of what is called the 'fixed point theorem' which we shall discuss in Section 5.7. When the wind blows on the surface of the earth (S^2), there is at least one place where the wind is not blowing.

Referring to the above examples, we will now give a number of definitions for transformation groups, and investigate the fundamental properties of those groups. We then consider transformation groups with additional structure, such as topology, C^∞ structure, and algebraic structure.

Definition 1.1 Let G be a group and X a set. By a *G-action* we mean a map

$$\varphi: G \times X \to X$$

such that:

(1) $\varphi(e, x) = x$ for all $x \in X$, where e is the identity of G;

(2) $\varphi(g_2, \varphi(g_1, x)) = \varphi(g_2 g_1, x)$ for all $g_1, g_2 \in G$ and $x \in X$.

The triple (X, G, φ) is called a *transformation group* or *G-action* on X, and X is called a *G-set*.

The above two conditions are natural requirements. Condition (1) says that the identity element e acts trivially. Condition (2) says that action by g_1, then g_2 coincides with action by the element $g_2 g_1$.

For $g \in G$, let

$$\varphi_g : X \to X$$

be the map defined by $\varphi_g(x) = \varphi(g, x)$. φ_g is called a *transformation by g*. Each φ_g is a bijective map of X, because it follows from conditions (1) and (2) that

$$\varphi_{g^{-1}} \cdot \varphi_g = \varphi_g \cdot \varphi_{g^{-1}} = \mathrm{id}$$

where id denotes the identity map on the set X. When φ is understood from the context, we shall often use the notation $g \cdot x$ or gx for $\varphi(g, x)$. If we employ this notation, it is easy to see that the above conditions become (1) $ex = x$, and (2) $g_2(g_1 x) = (g_2 g_1)x$. Moreover, we shall often suppress G and φ in the symbol (X, G, φ), and simply write X.

We now give some examples of G-sets.

Example 7 Let G be a group and H a subgroup of G. Let X be the set G/H of left cosets gH of H in G. Then the map

$$\varphi : G \times G/H \to G/H$$

defined by $\varphi(g', gH) = g'gH$ defines a G-action.

As mentioned before, we will put mathematical structures on X, G, and φ gradually. We give several such examples but since we shall discuss them precisely later we will not worry about details now. In the following, we always assume that φ satisfies conditions (1) and (2).

Example 8 Let G be a group, and X be a group, ring, or field. Then we may often define an action φ so that each φ_g is an endomorphism of the group, ring, or field (depending on the structure of X). In this case we will find that each φ_g is in fact an isomorphism.

Example 9 Let G be a topological group, X a topological space, and φ a continuous map. In this case we will find that each φ_g is a homeomorphism.

Example 10 Let G be a Lie group, X a C^∞ manifold, and φ a C^∞ map. Then each φ_g is a diffeomorphism.

In this book, we shall deal mainly with the structures of Examples 9 and 10.

We now introduce some fundamental terminology. Given a transformation group (X, G, φ), we give the following definition, using the notation gx instead of $\varphi(g, x)$.

(1) For $x \in X$, the subset

$$G(x) = \{gx \in X \mid g \in G\}$$

is called the *orbit* of x.

(2) For $x \in X$, the set

$$G_x = \{g \in G \mid gx = x\}$$

of elements of G leaving x fixed is called the *isotropy group at* x.

(3) The set

$$X^G = \{x \in X \mid G_x = G\}$$

is called the *fixed point set.*

We next investigate the fundamental properties of the objects defined above. In the rest of this section, we assume that a transformation group (X, G, φ) is given.

Proposition 1.1 *The isotropy group G_x is a subgroup of G.*

Proof From condition (1) of Definition 1.1, the identity e of G belongs to G_x. Given any two elements g_1, g_2 of G_x, we have, by (2),

$$(g_2 g_1)x = g_2(g_1 x) = g_2 x = x.$$

Hence $g_2 g_1 \in G_x$. For any $g \in G_x$, action by g^{-1} on both sides of $x = gx$ produces

$$g^{-1}x = g^{-1}(gx) = (g^{-1}g)x = ex = x.$$

Hence $g^{-1} \in G_x$. □

Proposition 1.2 *For $x_1, x_2 \in X$, either $G(x_1) = G(x_2)$ or $G(x_1) \cap G(x_2) = \varnothing$.*

Proof We have only to prove that $G(x_1) = G(x_2)$ if $G(x_1) \cap G(x_2) \neq \varnothing$. If there is an $x \in G(x_1) \cap G(x_2)$, then there exist $g_i \in G$ $(i = 1, 2)$ such that

$$g_1 x_1 = x = g_2 x_2.$$

For any $g \in G$, we have

$$gx_2 = (gg_2^{-1}g_2)x_2 = (gg_2^{-1})(g_2x_2)$$
$$= (gg_2^{-1})(g_1x_1) = (gg_2^{-1}g_1)x_1,$$

hence $G(x_2) \subset G(x_1)$. Since we have $G(x_2) \supset G(x_1)$ by a similar argument, then $G(x_1) = G(x_2)$. □

There is a fundamental relation between isotropy groups at two points on the same orbit.

Proposition 1.3 *For any $g \in G$ and $x \in X$, we have*

$$G_{gx} = gG_xg^{-1}.$$

Proof For any $g' \in G_{gx}$, we have $g'(gx) = gx$ by definition. Acting by g^{-1} on both sides of this equation, we have:

left side: $g^{-1}(g'(gx)) = g^{-1}((g'g)x) = (g^{-1}g'g)x,$

right side: $g^{-1}(gx) = (g^{-1}g)x = ex = x.$

Therefore, $g^{-1}g'g \in G_x$ holds. Hence we have $g' \in gG_xg^{-1}$. It follows that

$$G_{gx} \subset gG_xg^{-1}.$$

The proof of the reverse inclusion is similar and is left as an exercise for the reader. □

This proposition means that all the points translated by G of a given point have conjugate isotropy groups, and that all conjugates appear.

Given two transformation groups (X, G, φ) and (Y, G, ψ) with a common group G, a map $f: X \to Y$ is called a *G-map* or an *equivariant map* if the following equation holds for any $g \in G$ and $x \in X$:

$$f(\varphi(g, x)) = \psi(g, f(x)).$$

If we use the brief notation, the above becomes

$$f(gx) = gf(x).$$

Although this is easy to understand, it should not be forgotten that the action gx is given by φ and the action $gf(x)$ is given by ψ.

A G-map $f: X \to Y$ which is also bijective is called a *G-isomorphism.* When there exists a G-isomorphism $f: X \to Y$, we say that X and Y are *G-isomorphic* and denote G-isomorphism by $X \cong Y$.

A subset Y of a G-set X is said to be *G-invariant* if $gY \subset Y$ for any $g \in G$, where

$$gY = \{gy \mid y \in Y\}.$$

Lemma 1.4 *A G-invariant subset Y of a G-set X satisfies $gY = Y$ for any $g \in G$, and hence the G-action on X induces a G-action on Y.*

Proof The proof is left as an exercise.

Theorem 1.5 *Let X be a G-set. For any $x \in X$, $G(x)$ is a G-invariant subset of X, and*

$$G(x) \cong G/G_x.$$

Here the right-hand side is the G-set of Example 7.

Proof It is obvious that $G(x)$ is a G-invariant subset of X. We define a map

$$f\colon G/G_x \to G(x)$$

by

$$f(gG_x) = gx.$$

First we show that this definition is independent of the choice of coset representatives. Suppose that two elements g, g' of G satisfy $gG_x = g'G_x$. Then $G_x = g^{-1}g'G_x$; hence we have $g^{-1}g' \in G_x$. It follows from the definition of G_x that $(g^{-1}g')x = x$. By the properties of G-actions, we have $g'x = gx$ and

$$f(gG_x) = gx = g'x = f(g'G_x),$$

which shows that f is well defined.

Next we show that f is surjective. For any $y \in G(x)$, there exists $g \in G$ with $y = gx$. Hence $f(gG_x) = gx = y$, which means that f is surjective.

We now show that f is injective. Suppose that $gx = g'x$. This holds if and only if $g^{-1}g' \in G_x$, that is $G_x = g^{-1}g'G_x$. Hence we can conclude that $gG_x = g'G_x$, so f is injective.

Finally, we show that f is a G-map. Since we have, for all $g, g' \in G$,

$$f(g'(gG_x)) = f((g'g)G_x) = (g'g)x = g'(gx) = g'f(gG_x),$$

then f is a G-map. □

A transformation group (X, G, φ) is said to be:

(1) *trivial* if $G_x = G$ for all $x \in X$;

(2) *free* if $G_x = \{e\}$ for all $x \in X$;

(3) *semi-free* if $G_x = G$ or $\{e\}$ for all $x \in X$;

(4) *transitive* if there is precisely one orbit;

(5) *effective* if $\bigcap_{x \in X} G_x = \{e\}$.

1.2 Topological groups

In this section, we present some fundamental properties of topological groups. A topological group is a set with group structure as well as topological structure. These two structures are not, however, independent, but the group operations are continuous with respect to the topology. We now define a topological group precisely.

Definition 1.2 A set G is called a *topological group* if G satisfies the following conditions.

(1) G is a Hausdorff space.

(2) G is a group.

(3) The maps $\alpha: G \times G \to G$ and $\beta: G \to G$ defined by $\alpha(g, h) = gh$ and $\beta(g) = g^{-1}$ are continuous. Here $G \times G$ is equipped with the product topology.

Lemma 1.6 *Condition* (3) *in Definition* 1.2 *of a topological group is equivalent to the following condition.*

(3)′ *The map* $\gamma: G \times G \to G$ *defined by* $\gamma(g, h) = gh^{-1}$ *is continuous.*

Proof Since $\gamma(g, h) = \alpha(g, \beta(h))$, condition (3) implies condition (3)′. Conversely, assume (3)′. Then β is continuous since $\beta(g) = \gamma(e, g)$. Hence, using $\alpha(g, h) = \gamma(g, \beta(h))$, we know that α is continuous as well. □

Here are several straightforward examples of topological groups.

Example 1 Every group becomes a topological group when it is equipped with the discrete topology. Such a topological group with the discrete topology is called a *discrete group*.

Remark Since a topological group is a Hausdorff space by definition, a finite topological group is always a discrete group.

Example 2 The set $\mathbb{R}$ of real numbers and the set $\mathbb{C}$ of complex numbers are groups under addition, and they are also topological spaces equipped with the metric topologies. It is easily verified that the group operations are continuous. Thus $\mathbb{R}$ and $\mathbb{C}$ are topological groups.

Example 3 The set $GL(n, \mathbb{R})$ of non-singular $n \times n$ matrices over $\mathbb{R}$ and the set $GL(n, \mathbb{C})$ of non-singular $n \times n$ matrices over $\mathbb{C}$ are topological groups. Let $M(n, \mathbb{R})$ denote the set of all $n \times n$ matrices over $\mathbb{R}$. Identifying $M(n, \mathbb{R})$ with $\mathbb{R}^{n^2}$ naturally, we topologize $M(n, \mathbb{R})$ as the Cartesian product $\mathbb{R}^{n^2}$ of copies of $\mathbb{R}$ and give $GL(n, \mathbb{R})$ the relative topology of $M(n, \mathbb{R})$. Moreover $GL(n, \mathbb{R})$ is a group under multiplication. It is easy to see that $GL(n, \mathbb{R})$ satisfies condition (3) in Definition 1.2. Thus $GL(n, \mathbb{R})$ is a topological group. Similarly $GL(n, \mathbb{C})$ is a topological group. Later we will show that they are in fact Lie groups.

For $g \in G$ the *left translation map* $L_g: G \to G$, the *right translation map* $R_g: G \to G$, and the *inner automorphism* $I_g: G \to G$ are defined as follows:

$$L_g(h) = gh, \qquad R_g(h) = hg, \qquad I_g(h) = ghg^{-1}.$$

Proposition 1.7 *L_g, R_g, I_g, and β are homeomorphisms of G onto itself.*

Proof The map $l_g: G \to G \times G$ defined by $l_g(h) = (g, h)$ is continuous because of the product topology on $G \times G$. Since $L_g = \alpha \cdot l_g$, L_g is a continuous map. For given $h, k \in G$, it follows from

$$L_{gh}(k) = (gh)k = g(hk) = L_g \cdot L_h(k)$$

that $L_{gh} = L_g \cdot L_h$. Hence we have

$$L_g \cdot L_{g^{-1}} = L_{g^{-1}} \cdot L_g = L_e = \text{id}.$$

That is, $L_{g^{-1}} = (L_g)^{-1}$. Thus L_g is a homeomorphism. Similarly R_g is a homeomorphism. Since $I_g = R_{g^{-1}} \cdot L_g$ and $\beta^{-1} = \beta$, we see that these are also homeomorphisms. □

Corollary 1.8 *Let $\{U_\lambda\}$ be a neighbourhood basis at the identity element e of G. Then for any $g \in G$, $\{L_g(U_\lambda)\}$ is a neighbourhood basis at g.*

A subgroup H of a topological group G becomes naturally a topological group when equipped with the relative topology. As a subspace, if H is closed, open, or discrete in G, it is called a *closed subgroup*, an *open subgroup*, or a *discrete subgroup*, respectively. If the topological subgroup H is normal in G, we say that H is a *normal topological subgroup*. If the topological group H is commutative, we say that H is a *commutative topological group*.

Proposition 1.9 *If H is a subgroup of a topological group G, then $\bar{H}$, the closure of H, is also a subgroup. If H is normal or commutative, then so is $\bar{H}$.*

Proof Since the map γ in Lemma 1.6 (3)′ is continuous, we have

$$\gamma(\bar{H} \times \bar{H}) = \gamma(\overline{H \times H}) \subset \overline{\gamma(H \times H)} = \bar{H},$$

so that $\bar{H}$ is a subgroup of G. Suppose that H is a normal subgroup of G. Since I_g is a homeomorphism for any $g \in G$ by Proposition 1.7, we have

$$g\bar{H}g^{-1} = I_g(\bar{H}) = \overline{I_g(H)} = \bar{H},$$

so that $\bar{H}$ is a normal subgroup.

Let $f\colon G \times G \to G$ be the map defined by $f(g, h) = ghg^{-1}h^{-1}$. Then f is continuous, since f can be written as a composition of continuous maps. It is clear that for a subgroup H of G,

$$f(H \times H) = \{e\} \Leftrightarrow H \text{ is commutative.}$$

Now if H is a commutative subgroup, then we have

$$f(\bar{H} \times \bar{H}) = f(\overline{H \times H}) \subset \overline{f(H \times H)} = \overline{\{e\}} = \{e\},$$

so that $\bar{H}$ is a commutative subgroup. □

Proposition 1.10 *If H is a closed subgroup of a topological group G, then the normalizer $N(H)$ of H in G is also a closed subgroup.*

Proof For any $g \in \overline{N(H)}$, let $\{n_\lambda\}$ be a net in $N(H)$ converging to g. Then the net $\{n_\lambda h n_\lambda^{-1}\}$ converges to ghg^{-1} for any $h \in H$. Since $n_\lambda h n_\lambda^{-1} \in H$ and H is closed, ghg^{-1} belongs to H. Thus, we have shown that $gHg^{-1} \subset H$ for any $g \in \overline{N(H)}$. Since $\overline{N(H)}$ is a subgroup of G by Proposition 1.9, g^{-1} belongs to $\overline{N(H)}$ for any $g \in \overline{N(H)}$. Hence the inclusion $g^{-1}Hg \subset H$ holds as well. Thus $g \in N(H)$, and we have $\overline{N(H)} = N(H)$, as desired. □

Notation For subsets A and B of a topological group G, we put

$$AB = \{ab \mid a \in A, b \in B\},$$
$$A^{-1} = \{a^{-1} \mid a \in A\}.$$

Lemma 1.11 *Let A and O be subsets of a topological group G. If O is open, then AO and OA are open.*

Proof Note that $AO = \bigcup_{a \in A} aO$. Since $aO = L_a(O)$ and L_a is a homeomorphism, each aO is open, and hence AO is open. Similarly, OA is open as well. □

A subset A of a topological group G is called *symmetric* if $A = A^{-1}$ holds.

Lemma 1.12 *The symmetric open neighbourhoods of the identity e of a topological group G form an open neighbourhood basis at e.*

Proof Let U be an arbitrary open neighbourhood of e. Since β is a homeomorphism, and since $e \in U^{-1} = \beta(U)$, U^{-1} is also an open neighbourhood of e. If we set $V = U \cap U^{-1}$, then

$$V^{-1} = \{u^{-1} \mid u \in U, u^{-1} \in U\}$$
$$= \{u \mid u \in U, u^{-1} \in U\} = V,$$

so that V is a symmetric open neighbourhood, and satisfies $e \in V \subset U$. □

We now prepare some terminology and lemmas in point set topology which we will often use.

Let X be a topological space and Y a set. Given a surjection $f: X \to Y$, the *quotient topology* on Y induced by f is defined as follows. A subset O of Y is defined to be open if and only if $f^{-1}(O)$ is open in X. It is easy to see that this defines a topology on Y. Then Y endowed with this topology is called a *quotient space* or a *coset space*.

Let X and Y be topological spaces and $f: X \to Y$ a map. If $f(A)$ is an open (resp. closed) set for every open (resp. closed) set A of X, then f is called an *open* (resp. *closed*) map.

Lemma 1.13 *If $f_i: X_i \to Y_i$ are open maps for $i = 1, 2$, then $f_1 \times f_2: X_1 \times X_2 \to Y_1 \times Y_2$ is also an open map.*

Proof For any open set O of $X_1 \times X_2$, there exists a family $\{U_\lambda\}$ of open sets of X_1 and a family $\{V_\lambda\}$ of open sets of X_2 such that

$$O = \bigcup_\lambda U_\lambda \times V_\lambda.$$

Then we have

$$(f_1 \times f_2)(O) = \bigcup_\lambda f_1(U_\lambda) \times f_2(V_\lambda).$$

Since $f_1(U_\lambda)$ and $f_2(V_\lambda)$ are open sets of Y_1 and Y_2, respectively, $(f_1 \times f_2)(O)$ is an open set of $Y_1 \times Y_2$. □

Corollary 1.14 *The projection $p: X \times Y \to X$ is an open map.*

Proof Denote by $*$ the topological space consisting of one point and by p': $Y \to *$ the constant map. Identifying $X \times *$ with X, we have $p = \mathrm{id} \times p'$. Since both id and p' are open maps, it follows from Lemma 1.13 that p is also open. □

Example 4 Let X and Y be the set $\mathbb{R}$ of real numbers. Although both $\mathrm{id}: X \to X$ and $p': Y \to *$ are closed maps, the map

$$p = \mathrm{id} \times p': \mathbb{R} \times \mathbb{R} \to \mathbb{R}$$

is not a closed map, because the set

$$F = \{(x, y) \in \mathbb{R} \times \mathbb{R} \mid xy = 1\}$$

is a closed subset of $\mathbb{R} \times \mathbb{R}$, but $p(F) = \mathbb{R} - \{0\}$ is not a closed subset of $\mathbb{R}$. This means that the closed map versions of Lemma 1.13 and Corollary 1.14 do not hold in general.

Lemma 1.15 *Let X, Y, and Z be topological spaces and let*

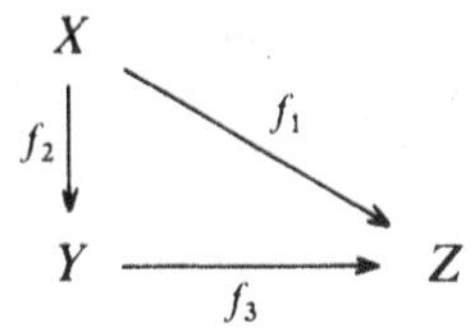

be a commutative diagram. If f_1 is a continuous map and if f_2 is a surjective open map, then f_3 is a continuous map.

Proof Let O be any open set of Z. Since f_2 is surjective, we have

$$f_3^{-1}(O) = f_2 \cdot f_2^{-1}(f_3^{-1}(O)) = f_2(f_1^{-1}(O)).$$

Since f_1 is continuous and f_2 is open, $f_3^{-1}(O)$ is an open set, which implies that f_3 is continuous. □

The following lemma is well known and will be used frequently.

Lemma 1.16 *Let X be a compact space and Y a Hausdorff space. Then a continuous map $f\colon X \to Y$ is closed. In particular, if f is bijective and continuous, f is a homeomorphism.*

Proof Since X is compact, any closed subset F of X is compact. Hence $f(F)$ is compact as well. Since Y is Hausdorff, $f(F)$ is a closed subset of Y. When f is bijective and continuous, this implies that f^{-1} is continuous, and hence f is a homeomorphism. □

Lemma 1.17 *In the commutative diagram of Lemma 1.15, we suppose that f_2 is surjective and Y is endowed with the quotient topology induced by the map f_2. If f_1 is continuous, then f_3 is continuous.*

Proof For any open set O of Z, the set

$$f_2^{-1}(f_3^{-1}(O)) = f_1^{-1}(O)$$

is open by assumption. It follows from the definition of the quotient topology that $f_3^{-1}(O)$ is open. □

Lemma 1.18 *In the commutative diagram of Lemma* 1.15, *we suppose that* f_2 *is a surjective continuous map. If* f_1 *is open (resp. closed), then* f_3 *is open (resp. closed).*

Proof Let O be any open (resp. closed) set of Y. Since f_2 is surjective, we have

$$f_3(O) = f_3 \cdot f_2(f_2^{-1}(O)) = f_1(f_2^{-1}(O)).$$

Hence $f_3(O)$ is open (resp. closed) by assumption. □

We now return to topological groups. Given a topological group G and a closed subgroup H, we consider the space G/H of left cosets gH of H in G, endowed with the quotient topology induced by the canonical projection $\pi\colon G \to G/H$.

Lemma 1.19 *Given a topological group G and its closed subgroup H, the quotient space G/H is a Hausdorff space and the projection map* $\pi\colon G \to G/H$ *is continuous and open.*

Proof The map π is continuous by definition. Since $\pi^{-1}(\pi(O)) = OH$ for $O \subset G$, it follows from Lemma 1.11 that $\pi^{-1}(\pi(O))$ is open if O is open. Hence $\pi(O)$ is open in G/H by definition. To show that G/H is Hausdorff, suppose that $g_1 H \neq g_2 H$ for $g_1, g_2 \in G$. Then $g_1^{-1} g_2 \notin H$. Since H is closed, there is a neighbourhood U of $g_1^{-1} g_2$ with $U \cap H = \varnothing$. Note that the map $f\colon G \times G \times G \to G$ defined by $f(g_1, g_2, g_3) = g_1 g_2 g_3$ is continuous since $f(g_1, g_2, g_3) = \alpha(g_1, \alpha(g_2, g_3))$. From continuity of f at the point $(e, g_1^{-1} g_2, e)$, there exists a neighbourhood V of e with $V g_1^{-1} g_2 V \subset U$. According to Lemma 1.12, there is a symmetric neighbourhood W of e with $W \subset V$. Since $W g_1^{-1} g_2 W \subset U$, we have $W g_1^{-1} g_2 W \cap H = \varnothing$. Thus $g_1^{-1} g_2 W \cap WH = \varnothing$ which implies that $g_2 W \cap g_1 WH = \varnothing$. Since H is a subgroup of G, we may conclude that $g_2 WH \cap g_1 WH = \varnothing$. Thus there exist disjoint open sets $\pi(g_1 W) = g_1 WH$ and $\pi(g_2 W) = g_2 WH$ in G/H containing $g_1 H$ and $g_2 H$, respectively. □

Proposition 1.20 *If H is a closed normal subgroup of a topological group G, then G/H is a topological group.*

Proof It follows from Lemma 1.19 that G/H is a Hausdorff space. In addition, G/H is a group since H is a normal subgroup of G. We now show that the group operations are continuous. Let $\bar{\gamma}\colon G/H \times G/H \to G/H$ be the map defined by $\bar{\gamma}(g_1 H, g_2 H) = g_1 g_2^{-1} H$. Since H is normal, this definition is independent of the choice of coset representatives. By Lemma 1.6, it suffices to show that $\bar{\gamma}$ is continuous. For this, consider the following commutative diagram

$$\begin{array}{ccc} G\times G & \xrightarrow{\ \gamma\ } & G \\ {\scriptstyle \pi\times\pi}\downarrow & & \downarrow{\scriptstyle \pi} \\ G/H\times G/H & \xrightarrow[\ \bar{\gamma}\ ]{} & G/H \end{array}$$

Clearly $\pi \times \pi$ is surjective. Since π is open, $\pi \times \pi$ is also open by Lemma 1.13. Of course $\pi \cdot \gamma$ is continuous, and hence continuity of $\bar{\gamma}$ follows from Lemma 1.15. □

Proposition 1.21 *The component G_0 of a topological group G containing the identity is a closed normal subgroup of G.*

Proof Since components are closed in general, it suffices to show that G_0 is a normal subgroup. As is well known, the product $G_0 \times G_0$ is connected and hence its continuous image $\gamma(G_0 \times G_0) = G_0 G_0^{-1}$ is also connected. Since $G_0 G_0^{-1}$ contains e, we have

$$G_0 G_0^{-1} \subset G_0,$$

which means that G_0 is a subgroup of G. According to Proposition 1.7, I_g is a homeomorphism for any $g \in G$. It follows that $gG_0g^{-1} = I_g(G_0)$ is a component of G and contains e, so we have $gG_0g^{-1} = G_0$. □

Corollary 1.22 *G/G_0 is a topological group.*

This follows by combining Propositions 1.20 and 1.21.

Let G and G' be topological groups. Then a map $f\colon G \to G'$ is called a *homomorphism* (of topological groups) when f is continuous as well as a homomorphism of groups. When f is a homeomorphism as well as an isomorphism of groups, f is called an *isomorphism* (of topological groups).

Proposition 1.23 *If $f\colon G \to G'$ is a homomorphism of topological groups, then* $\operatorname{Ker} f = \{g \in G \mid f(g) = e'$: the identity of $G'\}$ *is a closed normal subgroup of G. Suppose further that f is surjective. Then the induced map*

$$\bar{f}\colon G/\operatorname{Ker} f \to G'$$

is continuous as well as an isomorphism of groups. If G is compact or f is open, then $\bar{f}$ is an isomorphism of topological groups.

Proof Generally $\operatorname{Ker} f$ is a normal subgroup of G. Since the identity e' of G' is closed, it follows from continuity of f that $\operatorname{Ker} f = f^{-1}(e')$ is closed.

Consider the following commutative diagram

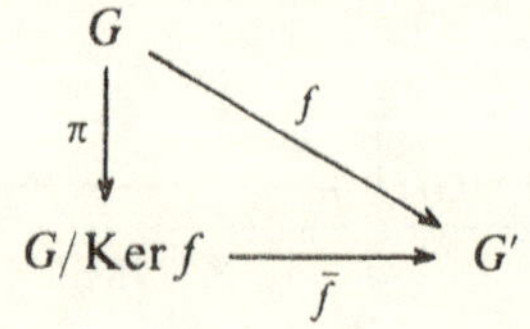

where π is the natural projection. Since $G/\mathrm{Ker}\, f$ is endowed with the quotient topology and f is continuous, $\bar{f}$ is continuous by Lemma 1.17. Furthermore, if f is surjective, then $\bar{f}$ is an isomorphism of groups. Suppose now that G is compact. Then its continuous image $G/\mathrm{Ker}\, f$ is also compact. This, together with the fact that G' is Hausdorff, implies that $\bar{f}$ is a closed map by Lemma 1.16. Hence $\bar{f}$ is a homeomorphism, and this gives an isomorphism of topological groups. If f is an open map, $\bar{f}$ is also an open map by Lemma 1.18. Hence $\bar{f}$ is a homeomorphism and this gives an isomorphism of topological groups in this case too. □

1.3 Examples of topological groups and semi-direct products

In this section we shall give examples of topological groups. Lie groups (which we shall deal with later) are typical examples of topological groups. In addition, we will also construct examples of topological groups which are not necessarily Lie groups. We then introduce the notions of direct products and semi-direct products which provide us with methods of constructing new topological groups out of old.

Example 1 As we mentioned in Example 3 in Section 1.2, $GL(n, \mathbb{R})$ and $GL(n, \mathbb{C})$ are topological groups. The following are all examples of topological subgroups of $GL(n, \mathbb{C})$:

$$\begin{aligned}
O(n) &= \{A \in GL(n, \mathbb{R}) \mid {}^tA A = I\},\\
SL(n, \mathbb{R}) &= \{A \in GL(n, \mathbb{R}) \mid \det A = 1\},\\
SO(n) &= O(n) \cap SL(n, \mathbb{R}),\\
U(n) &= \{A \in GL(n, \mathbb{C}) \mid {}^tA \bar{A} = I\},\\
SL(n, \mathbb{C}) &= \{A \in GL(n, \mathbb{C}) \mid \det A = 1\},\\
SU(n) &= U(n) \cap SL(n, \mathbb{C}).
\end{aligned}$$

In the above, we denote the transpose of A by ${}^tA$, the determinant of A by $\det A$, the complex conjugate of A by $\bar{A}$, and the unit matrix by I.

These examples are all closed subgroups of $GL(n, \mathbb{C})$. This will be shown using the fact that these subgroups are inverse images of single points under continuous maps (see Section 3.7).

We now construct examples of topological groups which are not necessarily Lie groups. Given two topological spaces X and Y, we denote by $\mathrm{Map}(X, Y)$ the set of continuous maps $f: X \to Y$. $\mathrm{Map}(X, Y)$ (or a subset $F \subset \mathrm{Map}(X, Y)$) is called a *mapping space.*

Given a compact subset C of X and an open subset O of Y, let

$$W(C, O) = \{f \in F \mid f(C) \subset O\}.$$

The *compact-open topology on* F has, as a sub-basis, the sets $W(C, O)$ for $C \subset X$ compact and $O \subset Y$ open. We always endow a mapping space with the compact-open topology. The following theorem will be used in Section 1.4 and is very helpful in understanding the topology of mapping spaces.

Theorem 1.24 *Let X, Y, and Z be topological spaces and F a subset of* $\mathrm{Map}(Y, Z)$. *For a map $f: X \to F$, we define a map $f': X \times Y \to Z$ by*

$$f'(x, y) = f(x)(y).$$

(1) *If f' is continuous, then f is continuous.*

(2) *When Y is a locally compact Hausdorff space, the converse of* (1) *holds.*

Proof (1) For any compact subset C of Y and any open set O of Z, we have only to prove that $f^{-1}(W(C, O))$ is open in X. For this, it suffices to show that for any point x of $f^{-1}(W(C, O))$ there exists an open neighbourhood U of x with $f(U) \subset W(C, O)$. Fix an arbitrary point x_0 of $f^{-1}(W(C, O))$. For $y \in C$, we have

$$f'(x_0, y) = f(x_0)(y) \in O.$$

Since f' is continuous by assumption, there exist an open neighbourhood U_y of x_0 and an open neighbourhood V_y of y with

$$f'(U_y \times V_y) \subset O.$$

(Since x_0 is fixed and y may vary, we write U_y. Note that this is not a misprint of U_{x_0}.)

If we consider this for each $y \in C$, we have trivially

$$\bigcup_{y \in C} V_y \supset C.$$

Since C is compact, there exists a finite set $\{y_1, \ldots, y_n\} \subset C$ with

$$\bigcup_{i=1}^{n} V_{y_i} \supset C.$$

We now put

$$U = \bigcap_{i=1}^{n} U_{y_i}, \qquad V = \bigcup_{i=1}^{n} V_{y_i}.$$

Then U is an open neighbourhood of x_0 and V is an open set of Y containing C.

In the following we show that

$$f'(U \times V) \subset O.$$

For any point $(x', y') \in U \times V$, choose i with $y' \in V_{y_i}$. Since $x' \in U \subset U_{y_i}$, we have

$$f'(x', y') \in f'(U_{y_i} \times V_{y_i}) \subset O.$$

Hence

$$f'(U \times C) \subset f'(U \times V) \subset O,$$

and

$$f(U) \subset W(C, O).$$

Therefore f is continuous.

(2) Suppose that f is continuous. For any non-empty open set O of Z, fix an arbitrary point (x, y) of $f'^{-1}(O)$. Note that $f(x)(y) \in O$. Since $f(x)$ is a continuous map from Y to Z, there exists an open set V containing y with

$$f(x)(V) \subset O.$$

By the assumption that Y is a locally compact Hausdorff space, there exists an open subset V' and a compact subset C with

$$y \in V' \subset C \subset V.$$

Hence we have

$$f(x)(C) \subset f(x)(V) \subset O,$$

and $f(x) \in W(C, O)$. Since f is continuous, there exists an open neighbourhood U of x with

$$f(U) \subset W(C, O).$$

Therefore we have

$$f'(U \times V') \subset f'(U \times C) \subset O,$$

which implies that f' is continuous. □

Lemma 1.25 *If Y is Hausdorff, then* Map(X, Y) *is Hausdorff. Hence any subspace of* Map(X, Y) *is Hausdorff as well.*

Proof Let f_1 and f_2 be different elements of Map(X, Y). Then there exists an element x of X with $f_1(x) \neq f_2(x)$. Since Y is Hausdorff, there are open sets O_1 and O_2 of Y such that

$$f_1(x) \in O_1, \qquad f_2(x) \in O_2, \qquad O_1 \cap O_2 = \varnothing.$$

Note that the one-point set $\{x\}$ is compact, and $W(\{x\}, O_1)$ and $W(\{x\}, O_2)$ are open sets of Map(X, Y) containing f_1 and f_2, respectively. Obviously

$$W(\{x\}, O_1) \cap W(\{x\}, O_2) = \varnothing,$$

which implies that Map(X, Y) is Hausdorff. □

For a topological space X, Homeo(X) denotes the set of all homeomorphisms of X onto itself. Homeo(X) is a group under composition. Since Homeo(X) is a subset of Map(X,X), we give Homeo(X) the relative topology.

Theorem 1.26 *If X is a compact Hausdorff space, then* Homeo(X) *is a topological group. Hence any subgroup of* Homeo(X) *is also a topological group.*

Proof Since X is Hausdorff, it follows from Lemma 1.25 that Map(X, X) is Hausdorff. Hence the subspace Homeo(X) of Map(X, X) is also Hausdorff. As we mentioned above, the group operation

$$\alpha\colon \mathrm{Homeo}(X) \times \mathrm{Homeo}(X) \to \mathrm{Homeo}(X)$$

is defined by $\alpha(f_1, f_2) = f_1 \cdot f_2$.

Taking any $W(C, O)$ containing $f_1 \cdot f_2$, we have $f_1 \cdot f_2(C) \subset O$. Here C is a compact set and O is an open set. Then $f_2(C)$ is compact, $f_1^{-1}(O)$ is open, and we have

$$f_2(C) \subset f_1^{-1}(O).$$

Since a compact Hausdorff space is normal, there exists an open set O' and a closed set C' with

$$f_2(C) \subset O' \subset C' \subset f_1^{-1}(O).$$

Since X is compact, its closed subset C' is compact, and we have

$$f_1 \in W(C', O) \qquad \text{and} \qquad f_2 \in W(C, O').$$

Now for any $f'_1 \in W(C', O)$ and for any $f'_2 \in W(C, O')$, we have

$$f'_1 \cdot f'_2(C) \subset f'_1(O') \subset f'_1(C') \subset O,$$

and hence $f'_1 \cdot f'_2 \in W(C, O)$. Thus we have shown that

$$\alpha(W(C', O) \times W(C, O')) \subset W(C, O).$$

It follows from the definition of a topology that an element of an open basis involving $f_1 \cdot f_2$ has the form

$$\bigcap_{i=1}^{n} W(C_i, O_i).$$

For each i, we choose C_i' and O_i' as above. Then

$$\left(\bigcap_{i=1}^{n} W(C_i', O_i)\right) \times \left(\bigcap_{i=1}^{n} W(C_i, O_i')\right)$$

is an open set of $\mathrm{Homeo}(X) \times \mathrm{Homeo}(X)$ containing (f_1, f_2), and clearly we have

$$\alpha\left(\left(\bigcap_{i=1}^{n} W(C_i', O_i)\right) \times \left(\bigcap_{i=1}^{n} W(C_i, O_i')\right)\right) \subset \bigcap_{i=1}^{n} W(C_i, O_i).$$

Thus we have shown that α is continuous.

Next we will show that the map β: $\mathrm{Homeo}(X) \to \mathrm{Homeo}(X)$ defined by $\beta(f) = f^{-1}$ is continuous. For a subset A of X, we denote by A^c the complement of A with respect to X. Note that for a compact subset C of X and for an open set O of X, C^c is open and O^c is closed. It follows that O^c is compact and we can consider $W(O^c, C^c)$. Then one can easily show that

$$\beta(W(C, O)) = W(O^c, C^c).$$

Since β is bijective and $\beta^{-1} = \beta$, we have

$$\begin{aligned} \beta^{-1}\left(\bigcap_{i=1}^{n} W(C_i, O_i)\right) &= \beta\left(\bigcap_{i=1}^{n} W(C_i, O_i)\right) \\ &= \bigcap_{i=1}^{n} \beta(W(C_i, O_i)) \\ &= \bigcap_{i=1}^{n} W(O_i^c, C_i^c), \end{aligned}$$

which implies that β is continuous. □

Next we investigate the structure of topological groups precisely. The reader might skip the rest of this section, returning to it when necessary.

Given two topological groups G_1 and G_2, we endow the Cartesian product $G_1 \times G_2$ with the product topology as well as the structure of the direct product of groups G_1 and G_2. As is easily shown, $G_1 \times G_2$ with this structure becomes a topological group, which is called the *direct product* of topological groups G_1 and G_2.

The semi-direct product appearing in the title of this section is a generalization of the notion of the direct product and is, so to speak, a twisted product.

Let N and G be topological groups. A map $f: N \to N$ is called an *automorphism* if f is bijective, and both f and f^{-1} are homomorphisms of topological groups. Then the set

$$\mathrm{Aut}(N) = \{f: N \to N \mid \text{automorphism}\}$$

is a group under composition. Since $\mathrm{Aut}(N)$ is a mapping space, it carries the compact-open topology. However, we do not use it now, but introduce the idea that a map $\theta\colon G \to \mathrm{Aut}(N)$ is continuous as follows. Given a map $\theta\colon G \to \mathrm{Aut}(N)$, we define a map $\theta'\colon G \times N \to N$ by

$$\theta'(g, n) = \theta(g)(n) \qquad \text{for} \qquad g \in G, n \in N.$$

If θ' is continuous, we say that θ is continuous. Refer to Theorem 1.24 for the relation between this continuity and that of θ with respect to the compact-open topology on $\mathrm{Aut}(N)$.

Given two topological groups N and G and a continuous homomorphism

$$\theta\colon G \to \mathrm{Aut}(N),$$

a new topological group

$$N \underset{\theta}{\rtimes} G \qquad (\text{or briefly } N \rtimes G)$$

is defined as follows.

First as a topological space, $N \underset{\theta}{\rtimes} G$ is given by the product space $N \times G$. Accordingly any element of $N \underset{\theta}{\rtimes} G$ can be written as (n, g) for $n \in N$, $g \in G$. Then the group operation $\circ$ is defined by $(n, g) \circ (n', g') = (n\theta(g)(n'), gg')$ for $n, n' \in N$, $g, g' \in G$ where $\theta(g)(n') \in N$, and $n\theta(g)(n')$ and gg' denote the multiplications in N and G, respectively. We now denote the identity element of N (resp. G) by 1 (resp. e). Then one easily verifies that $N \underset{\theta}{\rtimes} G$ is a group such that $(1, e)$ is the identity element and that the inverse element of (n, g) is given by

$$(n, g)^{-1} = (\theta(g^{-1})(n^{-1}), g^{-1}).$$

We can safely leave its proof to the reader.

Since both N and G are Hausdorff, $N \underset{\theta}{\rtimes} G$ is also Hausdorff. Since the map of multiplication

$$\alpha\colon \left(N \underset{\theta}{\rtimes} G\right) \times \left(N \underset{\theta}{\rtimes} G\right) \to N \underset{\theta}{\rtimes} G$$

is given by $((n, g), (n', g')) \to (n\theta(g)(n'), gg')$, it is continuous by the definition of continuity of θ. Similarly it is easily seen that the map $\beta\colon N \underset{\theta}{\rtimes} G \to N \underset{\theta}{\rtimes} G$ defined by $\beta(n, g) = (n, g)^{-1}$ is also continuous.

Thus $N \underset{\theta}{\rtimes} G$ becomes a topological group, which is called the *semi-direct product* of topological groups N and G with respect to the continuous homomorphism θ.

Next we consider when a given topological group can be written in the form of the semi-direct product. Let us consider a sequence

$$\cdots \longrightarrow G_{i-1} \xrightarrow{f_{i-1}} G_i \xrightarrow{f_i} G_{i+1} \longrightarrow \cdots$$

of topological groups, where f_i are homomorphisms of topological groups. Then the sequence is said to be *exact at* G_i if

$$\mathrm{Im}(f_{i-1}) = \mathrm{Ker}(f_i).$$

Here $\mathrm{Im}(f_{i-1}) = \{f_{i-1}(g_{i-1}) | g_{i-1} \in G_{i-1}\}$ and $\mathrm{Ker}(f_i) = \{g_i \in G_i | f_i(g_i) = e_{i+1}$ (the identity of $G_{i+1})\}$. If the sequence is exact at every G_i, it is called an *exact sequence.*

We now consider a short exact sequence

$$1 \longrightarrow N \xrightarrow{\ i\ } A \xrightarrow{\ j\ } G \longrightarrow 1$$

of topological groups whose end groups are trivial. A *splitting* for $j: A \to G$ is a homomorphism $s: G \to A$ such that $j \cdot s$ is the identity on G.

The following proposition gives a necessary and sufficient condition for a topological group to be written in the form of a semi-direct product.

Proposition 1.27 *Let*

$$1 \longrightarrow N \xrightarrow{\ i\ } A \xrightarrow{\ i\ } G \longrightarrow 1$$

be a short exact sequence of topological groups such that the topology on N coincides with the relative topology of A via the inclusion i: N → A. Then there exists a splitting s: G → A for j: A → G if and only if there exist a continuous homomorphism $\theta: G \to \mathrm{Aut}(N)$ *and an isomorphism* $f: A \to N \rtimes_\theta G$ *such that* $f \cdot i(n) = (n, e)$ and $j \cdot f^{-1}(n, g) = g$.

The procedures of making θ from s and of making s from θ are converse to one another.

Proof ⇒ Given a splitting $s: G \to A$, define a map $\theta: G \to \mathrm{Aut}(N)$ by

$$\theta(g)(n) = i^{-1}(s(g)i(n)s(g^{-1})).$$

First we show that this is well defined. Since

$$\begin{aligned} j(s(g)i(n)s(g^{-1})) &= (j \cdot s(g))(j \cdot i(n))(j \cdot s(g^{-1})) \\ &= geg^{-1} = e, \end{aligned}$$

we have

$$s(g)i(n)s(g^{-1}) \in \mathrm{Ker}(j) = \mathrm{Im}(i).$$

Since i is injective by assumption, i^{-1} is defined on the set $\mathrm{Im}(i)$. It follows that $\theta(g)(n)$ is well defined.

One easily verifies that $\theta(g) \in \mathrm{Aut}(N)$ and that θ is a homomorphism of groups. By assumption, the topology on N coincides with the relative topology of A. Hence to see that θ' is continuous, it suffices to show that the composition

$$G \times N \xrightarrow{\ \theta'\ } N \xrightarrow{\ i\ } A$$

is continuous. Since this map is given by

$$(g, n) \mapsto \theta(g)(n) \mapsto i \cdot \theta(g)(n) = s(g)i(n)s(g^{-1}),$$

it is continuous as desired.

Thus we have shown that θ is a continuous homomorphism. Accordingly we can consider the semi-direct product $N \underset{\theta}{\rtimes} G$.

Claim A is isomorphic to $N \underset{\theta}{\rtimes} G$.

For this, we first define a map $f: A \to N \underset{\theta}{\rtimes} G$ by

$$a \mapsto (i^{-1}(as \cdot j(a^{-1})),\ j(a)).$$

As in the case of θ, it is easily seen that f is well defined and continuous. We now show that f is a homomorphism of groups. For any $a, b \in A$, we compute

$$\begin{aligned} f(a) \circ f(b) &= (i^{-1}(as \cdot j(a^{-1})), j(a)) \circ (i^{-1}(bs \cdot j(b^{-1})), j(b)) \\ &= (i^{-1}(as \cdot j(a^{-1}))\theta(j(a))(i^{-1}(bs \cdot j(b^{-1}))),\ j(a)j(b)) \\ &= (i^{-1}(as \cdot j(a^{-1}))i^{-1}(s \cdot j(a)i(i^{-1}(bs \cdot j(b^{-1})))s(j(a)^{-1})), j(ab)) \\ &= (i^{-1}(as \cdot j(a^{-1}))i^{-1}(s \cdot j(a)bs \cdot j(b^{-1})s(j(a)^{-1})), j(ab)) \\ &= (i^{-1}(as \cdot j(a^{-1})s \cdot j(a)bs \cdot j(b^{-1})s(j(a)^{-1})), j(ab)) \\ &= (i^{-1}(abs \cdot j((ab)^{-1})), j(ab)) \\ &= f(ab). \end{aligned}$$

On the other hand, we define a map $f': N \underset{\theta}{\rtimes} G \to A$ of opposite direction by

$$(n, g) \mapsto i(n)s(g).$$

Then it is easy to see that f' is continuous and satisfies

$$f' \cdot f = \text{id}, \qquad f . f' = \text{id}.$$

Hence $f' = f^{-1}$ and f is an isomorphism of topological groups. Through this isomorphism, i and j are expressed as

$$\begin{aligned} f \cdot i(n) &= (i^{-1}(i(n)s \cdot j(i(n)^{-1})),\ j \cdot i(n)) = (n, e), \\ j \cdot f'((n, g)) &= j(i(n)s(g)) = j \cdot s(g) = g. \end{aligned}$$

$\Leftarrow$ Suppose that we are given an isomorphism $f: A \to N \underset{\theta}{\rtimes} G$ satisfying $f \cdot i(n) = (n, e)$ and $j \cdot f^{-1}((n, g)) = g$. Let $s: G \to A$ be the map defined by

$$g \mapsto f^{-1}((1, g)).$$

Then for any $g, g' \in G$, we have

$$\begin{aligned} s(gg') &= f^{-1}((1, gg')) = f^{-1}((1, g) \circ (1, g')) \\ &= f^{-1}((1, g))f^{-1}((1, g')) = s(g)s(g'). \end{aligned}$$

Obviously s is continuous, hence s is a homomorphism of topological groups. On the other hand, we have

$$j \cdot s(g) = j \cdot f^{-1}((1, g)) = g$$

by assumption. Hence s is a splitting for j.

We now show that the procedures of making θ from s and of making s from θ are converse to one another.

Given a splitting s, we make θ from s, and then make s' from θ as above. Then we have

$$s'(g) = f^{-1}((1, g)) = i(1)s(g) = s(g).$$

Conversely, given a homomorphism θ first, we make s from θ, and then make θ' from s. Then we have

$$\begin{aligned} \theta'(g)(n) &= i^{-1}(s(g)i(n)s(g^{-1})) \\ &= i^{-1}(f^{-1}((1, g))f^{-1}((n, e))f^{-1}((1, g^{-1}))) \\ &= i^{-1} \cdot f^{-1}((1, g) \circ (n, e) \circ (1, g^{-1})) \\ &= i^{-1} \cdot f^{-1}((\theta(g)(n), g) \circ (1, g^{-1})) \\ &= i^{-1} \cdot f^{-1}((\theta(g)(n)\theta(g)(1), gg^{-1})) \\ &= i^{-1} \cdot f^{-1}((\theta(g)(n), e)) = \theta(g)(n). \end{aligned}$$

□

Corollary 1.28 *If $A \cong N \rtimes_{\theta} G$, then the quotient A/N is a topological group and $A/N \cong G$.*

Proof According to Proposition 1.27, N can be written as $\mathrm{Ker}(j)$ and hence N is a closed normal subgroup of A. It follows from Proposition 1.20 that A/N is a topological group. Let π: $A \to A/N$ be the projection map. Then $\pi \cdot s$: $G \to A/N$ gives an isomorphism of groups and is continuous. On the other hand, the inverse $(\pi \cdot s)^{-1}$: $A/N \to G$ is always continuous by Lemma 1.17. □

Now we give examples of applications of Proposition 1.27 and Corollary 1.28.

Example 2 Consider the following short exact sequences:

$$1 \longrightarrow SO(n) \xrightarrow{i} O(n) \xrightarrow{j} \mathbb{Z}_2 \longrightarrow 1$$

$$1 \longrightarrow SU(n) \xrightarrow{i} U(n) \xrightarrow{j} U(1) \longrightarrow 1$$

$$1 \longrightarrow SL(n, \mathbb{R}) \xrightarrow{i} GL(n, \mathbb{R}) \xrightarrow{j} \mathbb{R}^* \longrightarrow 1$$

$$1 \longrightarrow SL(n, \mathbb{C}) \xrightarrow{i} GL(n, \mathbb{C}) \xrightarrow{j} \mathbb{C}^* \longrightarrow 1.$$

Here $\mathbb{R}^* = \mathbb{R} - \{0\}$, $\mathbb{C}^* = \mathbb{C} - \{0\}$, the i are inclusion maps and $j = \det$. Obviously splittings $s: G \to A$ for j are all given by

$$s(x) = \begin{bmatrix} x & & & 0 \\ & 1 & & \\ & & \ddots & \\ 0 & & & 1 \end{bmatrix}$$

It follows from Proposition 1.27 that we have isomorphisms of topological groups:

$$O(n) \cong SO(n) \rtimes \mathbb{Z}_2,$$

$$U(n) \cong SU(n) \rtimes U(1),$$

$$GL(n, \mathbb{R}) \cong SL(n, \mathbb{R}) \rtimes \mathbb{R}^*,$$

$$GL(n, \mathbb{C}) = SL(n, \mathbb{C}) \rtimes \mathbb{C}^*.$$

Therefore we have isomorphisms of topological groups:

$$O(n)/SO(n) \cong \mathbb{Z}_2$$

$$U(n)/SU(n) \cong U(1),$$

$$GL(n, \mathbb{R})/SL(n, \mathbb{R}) \cong \mathbb{R}^*,$$

$$GL(n, \mathbb{C})/SL(n, \mathbb{C}) \cong \mathbb{C}^*.$$

1.4 Topological transformation groups

By a *topological transformation group* we mean a triple (G, X, φ), where G is a topological group, X is a topological space, and $\varphi: G \times X \to X$ is a continuous map satisfying the conditions (1) and (2) of Definition 1.1. The space X, together with a given action φ of G, is called a G-*space*.

Example 1 Any topological group G acts on itself by group operation.

Example 2 Let G be a topological group and H a subgroup of G. Without considering topology, G acts on G/H in the manner of Example 7 in Section 1.1. The action is just the $\bar{\alpha}$ in the following commutative diagram:

$$\begin{array}{ccc} G\times G & \xrightarrow{\quad\alpha\quad} & G \\ {\scriptstyle \mathrm{id}\times\pi}\downarrow & & \downarrow{\scriptstyle \pi} \\ G\times G/H & \xrightarrow[\quad\bar{\alpha}\quad]{} & G/H \end{array}$$

In Lemma 1.19, the assumption that H is a closed subgroup is employed only to show that G/H is a Hausdorff space. Hence without assuming H a closed subgroup, we have that π is surjective and open. It follows from Lemma 1.13 that $\mathrm{id}\times\pi$ is surjective and open. From the diagram above, continuity of $\bar{\alpha}$ follows from Lemma 1.15.

Example 3 Let X be a topological space and $\mathbb{R}$ the topological group of real numbers. Then an $\mathbb{R}$-action on X is called a *dynamical system* or a *1-parameter group of homeomorphisms* of X.

Lemma 1.29 *Let (X, G, φ) be a topological transformation group. Given $g\in G$, let $\varphi_g\colon X\to X$ be the map defined by $\varphi_g(x)=\varphi(g, x)$ as in Section 1.1. Then φ_g is a homeomorphism of X.*

Proof Given $g\in G$, let $i_g\colon X\to G\times X$ be the map defined by $i_g(x)=(g, x)$. By the definition of the product topology, i_g is continuous. Since $\varphi_g=\varphi\cdot i_g$, φ_g is continuous. As shown in Section 1.1, φ_g is bijective and $(\varphi_g)^{-1}=\varphi_{g^{-1}}$, and hence $(\varphi_g)^{-1}$ is also continuous. Hence φ_g is a homeomorphism of X. □

By Lemma 1.29, the correspondence $g\mapsto\varphi_g$ defines a map

$$\Phi\colon G\to \mathrm{Homeo}(X).$$

Although $\mathrm{Homeo}(X)$ is not necessarily a topological group in general, it is a group as well as a topological space.

Lemma 1.30 *Φ is a group homomorphism.*

Proof For any $g_1, g_2\in G$, we have

$$\Phi(g_1g_2)=\varphi_{g_1g_2}=\varphi_{g_1}\cdot\varphi_{g_2}=\Phi(g_1)\cdot\Phi(g_2).$$ □

Lemma 1.31 *Given a homomorphism* $\Phi: G \to \mathrm{Homeo}(X)$, *we define* $\varphi: G \times X \to X$ by $\varphi(g, x) = \Phi(g)(x)$. *Then* φ *is a* G*-action in the sense of Section* 1.1.

The proof is clear and left to the reader.

Proposition 1.32 *Under the above circumstances, if* φ *is continuous, then so is* Φ. *If* X *is a locally compact Hausdorff space, then* φ *is continuous if and only if* Φ *is continuous.*

Proof If we replace X by G and Y, Z by X in Theorem 1.24, then $\mathrm{Homeo}(X)$ is a subspace of $\mathrm{Map}(X, X)$ and Proposition 1.32 follows immediately from Theorem 1.24. □

Theorem 1.33 *Let* X *be a compact Hausdorff space. Then* $\mathrm{Homeo}(X)$ *is a topological group, and considering a continuous* G*-action* φ *on* X *is equivalent to considering a homomorphism* $\Phi: G \to \mathrm{Homeo}(X)$ *of topological groups.*

Proof The first half is nothing but Theorem 1.26. Since a compact Hausdorff space is a locally compact Hausdorff space, the equivalence of the continuity of φ and Φ is immediate from Proposition 1.32.

On the other hand, it follows from Lemmas 1.30 and 1.31 that φ is a G-action in the sense of Section 1.1 if and only if Φ is a homomorphism of groups. □

Given a continuous action $\varphi: G \times X \to X$, we define $\Phi: G \to \mathrm{Homeo}(X)$ as above. The kernel of this homomorphism Φ will be called the 'kernel of the action φ'. That is, we put

$$\mathrm{Ker}\,\varphi = \mathrm{Ker}\,\Phi.$$

Lemma 1.34 $\mathrm{Ker}\,\varphi$ *is a normal subgroup of* G.

Proof This follows directly from the fact that Φ is a homomorphism of groups. □

Lemma 1.35 $\mathrm{Ker}\,\varphi = \bigcap_{x \in X} G_x$.

The proof is left to the reader as an exercise.

Lemma 1.36 *Fix a point* x *of* X. *If we define a map* $i: G \to G \times X$ *by* $i(g) = (g, x)$, *then* i *is continuous and we have the following equality:*

$$G - G_x = i^{-1}(\varphi^{-1}(X - \{x\})).$$

Proof For any element g of $G - G_x$, we have

$$\varphi \cdot i(g) = \varphi(g, x) \neq x.$$

This means that $g \in i^{-1}(\varphi^{-1}(X - \{x\}))$. Thus

$$G - G_x \subset i^{-1}(\varphi^{-1}(X - \{x\})).$$

Conversely for any element g of $i^{-1}(\varphi^{-1}(X - \{x\}))$, we have

$$\varphi \cdot i(g) = \varphi(g, x) \in X - \{x\}.$$

Hence we have $\varphi(g, x) \neq x$, which means that $g \in G - G_x$. Thus

$$G - G_x \supset i^{-1}(\varphi^{-1}(X - \{x\})).$$ □

Theorem 1.37 *If X is a T_1-space (topological space such that each point is closed), then the isotropy group G_x at any point x of X is closed.*

Proof By assumption, $X - \{x\}$ is open. It follows from Lemma 1.36 that $G - G_x$ is open, and hence G_x is closed. □

Corollary 1.38 *If X is a T_1-space, then* Ker φ *is a closed subgroup of G.*

Proof Since G_x is closed by Theorem 1.37, Corollary 1.38 is immediate from Lemma 1.35. □

Corollary 1.39 *If X is a T_1-space, then $G/$*Ker φ *is a topological group.*

Proof Combining Lemma 1.34 and Corollary 1.38, we have that Ker φ is a closed normal subgroup of G. Hence Proposition 1.20 yields Corollary 1.39. □

Theorem 1.40 *If X is a T_1-space, a topological transformation group (X, G, φ) induces an effective transformation group $(X, G/\text{Ker}\,\varphi, \bar{\varphi})$.*
($\bar{\varphi}$ will be given in the proof.)

Proof Define $\bar{\varphi}: G/\text{Ker}\,\varphi \times X \to X$ by

$$\bar{\varphi}(g\,\text{Ker}\,\varphi, x) = \varphi(g, x).$$

It is easily verified that $\bar{\varphi}$ is well defined and satisfies the conditions (1) and (2) of group action (Definition 1.1). Hence it suffices to check continuity of $\bar{\varphi}$. In view of Lemma 1.19, the natural projection $\pi: G \to G/\text{Ker}\,\varphi$ is open. As a consequence, we have that

$$\pi \times \text{id}: G \times X \to G/\text{Ker}\,\varphi \times X$$

is also open by Lemma 1.13. We now apply Lemma 1.15 to the following commutative diagram:

$$\begin{array}{ccc} G \times X & \xrightarrow{\varphi} & X \\ & {}_{\pi \times \mathrm{id}}\searrow \quad \nearrow_{\bar{\varphi}} & \\ & G/\operatorname{Ker}\varphi \times X & \end{array}$$

and deduce that $\bar{\varphi}$ is continuous. Evidently the action $\bar{\varphi}$ is effective. □

Corollary 1.41 *If X is a T_1-space, then there exists a continuous injective homomorphism $\bar{\Phi}: G/\operatorname{Ker}\varphi \to \operatorname{Homeo}(X)$ such that the following diagram commutes:*

$$\begin{array}{ccc} G & \xrightarrow{\Phi} & \operatorname{Homeo}(X) \\ & {}_{\pi}\searrow \quad \nearrow_{\bar{\Phi}} & \\ & G/\operatorname{Ker}\varphi & \end{array}$$

(Here $\operatorname{Homeo}(X)$ is not necessarily a topological group.)

Let G be a topological group and H a closed subgroup of G. If we put

$$H_0 = \bigcap_{g \in G} gHg^{-1},$$

then H_0 is a closed normal subgroup of G. Moreover H_0 is the maximum subgroup among the normal subgroups of G contained in H.

Proposition 1.42 *Let $(G/H, G, \varphi)$ be the continuous G-action of Example 2. Then the equality $\operatorname{Ker}\varphi = H_0$ holds.*

Proof For any $h \in H_0$ and $g \in G$, we have

$$hgH = g(g^{-1}hg)H = gH.$$

Hence H_0 acts trivially on G/H. That is, $H_0 \subset \operatorname{Ker}\varphi$. Conversely, suppose that $g_0 gH = gH$ for all $g \in G$. Then $g_0 \in gHg^{-1}$ for all $g \in G$. That is, $g_0 \in H_0$, which implies that $H_0 \supset \operatorname{Ker}\varphi$. □

Corollary 1.43 *G/H_0 acts effectively on G/H.*

We now introduce notations similar to those in topological groups. Given a subset J of a topological group G and a subset A of a G-space X, we put

$$JA = \{gx \mid g \in J,\ x \in A\}.$$

Lemma 1.44

(1) *If A is open in X, then JA is open in X.*

(2) *If J is compact and A is closed in X, then JA is closed in X.*

(3) *If both J and A are compact, then JA is compact.*

Proof (1) If A is open, then $gA = \varphi_g(A)$ is open for every $g \in G$ by Lemma 1.29. Hence the set

$$JA = \bigcup_{g \in J} gA$$

is also open in X.

(2) Given a subset Y of X, the following equivalences are easy to verify:

$$\begin{aligned} Y \cap JA = \varnothing &\Leftrightarrow J^{-1} Y \cap A = \varnothing \\ &\Leftrightarrow J^{-1} \times Y \cap \varphi^{-1}(A) = \varnothing. \end{aligned}$$

Fix an arbitrary point x of $X - JA$. Then we have in particular

$$J^{-1} \times \{x\} \cap \varphi^{-1}(A) = \varnothing,$$

that is,

$$J^{-1} \times \{x\} \subset G \times X - \varphi^{-1}(A).$$

Since A is closed in X, $\varphi^{-1}(A)$ is closed in $G \times X$. Thus $G \times X - \varphi^{-1}(A)$ is open and contains the set $J^{-1} \times \{x\}$. Therefore for each point g of J^{-1}, there exist an open neighbourhood U_g of g in G and an open neighbourhood V_g of x in X such that

$$U_g \times V_g \subset G \times X - \varphi^{-1}(A).$$

Note that x is fixed, and both U_g and V_g vary according as $g \in J^{-1}$. Trivially we have

$$J^{-1} \subset \bigcup_{g \in J^{-1}} U_g.$$

Since J is compact, $J^{-1} = \beta(J)$ is also compact so that there exist $g_1, \ldots, g_n \in J^{-1}$ with

$$J^{-1} \subset \bigcup_{i=1}^{n} U_{g_i}.$$

We now put

$$U = \bigcup_{i=1}^{n} U_{g_i}, \qquad V = \bigcap_{i=1}^{n} V_{g_i},$$

which are open in G and X, respectively. Note that

$$J^{-1} \times V \subset U \times V \subset G \times X - \varphi^{-1}(A),$$

that is, we have

$$J^{-1} \times V \cap \varphi^{-1}(A) = \varnothing.$$

Therefore it follows again from the equivalences mentioned at the beginning that

$$V \cap JA = \emptyset.$$

Thus we have shown that given any $x \in X - JA$, there exists an open neighbourhood V of x in X with

$$x \in V \subset X - JA,$$

which means that JA is closed in X.

(3) If J and A are compact, then their Cartesian product $J \times A$ is also compact. Hence its image

$$JA = \varphi(J \times A)$$

by the continuous map φ is also compact. □

Note that Lemma 1.44(2) does not hold under the condition that J is only closed in G. The following not only provides us with an example, but will also be quoted occasionally throughout the book.

Example 4 Let S^1 denote the group of complex numbers of norm 1 and let T^2 be the torus $S^1 \times S^1$. Given an irrational number a, we define a map

$$\varphi: \mathbb{R} \times T^2 \to T^2$$

by

$$\varphi(r, (z_1, z_2)) = (z_1 \exp 2\pi i r, z_2 \exp 2\pi i a r)$$

for $r \in \mathbb{R}$ and $z_i \in S^1$. Then φ is a continuous $\mathbb{R}$-action on T^2. Note that the point $(1, 1)$ of T^2 is closed. It is not difficult to see that the orbit $\mathbb{R}(1, 1)$ of $(1, 1)$ is dense in T^2 and does not coincide with T^2. Hence $\mathbb{R}(1, 1)$ is not closed in T^2.

Proposition 1.45 *If $\varphi: G \times X \to X$ is a continuous G-action, then φ is an open map. If G is compact, φ is a closed map.*

Proof Let $O \subset G \times X$ be open. Given $g \in G$, put

$$O_g = \{(g, x) \in O \mid x \in X\}, \qquad O'_g = \{x \in X \mid (g, x) \in O\}.$$

Then O_g is the section of O through $g \in G$ and O'_g is the projection of O_g to X. These are illustrated in Fig. 1.1.

It follows from the definition of product topology that O'_g is open in X. Clearly we have

$$\varphi(O) = \varphi\left(\bigcup_{g \in G} O_g\right) = \varphi\left(\bigcup_{g \in G} \{g\} \times O'_g\right) = \bigcup_{g \in G} \varphi_g(O'_g).$$

Since φ_g is a homeomorphism by Lemma 1.29, $\varphi_g(O'_g)$ is open. Thus $\varphi(O)$ is an open set in X.

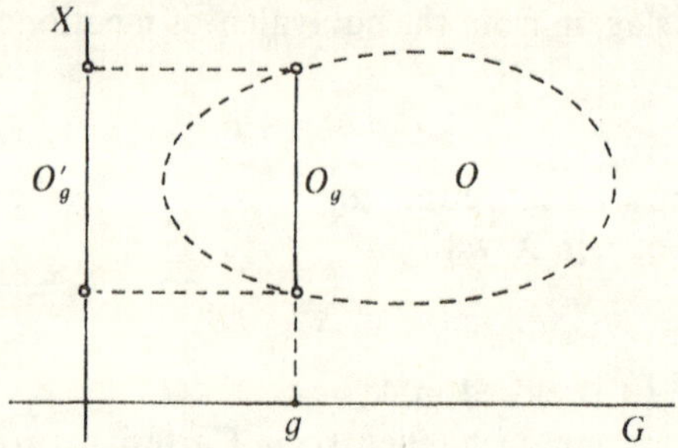

Fig. 1.1 O_g and O'_g.

Next we show the last half of the proposition. If we define a map

$$\Phi: G \times (G \times X) \to G \times X$$

by $(g, h, x) \mapsto (hg^{-1}, \varphi(g, x))$, then it is easy to see that Φ gives a G-action on $G \times X$. First we show that for any subset A of $G \times X$, the following equality holds:

$$\Phi(G \times A) = \varphi^{-1}(\varphi(A)). \tag{1.2}$$

For any $g \in G$ and $(h, x) \in A$, we have

$$\varphi \cdot \Phi(g, h, x) = \varphi(hg^{-1}, \varphi(g, x)) = \varphi(h, x) \in \varphi(A).$$

In consequence, we have $\Phi(G \times A) \subset \varphi^{-1}(\varphi(A))$. Conversely, for any $(h, x) \in \varphi^{-1}(\varphi(A))$, we have $\varphi(h, x) \in \varphi(A)$, and hence there exists an element (g, y) of A with

$$\varphi(h, x) = \varphi(g, y).$$

Accordingly we have $\varphi(h^{-1}g, y) = x$ and

$$\Phi(h^{-1}g, (g, y)) = (g(h^{-1}g)^{-1}, \varphi(h^{-1}g, y)) = (h, x).$$

Thus it is shown that the reverse inclusion

$$\Phi(G \times A) \supset \varphi^{-1}(\varphi(A))$$

holds.

Next, by a set theoretic argument, we have

$$\varphi^{-1}(\varphi(A)) = (\varphi^{-1}(\varphi(A)^c))^c \tag{1.3}$$

in general, where c denotes the operation of assigning complement. Since φ is surjective, we have

$$\varphi(A)^c = \varphi(\varphi^{-1}(\varphi(A)^c)). \tag{1.4}$$

We now suppose that G is compact and A is closed in $G \times X$. It follows from Lemma 1.44 that $\Phi(G \times A) = GA$ is closed in $G \times X$. Hence $\varphi^{-1}(\varphi(A))$ is also closed by (1.2). Accordingly it follows from (1.3) that $\varphi^{-1}(\varphi(A)^c)$ is open in

$G \times X$. As shown above, φ is an open map. It turns out that $\varphi(A)^c$ is open in X by (1.4). Thus we can conclude that $\varphi(A)$ is closed in X. □

Example 5 In Example 4, we put $A = G \times \{(1, 1)\}$. Then A is a closed set of $G \times T^2$. However we have

$$\varphi(A) \subsetneqq T^2 = \overline{\varphi(A)},$$

that is, $\varphi(A)$ is not closed in T^2. This example shows that the compactness assumption of G is indispensable in Proposition 1.45.

1.5 Fixed point sets

Given a transformation group (X, G, φ), its fixed point set X^G is one of the interesting objects in the study of transformation groups. In this section we generalize the notion of fixed point set and study its topological aspects.

Let (X, G, φ) be a topological transformation group. Given a subset J of G, we put

$$X^J = \{x \in X \mid J \subset G_x\}$$

which is called the *fixed point set of J* (or *J-fixed point set*). In particular, X^G is simply called the fixed point set (see Section 1.1). If J consists of a single element g of G, $X^{\{g\}}$ is simply denoted by X^g. We will be mainly concerned with the case where J is a subgroup of G. The reason will become clear soon.

The following theorem is fundamental.

Theorem 1.46 *Let (X, G, φ) be a topological transformation group. If X is Hausdorff, then X^J is closed in X for any $J \subset G$.*

Proof We first define a subset Δ of the direct product space $X \times X$ by

$$\Delta = \{(x, x) \mid x \in X\} \subset X \times X,$$

which is called the *diagonal*. As is well known, X is Hausdorff if and only if Δ is closed in $X \times X$. Given $g \in G$, we define a map

$$j_g: X \to X \times X$$

by $j_g(x) = (x, \varphi(g, x))$. Since j_g is continuous and since

$$j_g^{-1}(\Delta) = X^g,$$

X^g is closed in X. It follows from the equality

$$X^J = \bigcap_{g \in J} X^g$$

that X^J is closed in X. □

Lemma 1.47 *If X is Hausdorff, we have, for any subset J of G,*

$$X^J = X^{\bar{J}}.$$

Proof Since $J \subset \bar{J}$, we have trivially $X^J \supset X^{\bar{J}}$. Hence we have only to show that the reverse inclusion holds. Given $g \in \bar{J}$, there is a net $\{h_\nu\}$ in J converging to g. It follows from the continuity of φ that for any $x \in X^J$, we have

$$\varphi(g, x) = \lim_\nu \varphi(h_\nu, x) = \lim_\nu x = x.$$

Here the last equality follows from the fact that the limit of the net $\{x\}$ consisting solely of the constant point x is just x, since X is Hausdorff. Hence $x \in X^g$. Since g is an arbitrary element of $\bar{J}$, we have $x \in X^{\bar{J}}$. Thus we have shown that the reverse inclusion $X^J \subset X^{\bar{J}}$ holds. □

For a subset J of G, we denote by $\langle J \rangle$ the subgroup of G generated by J.

Lemma 1.48 $X^J = X^{\langle J \rangle}$.

Proof Trivially we have $X^J \supset X^{\langle J \rangle}$. Any element g of $\langle J \rangle$ can be written as

$$g = h_1^{m_1} h_2^{m_2} \cdots h_r^{m_r} \qquad h_i \in J,\ m_i \in \mathbb{Z},$$

by definition. Then for any $x \in X^J$, we have evidently $h_i^{-1} x = x$. Hence we have $gx = x$ by induction, that is, $x \in X^g$. Since g is an arbitrary element of $\langle J \rangle$, x lies in $X^{\langle J \rangle}$. Thus we have shown that $X^J \subset X^{\langle J \rangle}$. □

Theorem 1.49 *Let (X, G, φ) be a topological transformation group. If X is Hausdorff then we have, for any subset J of G,*

$$X^J = X^{\overline{\langle J \rangle}}.$$

(That is, as far as a fixed point set is concerned, we have only to deal with the fixed point sets of closed subgroups.)

Proof This theorem follows by combining Lemmas 1.47 and 1.48, and Proposition 1.9. □

Let (X, G, φ) be a topological transformation group and H a subgroup of G. Then G does not act on X^H in general, because X^H is not necessarily G-invariant.

Lemma 1.50 *X^H is $N(H)$-invariant. Hence if H is a normal subgroup of G, then X^H is G-invariant.*

Proof Given any $n \in N(H)$ and $h \in H$, there exists $h' \in H$ with $hn = nh'$. Hence for any $x \in X^H$, we have

$$h(nx) = (hn)x = (nh')x = n(h'x) = nx.$$

It follows that $nx \in X^H$, that is, X^H is $N(H)$-invariant. □

Since H acts trivially on X^H, there is a natural action of the quotient group $N(H)/H$ on X^H. If H is a closed subgroup of G, then H is a closed normal subgroup of $N(H)$, hence $N(H)/H$ is a topological group by Proposition 1.20. Thus X^H becomes an $N(H)/H$-space in this case.

Let $\rho\colon G \to G'$ be a homomorphism of topological groups, X a G-space and X' a G'-space. A *ρ-equivariant map* $f\colon X \to X'$ is a map that satisfies

$$f(gx) = \rho(g)f(x) \qquad \text{for all } g \in G \text{ and } x \in X.$$

When $G = G'$ and $\rho = \text{id}$, a ρ-equivariant map is simply called a *G-map*. If ρ is an isomorphism and if f is a homeomorphism, then f is called a *ρ-equivariant homeomorphism*.

Now, let H and K be conjugate closed subgroups of G. That is, there exists $g_0 \in G$ with

$$K = g_0 H g_0^{-1}.$$

Then it is easy to see that

$$N(K) = g_0 N(H) g_0^{-1}.$$

Lemma 1.51 *Let H and K be closed subgroups of G with $K = g_0 H g_0^{-1}$ for some $g_0 \in G$. Then the map*

$$\rho\colon N(H)/H \to N(K)/K$$

defined by $\rho(nH) = (g_0 n g_0^{-1})K$ for $n \in N(H)$ gives an isomorphism of topological groups.

Proof First we show that ρ is well defined. Let $n, n' \in N(H)$ be such that $nH = n'H$. Then there exists $h \in H$ with $n' = nh$ and we have

$$\begin{aligned}\rho(n'H) &= (g_0 n' g_0^{-1})K = (g_0 nh g_0^{-1})(g_0 H g_0^{-1}) \\ &= g_0 nhH g_0^{-1} = g_0 nH g_0^{-1} = (g_0 n g_0^{-1})(g_0 H g_0^{-1}) \\ &= (g_0 n g_0^{-1})K = \rho(nH).\end{aligned}$$

Hence the definition of ρ is independent of the choice of coset representatives. Next we show its continuity. Since the map $I_{g_0}\colon G \to G$ is continuous by Proposition 1.7, its restriction

$$I_{g_0} \mid N(H)\colon N(H) \to N(K)$$

is also continuous. Consider the following commutative diagram

$$\begin{array}{ccc} N(H) & \xrightarrow{I_{g_0}|N(H)} & N(K) \\ \downarrow{\scriptstyle \pi} & & \downarrow{\scriptstyle \pi} \\ N(H)/H & \xrightarrow{\rho} & N(K)/K \end{array}$$

Since each π is a surjective open continuous map, it follows from Lemma 1.15 that ρ is continuous. Finally we show that ρ is a homomorphism of groups. For any $n, n' \in N(H)$, we compute as follows:

$$\begin{aligned} \rho(nH \cdot n'H) &= \rho(nn'H) = g_0 nn' g_0^{-1} K = g_0 nn' g_0^{-1} g_0 H g_0^{-1} \\ &= g_0 nn' H g_0^{-1} = g_0 n H n' H g_0^{-1} \\ &= (g_0 n g_0^{-1})(g_0 H g_0^{-1})(g_0 n' g_0^{-1})(g_0 H g_0^{-1}) \\ &= \rho(nH) \cdot \rho(n'H). \end{aligned}$$

Thus we have shown that ρ is a homomorphism of topological groups.

Conversely, if we define a map

$$\rho' \colon N(K)/K \to N(H)/H$$

by $\rho'(nK) = (g_0^{-1} n g_0)H$ for $n \in N(K)$, then we have similarly that ρ' is a homomorphism of topological groups. Moreover it is obvious that $\rho' \cdot \rho = \rho \cdot \rho' = \mathrm{id}$. Thus we see that ρ is an isomorphism of topological groups. □

Proposition 1.52 *Let G, H, K, g_0 and ρ be as in Lemma* 1.51 *and* (X, G, φ) *a topological transformation group. Then if we define a map $f \colon X \to X$ by $f(x) = g_0 x$, its restriction $f|X^H \colon X^H \to X^K$ gives a ρ-equivariant homeomorphism.*

Proof First we show that $f(X^H) \subset X^K$. Given $k \in K$, there exists $h \in H$ with $k = g_0 h g_0^{-1}$. Hence for any $x \in X^H$, we have

$$kf(x) = kg_0 x = g_0 h g_0^{-1} g_0 x = g_0 hx = g_0 x = f(x),$$

that is, $f(X^H) \subset X^K$. Conversely, if we define a map $f' \colon X \to X$ by $f'(x) = g_0^{-1} x$, we have similarly that $f'(X^K) \subset X^H$. Evidently, we have

$$f' \cdot f = f \cdot f' = \mathrm{id}$$

so that the map $f|X^H \colon X^H \to X^K$ is bijective. Moreover f and f' are continuous, hence $f|X^H$ is a homeomorphism.

Finally we show that $f|X^H$ is ρ-equivariant. Given any $n \in N(H)$ and $x \in X^H$, we compute

$$f(nHx) = g_0 nHx = (g_0 n g_0^{-1})(g_0 H g_0^{-1}) g_0 x$$
$$= (g_0 n g_0^{-1}) K g_0 x = \rho(nH) f(x),$$

that is, $f|X^H$ is ρ-equivariant. □

1.6 Orbits and orbit spaces

Let (X, G, φ) be a topological transformation group. If we do not take the topology into consideration, there is a G-isomorphism

$$G/G_x \cong G(x)$$

of G-sets by Theorem 1.5. Let us now consider whether this is a G-homeomorphism or not, where G/G_x is equipped with the quotient topology of G. If we define a map $f\colon G \to G(x)$ by

$$f(g) = \varphi(g, x),$$

then f is continuous. It follows from Theorem 1.5 that the following diagram

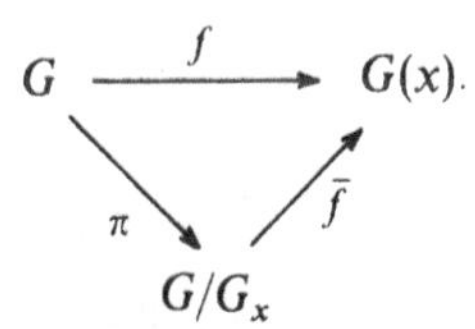

commutes and $\bar{f}$ is a bijective G-map. Since G/G_x is endowed with the quotient topology, $\bar{f}$ is continuous by Lemma 1.17. However $\bar{f}$ may not be a homeomorphism.

Proposition 1.53 *If G is compact and X is Hausdorff, then $\bar{f}\colon G/G_x \to G(x)$ is a G-homeomorphism.*

Proof We have only to prove that $\bar{f}^{-1}$ is continuous. Observe first that G/G_x is compact. Since X is Hausdorff, its subset $G(x)$ is also Hausdorff. Therefore the continuity of $\bar{f}^{-1}$ follows from Lemma 1.16. □

The following example shows that $\bar{f}$ is not in general a homeomorphism if G is not compact.

Example 1 In Example 4 of Section 1.4, the action is free so that we have $G/G_{(1,1)} = G = \mathbb{R}$, which is locally connected. On the other hand, the orbit

$G((1, 1))$ of $(1, 1)$ is not locally connected. Hence $G/G_{(1,1)}$ and $G((1, 1))$ are not homeomorphic. This is due to the lack of compactness of G.

Definition 1.3 For a G-space X, we consider a relation $\sim$ on X as follows:

$$x \sim y \Leftrightarrow \text{there exists an element } g \text{ of } G \text{ with } y = gx.$$

Then $\sim$ is an equivalence relation and we denote by X/G the quotient set $X/\sim$ with respect to the above relation. That is, X/G is the set whose elements are the orbits of G on X. Then X/G endowed with the quotient topology is called the *orbit space* of X.

For example, put $G = U(1)$ and $X = U(1) \times Y$. If $U(1)$ acts naturally on the $U(1)$ factor of X, then X/G is homeomorphic to Y, as illustrated in Fig. 1.2.

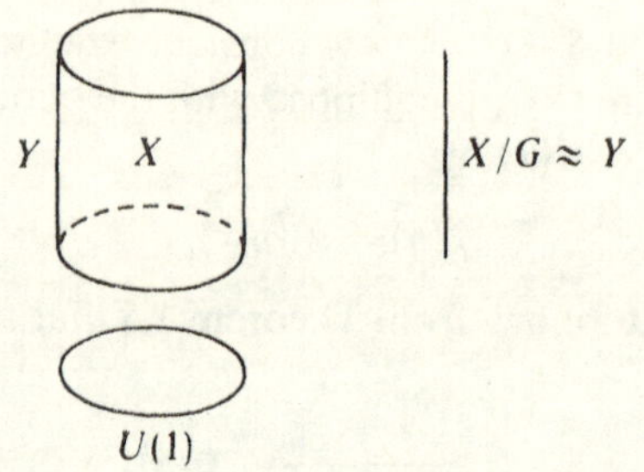

Fig. 1.2 X/G is homeomorphic to Y.

Example 2 Let G be a topological group and H a subgroup of G. Then an H-action on G, $\varphi: H \times G \to G$, is defined by $\varphi(h, g) = gh^{-1}$ for $h \in H$, $g \in G$. The orbit space of this action is nothing but the coset space G/H which is by now familiar.

Let us investigate the fundamental properties of the topology of orbit spaces.

Proposition 1.54 *The projection map $\pi: X \to X/G$ is a continuous open map.*

Proof It follows by definition of the quotient topology that the projection π is continuous. If $O \subset X$ is open, then GO is open by Lemma 1.44. Thus $\pi^{-1}(\pi(O)) = GO$ is open for O open in X and, by definition, this means that $\pi(O)$ is open in X/G. Thus $\pi: X \to X/G$ is a continuous open map. □

Generally a quotient space of a space satisfying the first (resp. second) countability axiom does not necessarily satisfy the first (resp. second) countability axiom. However, such relations hold with respect to a G-space and its orbit space. To see this, we first prepare the following lemma, which will also be used subsequently.

Lemma 1.55 *Let X and Y be topological spaces and $\pi\colon X \to Y$ be a surjective continuous open map.*

(1) *If X satisfies the first countability axiom, then so does Y.*

(2) *If X satisfies the second countability axiom, then so does Y.*

Proof Since the proof of (1) is similar to that of (2), we prove (2). Let $\{O_n\}$ be a countable open base for X. Then we show that $\{\pi(O_n)\}$ is a countable open base for Y. Since π is an open map, $\pi(O_n)$ is open in Y. If $O \subset Y$ is open, then $\pi^{-1}(O)$ can be written as

$$\pi^{-1}(O) = \bigcup_{n_i} O_{n_i}$$

by assumption, where $\{O_{n_i}\}$ is a subset of $\{O_n\}$. Since π is surjective, we have

$$O = \pi(\pi^{-1}(O)) = \pi\left(\bigcup_{n_i} O_{n_i}\right) = \bigcup_{n_i} \pi(O_{n_i}),$$

which means that $\{\pi(O_n)\}$ is a countable open base for Y. □

Corollary 1.56 *Let X be a G-space. If X satisfies the first (resp. second) countability axiom, then the orbit space X/G satisfies the first (resp. second) countability axiom as well.*

Proof It follows from Proposition 1.54 that the projection $\pi\colon X \to X/G$ is a surjective continuous open map. Hence Corollary 1.56 is immediate from Lemma 1.55. □

Example 3 Let us consider the orbit space of Example 4 in Section 1.4. Although the orbit space $T^2/\mathbb{R}$ is an infinite set, it has the trivial topology. To see this, let O be a non-empty open set of $T^2/\mathbb{R}$. Then $\pi^{-1}(O)$ is a non-empty open set of T^2. Since every orbit is dense in T^2, it intersects $\pi^{-1}(O)$. Since $\pi^{-1}(O)$ is the union of some orbits, this means that every orbit is included in $\pi^{-1}(O)$. Hence we can conclude that $\pi^{-1}(O) = T^2$. It follows that $O = T^2/\mathbb{R}$. Thus an open set of $T^2/\mathbb{R}$ is either empty or $T^2/\mathbb{R}$. That is, $T^2/\mathbb{R}$ has the trivial topology.

This example shows that the topology of the orbit space is not so interesting in general. Accordingly we assume that G is compact, so that the orbit space has reasonably desirable properties, as we see in the following.

First we shall prepare a well-known lemma.

Let X and Y be topological spaces. A map $f: X \to Y$ is called *proper*, if $f^{-1}(C)$ is compact for every compact set $C \subset Y$.

Lemma 1.57 *Let X and Y be topological spaces and $f: X \to Y$ be a closed map. If $f^{-1}(y)$ is compact for each $y \in Y$, then f is proper.*

Proof Given a compact set $C \subset Y$, let $\{U_s \mid s \in S\}$ be an arbitrary open covering of $f^{-1}(C)$. For each $y \in C$, $\{U_s \mid s \in S\}$ is also an open covering of $f^{-1}(y)$ so that there is a finite subset $S_y \subset S$ of indices with

$$f^{-1}(y) \subset \cup \{U_s \mid s \in S_y\}$$

since $f^{-1}(y)$ is compact. Now we put

$$U_y = \cup \{U_s \mid s \in S_y\}, \qquad V_y = Y - f(X - U_y).$$

Then $y \in V_y$ and V_y is an open set since f is a closed map. We now claim that

$$f^{-1}(V_y) \subset U_y,$$

because for each $x \in f^{-1}(V_y)$, we have

$$f(x) \in V_y = Y - f(X - U_y),$$

and hence $x \notin X - U_y$, which means that $x \in U_y$.

Considering the above for each $y \in C$, we have an open covering

$$C \subset \bigcup_{y \in C} V_y.$$

Since C is compact, there is a finite subset $\{y_1, \ldots, y_n\}$ of C with

$$C \subset \bigcup_{i=1}^{n} V_{y_i}.$$

Thus we get a finite subcovering

$$f^{-1}(C) \subset \bigcup_{i=1}^{n} f^{-1}(V_{y_i}) \subset \bigcup_{i=1}^{n} U_{y_i} = \cup \left\{ U_s \mid s \in \bigcup_{i=1}^{n} S_{y_i} \right\},$$

which means that $f^{-1}(C)$ is compact. □

Proposition 1.58 *Let G be a compact topological group and X a G-space. Then the following hold.*

(1) *$\pi: X \to X/G$ is a closed map.*

(2) *If X is Hausdorff, then so is the orbit space X/G.*

(3) *$\pi: X \to X/G$ is proper.*

(4) *X is compact $\Leftrightarrow X/G$ is compact.*

(5) *X is locally compact $\Leftrightarrow X/G$ is locally compact.*

Proof (1) Given a subset $A \subset X$, we remark first that

$$GA = \pi^{-1}(\pi(A)) = (\pi^{-1}(\pi(A)^c))^c.$$

If A is closed in X, then GA is closed by Lemma 1.44, so that $\pi^{-1}(\pi(A)^c)$ is open. It follows by definition of the quotient topology that $\pi(A)^c$ is open as well. Hence $\pi(A)$ is closed in X/G.

(2) Any two different points of X/G can be written as $\pi(x), \pi(y)$ for some $x, y \in X$ with

$$G(x) \cap G(y) = \varnothing.$$

Since G is compact, $G(x)$ and $G(y)$ are also compact. If X is Hausdorff, it follows from the general theory of point set topology that there exist open sets U and V of X with

$$G(x) \subset U, \qquad G(y) \subset V, \qquad U \cap V = \varnothing.$$

Then we have

$$\overline{U} \cap G(y) \subset \overline{U} \cap V = \varnothing,$$

which means that

$$\pi(y) \notin \pi(\overline{U}).$$

Thus we have

$$\pi(x) \in \pi(U), \qquad \pi(y) \in \pi(\overline{U})^c, \qquad \pi(U) \cap \pi(\overline{U})^c = \varnothing.$$

In view of Proposition 1.54, we see that $\pi(U)$ is open in X/G. On the other hand, it follows from (1) above that $\pi(\overline{U})$ is closed, that is, $\pi(\overline{U})^c$ is open. Thus $\pi(U)$ and $\pi(\overline{U})^c$ are disjoint open sets separating $\pi(x)$ and $\pi(y)$.

(3) On account of (1) above, the projection $\pi: X \to X/G$ is closed. Moreover each orbit $G(x)$ $(= \pi^{-1}(\pi(x)))$ is compact. Hence the assumption of Lemma 1.57 is satisfied so that Lemma 1.57 yields (3).

(4) $\Rightarrow$ If X is compact, its continuous image X/G is also compact.

$\Leftarrow$ If X/G is compact, $\pi^{-1}(X/G) = X$ is also compact by (3) above.

(5) $\Rightarrow$ Given $y \in X/G$, choose an arbitrary point $x \in \pi^{-1}(y)$. If X is locally compact, then there exist an open neighbourhood U of x and a compact set C including U. Then

$$y = \pi(x) \in \pi(U) \subset \pi(C).$$

Since $\pi(U)$ is open and $\pi(C)$ is compact, X/G is locally compact.

$\Leftarrow$ Suppose that X/G is locally compact. Then for any $x \in X$, there exist an open neighbourhood U of $\pi(x)$ and a compact set C including U. It follows that

$$x \in \pi^{-1}(U) \subset \pi^{-1}(C).$$

Since $\pi^{-1}(U)$ is open and $\pi^{-1}(C)$ is compact by (3), X is locally compact. □

Proposition 1.59 *Let H be a closed normal subgroup of a topological group G and X a G-space. Then the G-action on X induces a G/H-action on the orbit space X/H and there is a natural homeomorphism*

$$X/G \to (X/H)/(G/H).$$

Proof Since H is a closed normal subgroup of G, it follows from Proposition 1.20 that G/H is a topological group. Evidently the map $\varphi\colon G \times X \to X$ of the G-action induces a map $\bar{\varphi}\colon G/H \times X/H \to X/H$ so that we have the following commutative diagram:

$$\begin{array}{ccc} G \times X & \xrightarrow{\varphi} & X \\ \Big\downarrow{\scriptstyle \pi_1 \times \pi_2} & & \Big\downarrow{\scriptstyle \pi_2} \\ G/H \times X/H & \xrightarrow{\bar{\varphi}} & X/H \end{array}$$

Here π_i $(i = 1, 2)$ denote the natural projection maps. In view of Lemma 1.19 and Proposition 1.54, both π_i are surjective continuous open maps. Hence $\pi_1 \times \pi_2$ is also a surjective continuous open map by Lemma 1.13. We now apply Lemma 1.15 to the diagram above and deduce that $\bar{\varphi}$ is continuous. Hence $\bar{\varphi}$ gives a continuous G/H-action on X/H.

Next consider the following commutative diagram:

$$\begin{array}{ccc} X & \xrightarrow{\pi_2} & X/H \\ \Big\downarrow{\scriptstyle \pi_3} & & \Big\downarrow{\scriptstyle \pi_4} \\ X/G & \xrightarrow{f} & (X/H)/(G/H) \end{array}$$

Here π_i are all projection maps and f is given by $f(G(x)) = (G/H)(H(x))$. Obviously f is bijective. Since the π_i are all surjective continuous open maps, it follows from Lemma 1.15 that f is continuous. Finally we show that f^{-1} is continuous. Let O be an open set of X/G. It follows from the commutative diagram above and the surjectivity of π_3 that

$$f(O) = f \cdot \pi_3(\pi_3^{-1}(O)) = \pi_4 \cdot \pi_2(\pi_3^{-1}(O)).$$

Since both π_2 and π_4 are open maps, $f(O)$ is open, that is, f^{-1} is continuous. □

Proposition 1.60 *Let H be a closed subgroup of a topological group G and $N(H)$ a normalizer of H in G. Then the map*

$$\bar{\varphi}\colon N(H)/H \times G/H \to G/H$$

defined by $\bar{\varphi}(nH, gH) = gn^{-1}H$ *for* $n \in N(H)$, $g \in G$ *gives a free* $N(H)/H$-*action on* G/H *and its orbit space is canonically homeomorphic to* $G/N(H)$.

Proof Observe that H is a closed normal subgroup of $N(H)$ and that $N(H)$ acts on G by

$$\varphi(n, g) = gn^{-1} \quad \text{for} \quad n \in N(H), g \in G.$$

As remarked in Example 2, G/H is nothing but the orbit space of the restricted H-action of this $N(H)$-action. Accordingly replacing X with G, G with $N(H)$ and H with the same H in Proposition 1.59, we have that the $N(H)$-action φ on G induces the $N(H)/H$-action $\bar{\varphi}$ on G/H which is given by

$$\bar{\varphi}(nH, gH) = gH(nH)^{-1} = gn^{-1}H.$$

Next we show that the action $\bar{\varphi}$ is free. Suppose now that some $g \in G$ and $n \in N(H)$ satisfy

$$\bar{\varphi}(nH, gH) = gH.$$

Then we have

$$gH = \bar{\varphi}(nH, gH) = gn^{-1}H,$$

hence $H = n^{-1}H$. Consequently n lies in H, which means that the action $\bar{\varphi}$ is free. Furthermore it follows from Proposition 1.59 that the orbit space of this action is canonically homeomorphic to $G/N(H)$. □

1.7 Homogeneous spaces and equivariant maps between them

A G-space X is said to be a *homogeneous space* if the G-action on X is transitive. Namely we say X is a homogeneous space if there is precisely one orbit.

Example 1 A left coset space G/H of a topological group G by a subgroup H together with the action of G by left translation is a typical example of a homogeneous space (see Example 2 in Section 1.4).

Proposition 1.61 *If* G *is a compact topological group and if* X *is a Hausdorff homogeneous space of* G, *then* X *is* G-*homeomorphic to the homogeneous space* G/G_x *of Example* 1 *above for any* $x \in X$.

Proof Since $G(x) = X$ by assumption, the proposition is nothing but Proposition 1.53. □

Remark Let X be the orbit $G((1, 1))$ in Example 1 in Section 1.6. Although X is a transitive G-space, X is not G-homeomorphic to $G/G_{(1,1)}$ $(=G)$. That is, a homogeneous space is not necessarily G-homeomorphic to G/G_x in general. Since, however, we often assume that G is compact and X is Hausdorff, a homogeneous space usually has the form G/H.

In the rest of this section, we assume that G is compact and spaces are Hausdorff.

Given Hausdorff G-spaces X and Y, what kind of G-maps $f\colon X \to Y$ are there? Since both X and Y can be written as the union of orbits, the investigation of a G-map between homogeneous spaces will provide us with a necessary condition that a G-map should satisfy locally. That is to say, a G-map $f\colon X \to Y$ satisfies

$$f(G(x)) = G(f(x))$$

for any $x \in X$ so that f is locally a G-map

$$f \mid G(x)\colon G(x) \cong G/G_x \to G(f(x)) \cong G/G_{f(x)},$$

where $\cong$ stands for a G-homeomorphism.

First we prepare a lemma.

Lemma 1.62 *Let G be a compact group and H and K be closed subgroups of G. Then the following hold.*

(1) *There exists a G-map $f\colon G/H \to G/K$ if and only if H is conjugate to a subgroup of K.*

(2) *If there exists $a \in G$ with $aHa^{-1} \subset K$, then the map*

$$f_a\colon G/H \to G/K$$

defined by $f_a(gH) = ga^{-1}K$ is a well-defined G-map and every G-map has this form.

(3) $f_a = f_b$ *if and only if* $ab^{-1} \in K$.

Proof (1), (2) Suppose that there exists a G-map $f\colon G/H \to G/K$. Then there is a $a \in G$ anyway with

$$f(H) = a^{-1}K.$$

The technical reason that we write $a^{-1}K$ instead of aK will become clear subsequently. Since f is a G-map, it must satisfy

$$f(gH) = gf(H) = ga^{-1}K \qquad \text{for all } g \in G.$$

Since the above equality makes sense, the following condition is satisfied

$$ga^{-1}K = f(gH) = f(ghH) = gha^{-1}K \qquad \text{for all } h \in H.$$

Hence we have $K = aha^{-1}K$, that is, $aha^{-1} \in K$ for all $h \in H$. Thus we have

$$aHa^{-1} \subset K.$$

Conversely, suppose that $a \in G$ satisfies $aHa^{-1} \subset K$. Then we can pursue the above procedure conversely so that a G-map f_a: $G/H \to G/K$ is well defined by putting

$$f_a(gH) = ga^{-1}K \qquad \text{for } g \in G.$$

The only thing that we have to check is continuity of f_a. It follows from the definition of f_a that the following diagram commutes:

$$\begin{array}{ccc} G & \xrightarrow{R_{a^{-1}}} & G \\ \Big\downarrow{\scriptstyle \pi} & & \Big\downarrow{\scriptstyle \pi} \\ G/H & \xrightarrow{f_a} & G/K \end{array}$$

Since the π are surjective continuous open maps, the continuity of f_a is immediate from Lemma 1.15.

(3) The result follows from the following equivalences:

$$\begin{aligned} ga^{-1}K = f_a(gH) = f_b(gH) = gb^{-1}K &\Leftrightarrow a^{-1}K = b^{-1}K \\ &\Leftrightarrow ab^{-1}K = K \\ &\Leftrightarrow ab^{-1} \in K. \end{aligned}$$ □

Given closed subgroups H and K of a compact group G, we put

$$N(H, K) = \{a \in G \mid aHa^{-1} \subset K\}.$$

Lemma 1.63 *$N(H, K)$ is a closed subset of G, and hence a compact set.*

Proof Given an arbitrary element b of the closure $\overline{N(H, K)}$ of $N(H, K)$, let $\{a_\lambda\}$ be a net in $N(H, K)$ converging to b. For any $h \in H$, it follows from continuity of the product that

$$\lim_\lambda a_\lambda h a_\lambda^{-1} = bhb^{-1}.$$

Since $a_\lambda h a_\lambda^{-1} \in K$ and K is closed, we can assert $bhb^{-1} \in K$.

Since this is true for arbitrary $h \in H$, we have $bHb^{-1} \subset K$, which means that $b \in N(H, K)$ by definition. Hence we have

$$\overline{N(H, K)} \subset N(H, K).$$

Thus $N(H, K)$ is a closed subset of G, and hence a compact set as well. □

Lemma 1.64 *If we define a K-action on G by $(k, g) \mapsto kg$, then $N(H, K)$ is a K-invariant subset of G.*

Proof For any $k \in K$ and $a \in N(H, K)$, we have

$$kaH(ka)^{-1} = k(aHa^{-1})k^{-1} \subset kKk^{-1} = K,$$

that is, $ka \in N(H, K)$. □

Lemma 1.65 *The orbit space $N(H, K)/K$ of the K-action on $N(H, K)$ in Lemma 1.64 is a compact Hausdorff space.*

Proof Since K is a closed subgroup of a compact group G, K is compact. Since G is a compact Hausdorff space, its closed subset $N(H, K)$ is also a compact Hausdorff space. Hence the orbit space $N(H, K)/K$ is a compact Hausdorff space by Proposition 1.58. □

We denote by

$$\mathrm{Map}^G(G/H, G/K)$$

the space of G-maps $G/H \to G/K$ endowed with the compact-open topology. That is, $\mathrm{Map}^G(G/H, G/K)$ has the relative topology of the mapping space $\mathrm{Map}(G/H, G/K)$.

Proposition 1.66 *Consider the K-action on $N(H, K)$ in Lemma 1.64. The orbit of $a \in N(H, K)$ under K is denoted by $K(a)$ as in Section 1.1. Then the map*

$$F: N(H, K)/K \to \mathrm{Map}^G(G/H, G/K)$$

defined by $F(K(a)) = f_a$ is a homeomorphism. Here $f_a \in \mathrm{Map}^G(G/H, G/K)$ is given by

$$f_a(gH) = ga^{-1}K.$$

Proof First we show that F is well defined. For any $k \in K$, $a \in N(H, K)$ and $g \in G$, we have

$$f_{ka}(gH) = g(ka)^{-1}K = ga^{-1}k^{-1}K = ga^{-1}K = f_a(gH).$$

That is, $f_{ka} = f_a$. This means that F is well defined. Moreover it is an immediate consequence of Lemma 1.62 that F is bijective. Hence it suffices to show that both F and F^{-1} are continuous. To see this, consider the map

$$F': N(H, K)/K \times G/H \to G/K$$

defined by

$$F'(K(a), gH) = ga^{-1}K$$

and the following commutative diagram

$$\begin{array}{ccc} N(H, K) \times G & \xrightarrow{F_1} & G \\ \downarrow{\scriptstyle \pi \times \pi} & & \downarrow{\scriptstyle \pi} \\ N(H, K)/K \times G/H & \xrightarrow{F'} & G/K \end{array}$$

Here F_1 is the map defined by $F_1(a, g) = ga^{-1}$, which is obviously continuous. Since π is continuous and $\pi \times \pi$ is a surjective open map, it follows from Lemma 1.15 that F' is continuous. We now apply Theorem 1.24 to deduce that F is continuous.

Next we show that F^{-1} is continuous. Since K is a closed subgroup of G, G/K is a Hausdorff space by Lemma 1.19. Accordingly $\mathrm{Map}(G/H, G/K)$ and its subspace $\mathrm{Map}^G(G/H, G/K)$ are also Hausdorff by Lemma 1.25. In view of Lemma 1.65, $N(H, K)/K$ is compact. Hence it follows from Lemma 1.16 that F is a closed map, that is, F^{-1} is continuous. □

Proposition 1.67 *The correspondence $K(a) \mapsto a^{-1}K$ gives a homeomorphism*

$$F: N(H, K)/K \to (G/K)^H$$

where $(G/K)^H$ denotes the fixed point set of H on G/K of Example 2 in Section 1.4.

Proof First we show that if $a \in N(H, K)$, then $a^{-1}K$ lies in $(G/K)^H$. Since $aha^{-1} \in K$ for $a \in N(H, K)$ and $h \in H$, we have $aha^{-1}K = K$. Hence $ha^{-1}K = a^{-1}K$ for all $h \in H$, which means that $a^{-1}K \in (G/K)^H$.

For $a, b \in N(H, K)$, we have

$$K(a) = K(b) \Leftrightarrow \text{there exists } k \in K \text{ with } b = ka$$
$$\Leftrightarrow a^{-1}K = b^{-1}K.$$

This means that F is well defined and injective.

Observe now that for $g \in G$,

$$gK \in (G/K)^H \Leftrightarrow g^{-1} \in N(H, K).$$

Thus, given $g \in G$ with $gK \in (G/K)^H$, we have $g^{-1} \in N(H, K)$ and $F(K(g^{-1})) = gK$, that is, F is surjective.

Next we show that F is continuous. To see this, consider the following commutative diagram:

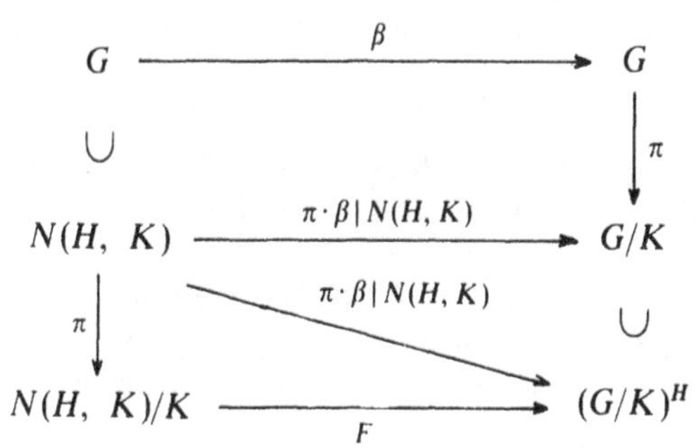

Since β (see Section 1.2) and π are continuous, $\pi \cdot \beta | N(H, K)$ is also continuous. Since $\pi: N(H, K) \to N(H, K)/K$ is a surjective open map, it follows from Lemma 1.15 that F is continuous.

Finally we show that F^{-1} is continuous. In view of Lemma 1.19, G/K is Hausdorff so that its subspace $(G/K)^H$ is also Hausdorff. On the other hand, since $N(H, K)/K$ is compact by Lemma 1.65, it follows from Lemma 1.16 that F is a closed map, that is, F^{-1} is continuous. □

Corollary 1.68 *The three spaces* $N(H, K)/K$, $\mathrm{Map}^G(G/H, G/K)$, *and* $(G/K)^H$ *are homeomorphic to each other.*

Lemma 1.69 *Given an element* g *of a compact group* G, *we put* $A = \{g^n \mid n = 0, 1, 2, \ldots.\}$. *Then the closure* $\bar{A}$ *of* A *is a subgroup of* G.

Proof Put $B = \{g^n \mid n \in \mathbb{Z}\}$ which is a subgroup of G. Then $\bar{B}$ is also a subgroup by Proposition 1.9. Suppose first that the identity e is isolated in $\bar{B}$. Then it follows from Corollary 1.8 that any point of $\bar{B}$ is isolated in $\bar{B}$. That is, $\bar{B}$ is a discrete subgroup of G. Since $\bar{B}$ is closed in a compact group, it is compact. Thus $\bar{B}$ is necessarily finite so that there is a positive integer n with $g^n = e$. That is, A is a finite cyclic group. Since $\bar{A} = A$ in this case, $\bar{A}$ is a subgroup of G.

Suppose next that e is not isolated in $\bar{B}$. Then for any symmetric open neighbourhood U of e in G, there are infinitely many integers n with $g^n \in U$, because if the set

$$\{n \in \mathbb{Z} \mid g^n \in U\}$$

is finite, then the set $(U - B) \cup \{e\}$ is also an open neighbourhood of e. As a consequence, e is isolated in B and hence in $\bar{B}$, which contradicts the assumption. If $g^n \in U$, then $g^{-n} \in U^{-1} = U$ since U is symmetric. After all, there are infinitely many positive integers n with $g^n \in U$.

We now fix an arbitrary positive integer i. By the observation above, there is a positive integer n such that $n > i$ and $g^n \in U$. Then we have

$$g^{n-i} \in g^{-i} U \cap A,$$

since $n - i > 0$. In particular, $g^{-i} U \cap A \neq \varnothing$. Since $g^{-i} U$ form an open neighbourhood base at g^{-i}, we have that $g^{-i} \in \bar{A}$. Thus we have

$$A \subset B \subset \bar{A} \subset \bar{B}.$$

Since $\bar{A}$ is a closed set including B, we have $\bar{A} \supset \bar{B}$. After all, $\bar{A}$ coincides with $\bar{B}$ which is a subgroup of G as we remarked at the beginning. □

Remark In Lemma 1.69, the compactness assumption of G is indispensable. For instance, let G be the additive group $\mathbb{R}$ of real numbers and A be $\{0, 1, 2, \ldots\}$. Then $\bar{A} = A$ and A is not a subgroup of $\mathbb{R}$.

Corollary 1.70 *For a closed subgroup* H *of a compact group* G, *we have*

$$N(H, H) = N(H).$$

Proof The inclusion $N(H) \subset N(H, H)$ is obvious. Conversely, given $g \in N(H, H)$, we put

$$A = \{g^n \mid n = 0, 1, 2, \ldots\}.$$

We now consider the map

$$\psi: G \times G \to G$$

defined by

$$\psi(g_1, g_2) = g_1 g_2 g_1^{-1}.$$

Then it follows from the assumption that

$$\psi(A \times H) \subset H.$$

Since ψ is continuous, we have

$$\psi(\bar{A} \times H) = \psi(\bar{A} \times \bar{H}) = \psi(\overline{A \times H}) \subset \bar{H} = H.$$

Note that $g^{-1} \in \bar{A}$ by Lemma 1.69 so that we have

$$g^{-1}Hg = \psi(g^{-1}, H) \subset H,$$

that is, $H \subset gHg^{-1}$. On the other hand, $gHg^{-1} \subset H$ by the assumption $g \in N(H, H)$. Consequently we have $gHg^{-1} = H$, that is, $g \in N(H)$. Thus the reverse inclusion $N(H) \supset N(H, H)$ holds. □

The group of all G-homeomorphisms of G/H onto itself with the compact-open topology is denoted by $\mathrm{Homeo}^G(G/H)$.

Proposition 1.71 *If G is a compact group and H is a closed subgroup of G, then $\mathrm{Homeo}^G(G/H)$ coincides with $\mathrm{Map}^G(G/H, G/H)$ and both are topological subgroups of the topological group $\mathrm{Homeo}(G/H)$ in Theorem 1.26.*

Proof Obviously, the inclusion

$$\mathrm{Homeo}^G(G/H) \subset \mathrm{Map}^G(G/H, G/H)$$

holds. Conversely, for any $f \in \mathrm{Map}^G(G/H, G/H)$, there is $a \in N(H, H)$ with $f = f_a$ by Lemma 1.62. Since $N(H, H) = N(H)$ by Corollary 1.70, a^{-1} is also an element of $N(H, H)$ and hence $f_{a^{-1}} \in \mathrm{Map}^G(G/H, G/H)$. Evidently we have

$$f_{a^{-1}} \cdot f_a = f_a \cdot f_{a^{-1}} = \mathrm{id}.$$

As a consequence $f = f_a$ lies in $\mathrm{Homeo}^G(G/H)$ so that the reverse inclusion

$$\mathrm{Homeo}^G(G/H) \supset \mathrm{Map}^G(G/H, G/H)$$

holds. Since both $\mathrm{Homeo}^G(G/H)$ and $\mathrm{Homeo}(G/H)$ are endowed with the compact-open topology, $\mathrm{Homeo}^G(G/H)$ is a topological subgroup of $\mathrm{Homeo}(G/H)$. □

Corollary 1.72 *Let G be a compact group, and H and K be closed subgroups of G. If there exist G-maps*

$$f: G/H \to G/K, \qquad f': G/K \to G/H,$$

then both f and f' are G-homeomorphisms, and H and K are conjugate in G.

Proof It follows from Proposition 1.71 that

$$f' \cdot f \in \mathrm{Map}^G(G/H, G/H) = \mathrm{Homeo}^G(G/H)$$
$$f \cdot f' \in \mathrm{Map}^G(G/K, G/K) = \mathrm{Homeo}^G(G/K).$$

Hence both f and f' are bijective G-maps. To see that f is a G-homeomorphism, it suffices to show that f^{-1} is continuous. Since $f' \cdot f \in \mathrm{Homeo}^G(G/H)$, $(f' \cdot f)^{-1}$ is continuous. Accordingly the equality $f^{-1} = (f' \cdot f)^{-1} \cdot f'$ implies that f^{-1} is also continuous. That is, f is a G-homeomorphism. Similarly f' is a G-homeomorphism.

Next we show that H and K are conjugate. It follows from Lemma 1.62 (1) that there are $a, b \in G$ with

$$aHa^{-1} \subset K, \qquad bKb^{-1} \subset H.$$

Combining these, we have $baH(ba)^{-1} \subset H$. We now apply Corollary 1.70 to deduce $baH(ba)^{-1} = H$ so that $aHa^{-1} = b^{-1}Hb$. Since $aHa^{-1} \subset K \subset b^{-1}Hb$, we can conclude that $aHa^{-1} = K$. □

Theorem 1.73 *Let G be a compact group and H a closed subgroup of G. Then the map*

$$F: N(H)/H \to \mathrm{Homeo}^G(G/H)$$

defined by $H(a)\ (= Ha = aH) \mapsto f_a$ *gives an isomorphism of topological groups where* f_a *is the map defined by* $f_a(gH) = ga^{-1}H$.

Proof By combining Propositions 1.66 and 1.71 and Corollary 1.70, we see that F is a homeomorphism. Accordingly it suffices to show that F is a homomorphism of groups. Given $a, b \in N(H)$ and $g \in G$, we compute

$$f_{ab}(gH) = g(ab)^{-1}H = gb^{-1}a^{-1}H$$
$$= f_a(gb^{-1}H) = f_a \cdot f_b(gH),$$

that is, we have

$$f_{ab} = f_a \cdot f_b.$$

It turns out that

$$F(aHbH) = F(abH) = f_{ab} = f_a \cdot f_b = F(aH) \cdot F(bH),$$

that is, F is a homomorphism. □

Corollary 1.74 *If H is a closed subgroup of a compact group G, then the following four spaces are homeomorphic to each other:*

$$N(H)/H, \qquad \mathrm{Homeo}^G(G/H), \qquad \mathrm{Map}^G(G/H, G/H), \qquad (G/H)^H.$$

Corollary 1.75 *If H is a closed normal subgroup of a compact group G, then the following three spaces are isomorphic as topological groups:*

$$G/H, \qquad \mathrm{Homeo}^G(G/H), \qquad \mathrm{Map}^G(G/H, G/H).$$

1.8 Orbit types and isotropy types

In this section we assume that G is a compact topological group and that all spaces are Hausdorff. We consider the family F of all Hausdorff homogeneous spaces of G. We put a relation $\sim$ on F as follows: for $X, Y \in F$,

$$X \sim Y \Leftrightarrow \text{there exists a } G\text{-homeomorphism } f: X \to Y.$$

Then $\sim$ is an equivalence relation. Equivalence classes under this equivalence relation are called *G-orbit types* or simply *types*. We denote the equivalence class including X by type(X) or (X). Moreover (X) is often written as X for brevity. The set $F/\sim$ of equivalence classes is called the *set of G-orbit types*.

According to Proposition 1.61, for any Hausdorff homogeneous space X of G, there exists a closed subgroup H of G such that X is G-homeomorphic to G/H. That is, every type(X) is represented by a coset space G/H.

A partial ordering of the set $F/\sim$ of G-orbit types is given by

$$\mathrm{type}(X) \geq \mathrm{type}(Y) \Leftrightarrow \text{there is a } G\text{-map } f: X \to Y,$$

because if both

$$\mathrm{type}(X) \geq \mathrm{type}(Y) \qquad \text{and} \qquad \mathrm{type}(X) \leq \mathrm{type}(Y)$$

hold, then we have type$(X) =$ type(Y) by Corollary 1.72. In this partially ordered set, type(G) is a maximum and type(G/G) is a minimum.

When a homogeneous space X of G is equivalent to G/H, the conjugacy class (H) of H in G is called the *isotropy type* of X. Generally given a G-space X, a G-orbit which is equivalent to G/H is called an *orbit with the isotropy type* (H) or an *orbit with the orbit type* (G/H).

For closed subgroups H, K of G, an ordering $\leq$ is given by

$$(H) \leq (K) \Leftrightarrow H \text{ is conjugate to a subgroup of } K.$$

This gives a partial ordering of conjugacy classes of closed subgroups of G, because if both

$$(H) \leq (K) \qquad \text{and} \qquad (H) \geq (K)$$

hold, then we have $(H) = (K)$ as shown in the proof of Corollary 1.72.

Proposition 1.75 *The correspondence* $\text{type}(G/H) \mapsto (H)$ *gives an anti-isomorphism from the partially ordered set* $F/\sim$ *to the partially ordered set of conjugacy classes of closed subgroups of* G.

This proposition follows easily from Lemma 1.62.

Given a G-space X, we say that X has *n orbit types* when there are exactly n different orbit types in X.

Given a G-space X and a closed subgroup H of G, we put

$$X(H) = \{x \in X \mid (G_x) = (H)\}$$

which is called *the union of the orbits with isotropy type* (H). Furthermore, we put

$$X(\geq H) = \{x \in X \mid (G_x) \geq (H)\}$$

which is called *the union of the orbits with isotropy type greater than or equal to* (H).

Proposition 1.77 $GX^H = X(\geq H)$.

Proof For any $g \in G$ and $x \in X^H$, we have $G_{gx} = gG_xg^{-1}$ and $G_x \supset H$. Hence $(G_{gx}) \geq (H)$. That is, $gx \in X(\geq H)$. Thus $GX^H \subset X(\geq H)$ holds true.

Conversely, for any $x \in X(\geq H)$, there exists $g \in G$ with

$$G_{gx} = gG_xg^{-1} \supset H.$$

Hence $gx \in X^H$ and $x = g^{-1}(gx) \in GX^H$. Thus $GX^H \supset X(\geq H)$ holds. □

Corollary 1.78 *If* G *is compact and* X *is a Hausdorff* G*-space, then* $X(\geq H)$ *is closed in* X.

Proof Since X^H is closed in X by Theorem 1.46, it follows from Lemma 1.44 that GX^H is closed in X. Therefore this corollary is immediate from Proposition 1.77. □

Generally in an ordered set A, an element a of A is called *maximal* when there is no element greater than a.

Corollary 1.79 *Under the assumption of Corollary* 1.78, *if* (H) *is maximal among the isotropy types which appear in* X, *then* $X(H)$ *is closed.*

Remark $X(H)$ is not closed in X in general. For instance, consider the $\mathbb{Z}_2$ $(= \{\pm 1\})$ action on $X = S^1 = \{z \in \mathbb{C} \mid |z| = 1\}$ defined by $(-1, z) \mapsto \bar{z}$. Then we have

$$X(\{e\}) = S^1(\{+1\}) = S^1 - \{1, -1\}$$

which is not closed. This is because $\{+1\}$ is not maximal among the isotropy types. On the other hand, $\mathbb{Z}_2$ is maximal and we have

$$S^1(\mathbb{Z}_2) = \{1, -1\}$$

which is closed in S^1.

Given a G-space X, we get a decomposition

$$X = \bigcup_{(H)} X(H)$$

which is a disjoint union of subsets of X. Here (H) run over all the conjugacy classes of subgroups of G. As remarked above, each $X(H)$ is not closed in X in general and is not easy to deal with on a case-by-case basis. However, if we consider them in the following order, they become easy to handle. The following two lemmas will be employed when we consider smooth actions on manifolds and decompose manifolds with respect to orbit types (see Chapter 4).

Lemma 1.80 *Given finitely many conjugacy classes $(H_1), \ldots, (H_n)$ of closed subgroups of G, we can arrange them such that*

$$(H_i) \geq (H_j) \Rightarrow i \leq j.$$

Proof First we show that there is a maximal element. If (H_1) is not maximal, there is a class (H_{i_1}) which is greater than (H_1). If (H_{i_1}) is not maximal, there is a class (H_{i_2}) which is greater than (H_{i_1}). If we continue this argument, there is j such that no classes are greater than (H_{i_j}), since $\{(H_i)\}$ is finite. Thus (H_{i_j}) is maximal.

We now choose an arbitrary maximal element from $\{(H_i)\}$, which we denote by (K_1). Next we choose an arbitrary maximal element from $\{(H_i)\} - (K_1)$, which we denote by (K_2). If we choose (K_i) inductively as above, this arrangement clearly satisfies the required condition. □

Fig. 1.3 illustrates an example of arrangements.

For our purposes a partially ordered set P is represented by a diagram in general as follows. For $a, b \in P$, we say that *a covers b* if $a > b$, and $a > x \geq b$ implies $x = b$. Elements of P are denoted by dots and we arrange them such that if $a > b$, then a is situated above b. Furthermore we connect two dots a and b by a segment if and only if a covers b.

Lemma 1.81 *Let G be a compact group and X a Hausdorff G-space. Suppose that the set $\{(H_i)\}$ of isotropy types appearing in X is finite and arrange them as*

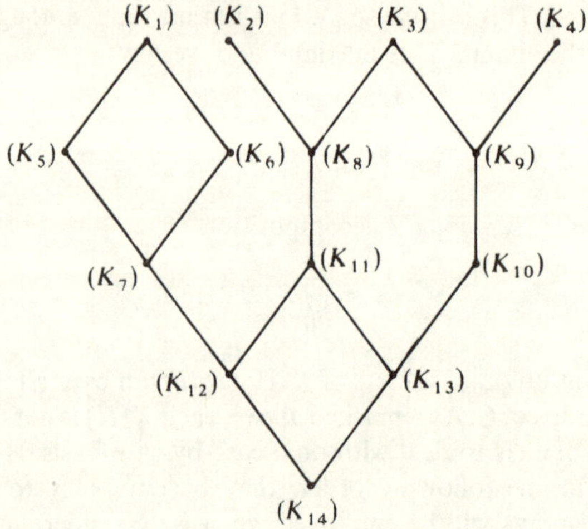

Fig. 1.3 An arrangement of finitely many conjugacy classes.

in Lemma 1.80. *If we put*

$$X_i = X(H_1) \cup \ldots \cup X(H_i),$$

then X_i *is closed in* X.

Proof By way of the arrangement, one easily verifies that

$$X_i = X(\geq H_1) \cup \ldots \cup X(\geq H_l).$$

Since each $X(\geq H_j)$ is closed by Corollary 1.78, X_i is closed in X as well. □

1.9 Induced transformation groups

Let G be a topological group and H a subgroup of G. Given an H-space X, we can construct naturally a G-space associated with X as follows. This construction is universal in some sense and corresponds to that of the induced representation in representation theory.

Construction Let $\varphi: H \times X \to X$ be an H-action on X. Then an H-action on $G \times X$ is defined by $(h, (g, x)) \mapsto (gh^{-1}, \varphi(h, x))$. The orbit space of this H-action is denoted by $G \underset{H}{\times} X$ and $\pi: G \times X \to G \underset{H}{\times} X$ denotes the projection. The H-orbit of (g, x) will be denoted by $[g, x]$. Define a G-action

$$\tilde{\varphi}: G \times G \underset{H}{\times} X \to G \underset{H}{\times} X$$

on $G \underset{H}{\times} X$ by putting

$$\tilde{\varphi}(g', [g, x]) = [g'g, x].$$

It is easy to see that this gives a continuous G-action.

The G-action $\tilde{\varphi}$ on $G \underset{H}{\times} X$ obtained in this way is called the *induced G-action* and the triple $(G \underset{H}{\times} X, G, \tilde{\varphi})$ is called the *induced transformation group* of (X, H, φ). The space $G \underset{H}{\times} X$, together with the action $\tilde{\varphi}$, is called the *induced G-space.*

We define a map $i_e: X \to G \times X$ by $i_e(x) = (e, x)$ and another map $i: X \to G \underset{H}{\times} X$ by the composition of the following maps

$$i: X \xrightarrow{i_e} G \times X \xrightarrow{\pi} G \underset{H}{\times} X.$$

Lemma 1.82 *If H is a compact subgroup of a topological group G, then i is an H-homeomorphism onto its image.*

Proof Since both i_e and π are continuous, i is also continuous. For any $h \in H$, we have

$$i(hx) = [e, hx] = [h, x] = h[e, x] = hi(x).$$

Hence i is an H-map. Next we show that i is injective. Suppose that $i(x) = i(y)$ for $x, y \in X$. It follows by definition that there exists an element h of H with $(eh^{-1}, hx) = (e, y)$. Hence $h = e$ and $x = y$.

Next we show that i is a closed map. Since G is Hausdorff, $i_e(X)$ is closed in $G \times X$. Accordingly for any closed set A of X, $i_e(A)$ is closed in $G \times X$, since $i_e(A)$ is closed in $i_e(X)$ in general. That is, i_e is a closed map. In view of Proposition 1.58, π is also closed, since H is compact. Thus we have shown that $i = \pi \cdot i_e$ is a closed map. □

Remark The projection $G \times X \to G$ induces a projection $p: G \underset{H}{\times} X \to G/H$. As we will see later, if G is a Lie group, then

$$p: G \underset{H}{\times} X \to G/H$$

gives a fibre bundle with fibre X. Hence, the space $G \underset{H}{\times} X$ is comparatively easy to deal with.

Theorem 1.83 (Universality) *The induced transformation group $(G \underset{H}{\times} X, G, \tilde{\varphi})$ has the following property. For any G-space Y and any H-map $f: X \to Y$,* *there is a unique G-map*

$$\tilde{f}: G \times_H X \to Y$$

such that the following diagram commutes:

$$\begin{array}{ccc} X & \xrightarrow{\quad i \quad} & G \times_H X \\ & \searrow^{f} & \downarrow \tilde{f} \\ & & Y \end{array}$$

Proof Define $\tilde{f}$ by putting

$$\tilde{f}([g, x]) = gf(x) \qquad \text{for} \qquad g \in G, x \in X.$$

First we show that $\tilde{f}$ is well defined. For any $h \in H$, we have

$$\begin{aligned} \tilde{f}([gh^{-1}, hx]) &= gh^{-1}f(hx) \\ &= gh^{-1}hf(x) \qquad \text{(since } f \text{ is an } H\text{-map)} \\ &= gf(x). \end{aligned}$$

That is, $\tilde{f}$ is defined independently of the choice of representatives. The commutativity of the diagram is seen by

$$\tilde{f} \cdot i(x) = \tilde{f}([e, x]) = ef(x) = f(x),$$

and $\tilde{f}$ is G-equivariant, since

$$\begin{aligned} \tilde{f}(g'[g, x]) = \tilde{f}([g'g, x]) &= (g'g)f(x) \\ &= g'(gf(x)) = g'\tilde{f}([g, x]). \end{aligned}$$

We now show that $\tilde{f}$ is continuous. If we denote by

$$\psi: G \times Y \to Y$$

the G-action on Y, the following diagram is commutative:

$$\begin{array}{ccc} G \times X & \xrightarrow{\quad \mathrm{id} \times f \quad} & G \times Y \\ \pi \downarrow & & \downarrow \psi \\ G \times_H X & \xrightarrow{\quad \tilde{f} \quad} & Y \end{array}$$

Since $\mathrm{id} \times f$ and ψ are continuous and π is a surjective open map, it follows from Lemma 1.15 that $\tilde{f}$ is continuous. Thus the existence of $\tilde{f}$ is shown.

Next we show the uniqueness of $\tilde{f}$. Let

$$\tilde{f}': G \underset{H}{\times} X \to Y$$

be another G-map with $\tilde{f}' \cdot i = f$. Then

$$\begin{aligned} \tilde{f}'([g, x]) &= \tilde{f}'(g[e, x]) \\ &= g\tilde{f}'([e, x]) \qquad \text{(since } \tilde{f}' \text{ is a } G\text{-map)} \\ &= g\tilde{f}' \cdot i(x) = gf(x) = \tilde{f}([g, x]). \end{aligned}$$

That is, $\tilde{f}' = \tilde{f}$ holds. □

Theorem 1.84 (Uniqueness) *Given a subgroup H of a topological group G and an H-space X, there exist a unique G-space $\tilde{X}$ and a unique injective H-map $i: X \to \tilde{X}$ having the following property.*

If Y is a G-space and $f: X \to Y$ is a continuous H-map, then there is a unique continuous G-map $\tilde{f}$ such that the following diagram commutes:

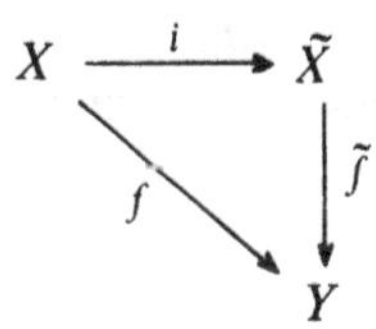

(1.5)

Proof Existence follows from Theorem 1.83 by putting

$$\tilde{X} = G \underset{H}{\times} X, \; i = i \; (i \text{ in Theorem 1.83}).$$

We now show uniqueness. Suppose that

$$i_1: X \to \tilde{X}_1 \qquad \text{and} \qquad i_2: X \to \tilde{X}_2$$

have the property (1.5). It follows from (1.5) that there are two unique G-maps

$$\tilde{i}_2: \tilde{X}_1 \to \tilde{X}_2 \qquad \text{and} \qquad \tilde{i}_1: \tilde{X}_2 \to \tilde{X}_1$$

such that the following diagrams commute:

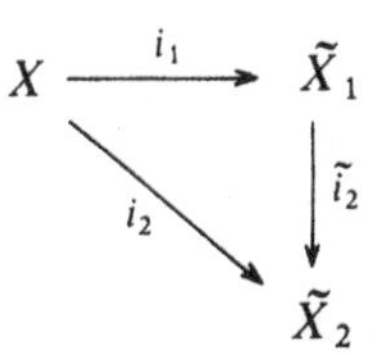

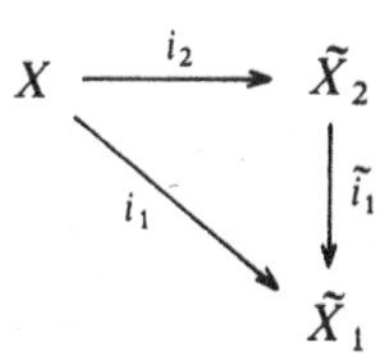

By combining these diagrams in two ways, we get the following commutative diagrams:

$$\begin{array}{ccc} X & \xrightarrow{i_1} & \tilde{X}_1 \\ & \searrow^{i_1} & \downarrow{\scriptstyle \tilde{i}_1 \cdot \tilde{i}_2} \\ & & \tilde{X}_1 \end{array} \qquad\qquad \begin{array}{ccc} X & \xrightarrow{i_2} & \tilde{X}_2 \\ & \searrow^{i_2} & \downarrow{\scriptstyle \tilde{i}_2 \cdot \tilde{i}_1} \\ & & \tilde{X}_2 \end{array}$$

If we replace both $\tilde{i}_1 \cdot \tilde{i}_2$ and $\tilde{i}_2 \cdot \tilde{i}_1$ with the identity map, then the diagrams above commute as well. It follows from the uniqueness of $\tilde{f}$ in property (1.5) that we have

$$\tilde{i}_1 \cdot \tilde{i}_2 = \mathrm{id}, \qquad \tilde{i}_2 \cdot \tilde{i}_1 = \mathrm{id}.$$

Hence both $\tilde{i}_1$ and $\tilde{i}_2$ are G-homeomorphisms. Moreover since i_1 corresponds to i_2 via the G-homeomorphism $\tilde{i}_2$, i is unique in this sense. □

Next we show a transitivity property.

Theorem 1.85 (Transitivity) *Let G be a topological group and H and K be subgroups of G with $H \subset K$. Given an H-space X, the map*

$$f: G \underset{K}{\times} \left(K \underset{H}{\times} X\right) \to G \underset{H}{\times} X$$

defined by

$$f([g, [k, x]]) = [gk, x] \qquad \text{for} \qquad g \in G, k \in K, x \in X$$

is a G-homeomorphism.

Proof It is easy to see that f is a well defined continuous G-map. Conversely, let

$$f': G \underset{H}{\times} X \to G \underset{K}{\times} \left(K \underset{H}{\times} X\right)$$

be the map defined by

$$f'([g, x]) = [g, [e, x]] \qquad \text{for} \qquad g \in G, x \in X.$$

We now show that f' is defined independently of the choice of representatives. If $[g, x] = [g', x']$, then there exists $h \in H$ with $g' = gh^{-1}$, $x' = hx$ by definition. Then

$$\begin{aligned} f'([g', x']) &= f'([gh^{-1}, hx]) = [gh^{-1}, [e, hx]] \\ &= [gh^{-1}, [h, x]] = [gh^{-1}, h[e, x]] \\ &= [g, [e, x]] = f'([g, x]). \end{aligned}$$

That is, f' is well defined. Although continuity of f' is almost clear, we show it for certainty's sake. To see this, consider the following commutative diagram:

$$\begin{array}{ccc}
G \times X & \xrightarrow{\tilde{f}'} & G \times K \times X \\
 & & \downarrow \pi \\
\Big\downarrow \pi & & G \times (K \times_H X) \\
 & & \downarrow \pi \\
G \times_H X & \xrightarrow{f'} & G \times_K (K \times_H X)
\end{array}$$

In the above, $\tilde{f}'$ is the continuous map defined by

$$\tilde{f}'(g, x) = (g, e, x) \qquad \text{for} \qquad g \in G,\ x \in X.$$

Since all the π are surjective continuous open maps, f' is continuous by Lemma 1.15. Moreover we have

$$\begin{aligned}
f \cdot f'([g, x]) &= f([g, [e, x]]) = [g, x], \\
f' \cdot f([g, [k, x]]) &= f'([gk, x]) = [gk, [e, x]] \\
&= [g, k[e, x]] = [g, [k, x]].
\end{aligned}$$

That is,

$$f \cdot f' = \text{id} \qquad \text{and} \qquad f' \cdot f = \text{id}.$$

Thus both f and f' are G-homeomorphisms. □

Corollary 1.86 *Let G be a topological group and $H_1, \ldots, H_n$ be subgroups of G with $H_1 \subset H_2 \subset \cdots \subset H_n$. Then for an H_1-space X, the map*

$$f \colon G \times_{H_n} \left(H_n \times_{H_{n-1}} \cdots \left(H_3 \times_{H_2} \left(H_2 \times_{H_1} X \right) \right) \cdots \right) \to G \times_{H_1} X$$

defined by

$$f([g, [h_n, [h_{n-1}, [\cdots [h_2, x]] \cdots]]]) = [g h_n h_{n-1} \cdots h_2, x]$$

for $g \in G$, $h_i \in H_i$, $x \in X$ is a G-homeomorphism.

Proof Theorem 1.85 covers the case $n = 2$ in this corollary. Hence the corollary is shown by using Theorem 1.85 inductively. □

We now sum up several straightforward remarks as lemmas.

Lemma 1.87 *If X is a G-space, then $G \times_G X$ and X are G-homeomorphic.*

Proof Define a map $f\colon G \times_G X \to X$ by putting $f([g, x]) = gx$. It is shown as usual that f is a well defined continuous G-map. Conversely, if we define a map $f'\colon X \to G \times_G X$ by putting $f'(x) = [e, x]$, then it is easily seen that f' is continuous with

$$f \cdot f' = \mathrm{id} \qquad \text{and} \qquad f' \cdot f = \mathrm{id}.$$

Thus f is a G-homeomorphism. □

Lemma 1.88 *If H is a subgroup of a topological group G and X is a space with trivial H-action, then $G \times_H X$ and $(G/H) \times X$ are G-homeomorphic. In particular if X is one point $*$ with trivial H-action, then $G \times_H *$ and G/H are G-homeomorphic.*

Proof Obvious.

Proposition 1.89 *For subgroups H, K of a topological group G with $H \subset K$, the map*

$$f\colon G \times_K (K/H) \to G/H$$

defined by

$$f([g, kH]) = gkH$$

is a G-homeomorphism.

Proof Applying Theorem 1.84 and Lemma 1.88 to the case $X = *$ (one point), we have

$$\begin{aligned} G \times_K K/H &\cong G \times_K \left(K \times_H * \right) \\ &\cong G \times_H * \cong G/H, \end{aligned}$$

where $\cong$ stands for a G-homeomorphism. It is shown from the definitions of the maps in Theorem 1.85 and Lemma 1.88 that the composition of the G-homeomorphisms above coincides with f. □

Proposition 1.90 *If H is a subgroup of a topological group G and X is an H-space, then the inclusion $i\colon X \to G \times_H X$ defined by $i(x) = [e, x]$ induces a homeomorphism*

$$\bar{i}\colon X/H \to \left(G \times_H X \right) \Big/ G.$$

Proof Since i is an H-map, each H-orbit $H(x)$ of X is mapped to the H-orbit $H([e, x])$ by i. Since $H([e, x])$ is involved in the G-orbit $G([e, x])$, we can put

$$\bar{i}(H(x)) = G([e, x]),$$

and the following diagram commutes:

$$\begin{array}{ccc} X & \xrightarrow{\ i\ } & G \times_H X \\ \downarrow{\scriptstyle \pi'} & & \downarrow{\scriptstyle \pi} \\ X/H & \xrightarrow{\ \bar{i}\ } & (G \times_H X)/G \end{array}$$

Note that i is continuous without assuming that H is compact in Lemma 1.82. Hence we see that $\bar{i}$ is continuous by applying Lemma 1.15 to the diagram above.

On the other hand, the projection $p: G \times X \to X$ induces a map

$$\bar{p}: G \times_H X \to X/H$$

which is continuous by Lemma 1.15. Since the map $\bar{p}$ is given by $\bar{p}([g, x]) = H(x)$, it has the following factorization:

$$\begin{array}{ccc} G \times_H X & \xrightarrow{\ \bar{p}\ } & X/H \\ & {\scriptstyle \pi}\searrow \quad \nearrow{\scriptstyle \bar{\bar{p}}} & \\ & (G \times_H X)/G & \end{array}$$

Hence, appealing to Lemma 1.15 again, we see that $\bar{\bar{p}}$ is continuous as well. Evidently we have

$$\bar{i} \cdot \bar{\bar{p}} = \text{id}, \qquad \bar{\bar{p}} \cdot \bar{i} = \text{id},$$

so that $\bar{i}$ is a homeomorphism. □

Proposition 1.91 *Let K be a subgroup of a topological group G and H a closed normal subgroup of K. Let X be a K-space such that the restricted H-action on X is trivial. Then the map*

$$f: G \times_K X \to (G/H) \times_{K/H} X$$

defined by

$$f([g,x]) = [gH, x]$$

is a G-homeomorphism.

Proof Since H is a closed normal subgroup of K by assumption, K/H is a topological group by Proposition 1.20. Note that $G \times_H X = (G/H) \times X$ by assumption. It follows from Proposition 1.59 that the map $f\colon G \times_K X \to (G/H) \times_{K/H} X$ is a homeomorphism. Moreover f is obviously a G-map. □

Corollary 1.92 *If H is a closed subgroup of a topological group G and X is a G-space, then the following two G-spaces are G-homeomorphic*:

$$G \times_{N(H)} X^H \cong (G/H) \times_{N(H)/H} X^H.$$

Proof Note that H is a closed normal subgroup of the normalizer $N(H)$ of H in G and X^H is an $N(H)$-space such that the restricted H-action on X^H is trivial. Hence if we put $K = N(H)$, the condition of Proposition 1.91 is satisfied so that Proposition 1.91 yields this corollary. □

Lemma 1.93 *Let G be a compact group and X a T_1, G-space. If X has only one orbit type (G/H), then we have $G_x = H$ for any $x \in X^H$.*

Proof Since X has only the orbit type (G/H), there exists $g \in G$ with $G_{gx} = H$. Hence H is a closed subgroup of G by Theorem 1.37. Since $x \in X^H$, we have

$$g^{-1}Hg = g^{-1}G_{gx}g = G_x \supset H.$$

Hence we can conclude that $G_x = H$ by Corollary 1.70. □

Corollary 1.94 *Under the assumption of Lemma* 1.93, *X^H is a free $N(H)/H$-space.*

Proof If $(nH)x = x$ for $x \in X^H$ and $n \in N(H)$, then $nx = x$. Hence $n \in H$ by Lemma 1.93. That is, $N(H)/H$ acts freely on X^H. □

Let G be a topological group and H a closed subgroup of G. For a G-space X, the inclusion $j\colon X^H \to X$ is an $N(H)$-map so that we get the induced G-space $G \times_{N(H)} X^H$ and the continuous G-map

$$\tilde{j}\colon G \times_{N(H)} X^H \to X$$

by Theorem 1.83.

Theorem 1.95 *Let G be a compact group and X a Hausdorff G-space. If X has only one orbit type (G/H), then the map*

$$\tilde{j}: G \underset{N(H)}{\times} X^H \to X$$

is a G-homeomorphism.

Proof It follows from the assumption that each G-orbit in X contains an H fixed point. Hence $\tilde{j}$ is surjective. Suppose next that

$$\tilde{j}([g, x]) = \tilde{j}([g', x']) \qquad \text{for} \qquad g, g' \in G, \qquad x, x' \in X^H.$$

Then $gx = g'x'$ so that $x' = g'^{-1}gx$. In view of Lemma 1.93, we have

$$H = G_{x'} = G_{g'^{-1}gx} = g'^{-1}gG_x(g'^{-1}g)^{-1} = g'^{-1}gH(g'^{-1}g)^{-1}$$

so that $g'^{-1}g \in N(H)$. Accordingly

$$[g', x'] = [g', g'^{-1}gx] = [g'(g'^{-1}g), x] = [g, x],$$

that is, $\tilde{j}$ is injective.

Finally we show that $\tilde{j}^{-1}$ is continuous. Recall that the map $\varphi: G \times X \to X$ of the G-action is closed by Proposition 1.45. Since X^H is closed in X by Theorem 1.46, $G \times X^H$ is closed in $G \times X$. It turns out that the restriction

$$\varphi \mid G \times X^H: G \times X^H \to X$$

of φ is also closed. Observe now that the following diagram

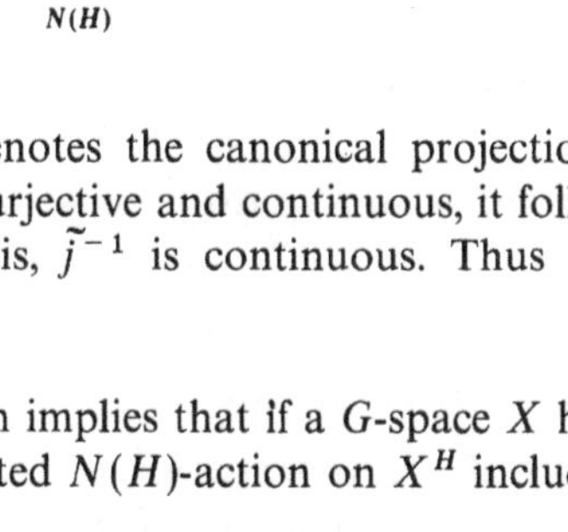

commutes where π denotes the canonical projection. Since $\varphi \mid G \times X^H$ is closed and since π is surjective and continuous, it follows from Lemma 1.18 that $\tilde{j}$ is closed, that is, $\tilde{j}^{-1}$ is continuous. Thus we see that $\tilde{j}$ is a G-homeomorphism. □

Remark This theorem implies that if a G-space X has only one orbit type (G/H), then the restricted $N(H)$-action on X^H includes all the information about the G-space X.

Corollary 1.96 *Under the assumption of Theorem* 1.95, *the inclusion map* $j: X^H \to X$ *induces a homeomorphism*:

$$X^H/N(H) \to X/G.$$

Proof It follows from Theorem 1.95 that the map

$$\bar{\tilde{j}}: \left(G \underset{N(H)}{\times} X^H \right) \Big/ G \to X/G$$

induced by $\tilde{j}$ is a homeomorphism. On the other hand, there is a homeomorphism

$$\tilde{i}: X^H/N(H) \to \left(G \underset{N(H)}{\times} X^H \right) \Big/ G$$

by Proposition 1.90. Hence the composition $\bar{\tilde{j}} \cdot \tilde{i}$ gives the desired homeomorphism. □

Corollary 1.97 *Under the assumption of Theorem* 1.95, *the map*

$$(G/H) \underset{N(H)/H}{\times} X^H \to X$$

defined by $[gH, x] \mapsto gx$ *is a G-homeomorphism.*

Proof We have only to combine Proposition 1.91 and Theorem 1.95. □

Proposition 1.98 *Let G be a topological group and H a closed subgroup of G. If X is a G-space, then* $G \underset{H}{\times} X$ *and* $(G/H) \times X$ *are G-homeomorphic. Here the G-action on* $(G/H) \times X$ *is given by the diagonal action.*

Proof Define two maps $f: G \underset{H}{\times} X \to (G/H) \times X$ and $f': (G/H) \times X \to G \underset{H}{\times} X$ by putting $f([g, x]) = (gH, gx)$ and $f'(gH, y) = [g, g^{-1}y]$. Then one easily verifies that these are well-defined G-maps with $f' \cdot f = f \cdot f' = \text{id}$. Continuity of them is shown by the usual argument so both f and f' are G-homeomorphisms. □

Exercises for Chapter 1

1. Determine the isotropy group at the point $(1, 0, \ldots, 0)$ of the natural $GL(n, \mathbb{R})$-action on $\mathbb{R}^n$.

2. Let G be a connected topological group and U a neighbourhood of the identity e in G. Then show that G is generated by U.

3. Show that the centre of a topological group G is closed in G. Also show that a discrete subgroup of a topological group G is closed in G.

4. Let G and G' be topological groups. Show that if a homomorphism $f: G \to G'$ of groups is continuous at a point, then f is globally continuous.

5. Let G be a connected topological group and H a discrete normal subgroup of G. Show that H is contained in the centre of G.

6. Let H be a closed subgroup of a topological group G. Show that if both H and G/H are connected, then so is G.

7. Let H be a normal subgroup of a topological group G. Show that, given a G-space X, the equality $X(\geq H) = X^H$ holds.

8. Given an element g of $S^1 = U(1)$, put $A_g = \{g^n \mid n = 0, 1, 2, \ldots\}$. Show that the closure $\bar{A}_g$ of A_g in S^1 is either a finite cyclic group or S^1.

9. Let S_n, A_n, and D_n denote the symmetric group, the alternating group, and the dihedral group of degree n, respectively. Show that these groups are related to the exact sequences of Example 2 in Section 1.3 in the following form.

$$\begin{array}{ccccccccc} 1 & \longrightarrow & SO(n) & \longrightarrow & O(n) & \longrightarrow & \mathbb{Z}_2 & \longrightarrow & 1 \\ & & \cup & & \cup & & \| & & \\ 1 & \longrightarrow & A_n & \longrightarrow & S_n & \longrightarrow & \mathbb{Z}_2 & \longrightarrow & 1 \end{array}$$

$$\begin{array}{ccccccccc} 1 & \longrightarrow & SO(2) & \longrightarrow & O(2) & \longrightarrow & \mathbb{Z}_2 & \longrightarrow & 1 \\ & & \cup & & \cup & & \| & & \\ 1 & \longrightarrow & \mathbb{Z}_n & \longrightarrow & D_n & \longrightarrow & \mathbb{Z}_2 & \longrightarrow & 1. \end{array}$$

Find splittings for these sequences such that the diagrams above commute and show that

$$S_n \cong A_n \rtimes \mathbb{Z}_2 \quad \text{and} \quad D_n \cong \mathbb{Z}_n \rtimes \mathbb{Z}_2.$$

10. Put $G = SL(2, \mathbb{R})$ and $\mathbb{H} = \{z \in \mathbb{C} \mid \operatorname{Im} z > 0\}$. Define a map $\varphi: G \times \mathbb{H} \to \mathbb{H}$ by

$$\varphi\left(\begin{bmatrix} a & b \\ c & d \end{bmatrix}, z\right) = \frac{az + b}{cz + d}.$$

Show that φ gives a transitive G-action and determine the isotropy group at the point i.

11. Show by example that the compactness assumption of G in Corollary 1.70 cannot be omitted.

2 Fibre bundles

Given two topological spaces X and Y, the construction of the Cartesian product $X \times Y$ is a method of constructing new topological spaces out of old. A generalization of this Cartesian product is the notion of fibre bundles, which have the Cartesian product structures locally. Its definition is given in terms of transformation groups; hence one may say that the theory of fibre bundles is an application of the theory of transformation groups.

As will be shown later, fibre bundles exist naturally as randomly placed geometric objects. They enable us to clearly understand the structures of transformation groups on manifolds as will be discussed in Chapter 4. Moreover, the theory of fibre bundles is applied not only to the theory of transformation groups but also to various other fields of mathematics, such as: Lie group theory; homogeneous spaces; covering spaces; real, complex, or algebraic vector bundles; characteristic classes; differential topology; and differential geometry. For instance, the concept of fibre bundles contributed essentially to the study of exotic differential structures on the 7-sphere due to Milnor, which foretold the dawn of differential topology.

2.1 Fibre bundles

First we give the strict definition.

Definition 2.1 (Coordinate bundle). Let K be a topological group acting effectively on a topological space F. Let E and X be Hausdorff spaces and $\pi: E \to X$ a continuous map. Suppose that there exist an open covering $\{U_\alpha\}$ $(\alpha \in \Lambda)$ of X and a homeomorphism $\varphi_\alpha: \pi^{-1}(U_\alpha) \to U_\alpha \times F$ for each $\alpha \in \Lambda$ having the following three properties.

(1) $p_\alpha \cdot \varphi_\alpha = \pi$ where $p_\alpha: U_\alpha \times F \to U_\alpha$ is the projection.

(2) Define

$$\varphi_{\alpha,x}: \pi^{-1}(x) \to F \qquad \text{for} \quad x \in U_\alpha$$

by $\varphi_{\alpha,x}(z) = p'_\alpha \cdot \varphi_\alpha(z)$ for $z \in \pi^{-1}(x)$ where $p'_\alpha: U_\alpha \times F \to F$ is another projection. Put

$$\theta_{\beta\alpha}(x) = \varphi_{\beta,x} \cdot \varphi_{\alpha,x}^{-1} \qquad \text{for} \quad x \in U_\alpha \cap U_\beta.$$

Then $\theta_{\beta\alpha}(x) \in K$.

(3) The map $\theta_{\beta\alpha}: U_\alpha \cap U_\beta \to K$ so obtained is continuous.

Then summing up the above information, the system $(E, \pi, X, F, K, U_\alpha, \varphi_\alpha)$ is called a *coordinate bundle.*

Remark The reason why we include the assumption that K acts effectively on F is that the statement $\theta_{\beta\alpha}(x) \in K$ in (2) makes sense. As shown in Section 1.4, any action can be regarded as an effective action of a quotient group. Hence we have only to deal with effective actions. Even if we do not assume effectiveness, it is possible to give a generalized definition of coordinate bundles to the sacrifice of order.

Definition 2.2 (Fibre bundles). Two coordinate bundles $(E, \pi, X, F, K, U_\alpha, \varphi_\alpha)$ and $(E, \pi, X, F, K, U'_\mu, \varphi'_\mu)$ which have in common E, π, X, F, and K are said to be *equivalent in the strict sense,* if $\bar{\theta}_{\mu\alpha}(x) = \varphi'_{\mu,x} \cdot \varphi_{\alpha,x}^{-1} \in K$ for $x \in U_\alpha \cap U'_\mu$ and $\bar{\theta}_{\mu\alpha}: U_\alpha \cap U'_\mu \to K$ is continuous. An equivalence class $\xi = (E, \pi, X, F, K)$ of coordinate bundles with respect to the above equivalence relation is called a *fibre bundle,* and E is called the *total space,* π the *projection,* X the *base space,* F the *fibre,* and K *the structure group.* Also, U_α of a coordinate bundle $(E, \pi, X, F, K, U_\alpha, \varphi_\alpha)$ is called the *coordinate neighbourhood,* φ_α the *coordinate function,* and $\theta_{\beta\alpha}$ the *transition function.* For each $x \in X$, $\pi^{-1}(x)$ is called the *fibre over* x and is often written as E_x.

Remark A coordinate bundle which is a representative of a fibre bundle ξ is often called a fibre bundle for simplicity or, briefly, a bundle and $\pi: E \to X$ or E is short for $\xi = (E, \pi, X, F, K)$.

Fibre bundles, especially when their structures are complicated, are a powerful tool. Let us begin with simple examples to illustrate the general idea of fibre bundles for the benefit of the reader.

Example 1 *The product bundle.* Let $E = X \times F$ and $\pi: X \times F \to X$ be the projection. Then a fibre bundle (E, π, X, F, K) is called a *product bundle* if $\mathrm{id}: E \to X \times F$ is a coordinate function.

Example 2 *The Möbius band.* Let I denote the closed interval $[-1, 1]$. Then the fibre F is given by I and the base space X is a circle obtained from I by identifying its ends. The total space E is obtained from the product $I \times F$ by matching the two ends with a twist, that is, the quotient space $(I \times F)/\sim$ of $I \times F$ with respect to the identification $(1, y) \sim (-1, -y)$. E is the Möbius band. The projection $I \times F \to I$ on the first factor carries over under this matching into a projection $\pi: E \to X$. Define two open sets U_α and U_β of X by

$$U_\alpha = (-\tfrac{2}{3}, \tfrac{2}{3}), \qquad U_\beta = \{[-1, -\tfrac{1}{3}) \cup (\tfrac{1}{3}, 1]\}/\sim$$

where $/\sim$ stands for the identification of -1 and 1. Then $\{U_\alpha, U_\beta\}$ is an open covering of X and satisfies

$$U_\alpha \cap U_\beta = (-\tfrac{2}{3}, -\tfrac{1}{3}) \cup (\tfrac{1}{3}, \tfrac{2}{3}).$$

We now denote the equivalence class of $(x, y) \in I \times F$ by $\{x, y\} \in E$, since the notation $[x, y]$ is already used for a closed interval. Moreover we denote the equivalence class of $x \in I$ by the same $x \in X$ for brevity. Then we define two maps

$$\varphi_\alpha : \pi^{-1}(U_\alpha) \to U_\alpha \times F \qquad \text{by} \qquad \varphi_\alpha = \mathrm{id}$$

and

$$\varphi_\beta : \pi^{-1}(U_\beta) \to U_\beta \times F \qquad \text{by}$$

$$\varphi_\beta(\{x, y\}) = \begin{cases} (x, y) & \text{for} \quad x \in (\tfrac{1}{3}, 1],\, y \in F \\ (x, -y) & \text{for} \quad x \in [-1, -\tfrac{1}{3}),\, y \in F. \end{cases}$$

One easily verifies that φ_α and φ_β give coordinate functions and satisfy

$$\theta_{\beta\alpha} | (\tfrac{1}{3}, \tfrac{2}{3}) = \mathrm{id}, \qquad \theta_{\beta\alpha} | (-\tfrac{2}{3}, -\tfrac{1}{3}) = -\mathrm{id}.$$

Next let K be the cyclic group $\mathbb{Z}_2$ of order 2 with the discrete topology and g its generator. Then K acts effectively on F by $g \cdot y = -y$ and $\theta_{\beta\alpha}(x) \in K$ for any $x \in U_\alpha \cap U_\beta$. Since $\theta_{\beta\alpha}$ is constant on each of the two components of $U_\alpha \cap U_\beta$, it is continuous. Thus we have shown that $\pi : E \to X$ is a fibre bundle. The construction is illustrated in Fig. 2.1.

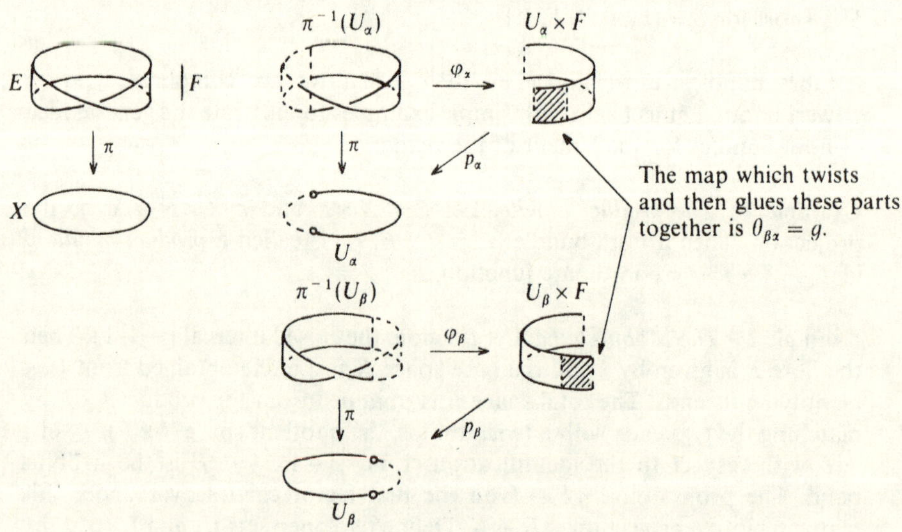

Fig. 2.1 Construction of the Möbius band.

Example 3 *The Klein bottle.* The preceding construction is modified by replacing the fibre by a circle. The base space X is the same circle as that of Example 2. Although the fibre F is also a circle, we express it as the unit circle $\{z \in \mathbb{C} \mid |z| = 1\}$ of the complex numbers $\mathbb{C}$. Let K be $\mathbb{Z}_2$ as above whose generator g acts on F by $g \cdot z = \bar{z}$ where $\bar{z}$ denotes the complex conjugate of z. The total space E is obtained from the product $I \times F$ by matching the two ends with a reflection, that is, the quotient space $(I \times F)/\sim$ of $I \times F$ with respect to the identification $(1, z) \sim (-1, \bar{z})$. Then E is the Klein bottle. The projection $I \times F \to I$ on the first factor carries over under this matching into a projection $\pi: E \to X$. The two open sets U_α and U_β are the same as above and we put $\varphi_\alpha = \mathrm{id}$ and

$$\varphi_\beta(\{x, z\}) = \begin{cases} (x, z) & \text{for } x \in (\tfrac{1}{3}, 1], z \in F \\ (x, \bar{z}) & \text{for } x \in [-1, -\tfrac{1}{3}), z \in F. \end{cases}$$

The rest of the construction is similar to Example 2.

Let $\xi = (E, \pi, X, F, K)$ and $\xi' = (E', \pi', X', F, K)$ be two fibre bundles having the same fibre and the same structure group. A continuous map $\bar{f}: E \to E'$ is called a *bundle map* from ξ to ξ' if the following two conditions are satisfied.

(1) There is a continuous map f: $X \to X'$ such that the following diagram commutes,

$$\begin{array}{ccc} E & \xrightarrow{\bar{f}} & E' \\ {\scriptstyle \pi}\downarrow & & \downarrow{\scriptstyle \pi'} \\ X & \xrightarrow{f} & X' \end{array}$$

(2) Let $\{U_\alpha, \varphi_\alpha\}$ and $\{V'_\mu, \varphi'_\mu\}$ be pairs of coordinate neighbourhoods and coordinate functions of ξ and ξ', respectively. For any α, μ with $U_\alpha \cap f^{-1}(V'_\mu) \neq \varnothing$, we put

$$f_{\mu\alpha}(x) = \varphi'_{\mu, f(x)} \cdot \bar{f} \cdot \varphi_{\alpha, x}^{-1} \qquad \textit{for} \quad x \in U_\alpha \cap f^{-1}(V'_\mu).$$

Then $f_{\mu\alpha}(x) \in K$ and the map

$$f_{\mu\alpha}: U_\alpha \cap f^{-1}(V'_\mu) \to K$$

is continuous.

If $X = X'$ and $f = \mathrm{id}$ in the above, $\bar{f}$ is called an *isomorphism*. When there exists an isomorphism between two fibre bundles ξ and ξ', we say that ξ and

ξ' are *isomorphic* (or *equivalent*) and write $\xi \cong \xi'$ or $E \cong E'$. A bundle that is isomorphic to a product bundle is said to be *trivial.*

Remark If f is a homeomorphism, $\bar{f}$ is also a homeomorphism and $\bar{f}^{-1}$ is a bundle map as well. The proof is left as an exercise.

A fibre bundle $\xi = (E, \pi, X, F, K)$ is called a *covering* if the following conditions are satisfied.

(1) E is connected;

(2) F is a discrete space.

Then we call π a *covering map* and E a *covering space* of X.

Proposition 2.1 *The projection* π: $E \to X$ *of a fibre bundle is an open map.*

Proof Let $\varphi_\alpha : \pi^{-1}(U_\alpha) \to U_\alpha \times F$ be a coordinate function. Since $O \cap \pi^{-1}(U_\alpha)$ is open in $\pi^{-1}(U_\alpha)$ for any open set O of E, $\varphi_\alpha(O \cap \pi^{-1}(U_\alpha))$ is open in $U_\alpha \times F$. Since the projection p_α: $U_\alpha \times F \to U_\alpha$ is an open map by Corollary 1.14 and since

$$\pi(O \cap \pi^{-1}(U_\alpha)) = p_\alpha \cdot \varphi_\alpha(O \cap \pi^{-1}(U_\alpha)),$$

$\pi(O \cap \pi^{-1}(U_\alpha))$ is open in U_α and hence in X. Observe now that

$$\pi(O) = \pi\left(O \cap \bigcup_\alpha \pi^{-1}(U_\alpha)\right) = \pi\left(\bigcup_\alpha (O \cap \pi^{-1}(U_\alpha))\right)$$
$$= \bigcup_\alpha \pi(O \cap \pi^{-1}(U_\alpha))$$

which means that $\pi(O)$ is open in X. □

2.2 G_1–G_2 bispaces and twisted products

Let G be a topological group and X a topological space. By a *right* G-action on X we mean a continuous map:

$$\varphi : X \times G \to X$$

such that:

(1) $\varphi(x, e) = x$ for $x \in X$;

(2) $\varphi(\varphi(x, g_1), g_2) = \varphi(x, g_1 g_2)$ for $x \in X, g_i \in G$.

The space X, together with a given right G-action φ, is called a *right* G-space. When φ is understood from the context, we shall often use the notation $x \cdot g$

or xg for $\varphi(x, g)$. When compared with this right G-action, the previous action is sometimes emphasized as a *left* G-action. Whenever we say simply a G-action without 'right' or 'left', we mean a left G-action.

Example 1 Let G be a topological group and H a subgroup of G. Then H acts on G from the right by the multiplication in G. Its orbit space is nothing but the homogeneous space G/H which has already appeared often.

We now look into the relation between a left G-action and a right G-action.

Lemma 2.2 *Given a left G-action $\varphi_L: G \times X \to X$, the map $\varphi_R: X \times G \to X$ defined by $\varphi_R(x, g) = \varphi_L(g^{-1}, x)$ for $g \in G$, $x \in X$ gives a right G-action. Conversely, given a right G-action $\varphi_R: X \times G \to X$, the map $\varphi_L: G \times X \to X$ defined by $\varphi_L(g, x) = \varphi_R(x, g^{-1})$ gives a left G-action.*

Proof Suppose that a left G-action φ_L is given. Then it is clear that the map φ_R defined as above is continuous and $\varphi_R(x, e) = x$. For $g_1, g_2 \in G$, we have

$$\begin{aligned}\varphi_R(x, g_1 g_2) &= \varphi_L((g_1 g_2)^{-1}, x)\\ &= \varphi_L(g_2^{-1} g_1^{-1}, x) = \varphi_L(g_2^{-1}, \varphi_L(g_1^{-1}, x))\\ &= \varphi_R(\varphi_L(g_1^{-1}, x), g_2)\\ &= \varphi_R(\varphi_R(x, g_1), g_2).\end{aligned}$$

That is, φ_R is a right G-action. The proof of the converse is similar. □

It follows from Lemma 2.2 that a left G-space can be regarded as a right G-space and vice versa; therefore either will do. However it is convenient to employ both terms. In particular, suppose that two groups G_1 and G_2 act on a space X such that both actions commute. That is, for $g_i \in G_i$ and $x \in X$, the equality $g_2 \cdot (g_1 \cdot x) = g_1 \cdot (g_2 \cdot x)$ holds. We now regard the G_2-action as a right G_2-action as in Lemma 2.2, that is, we put $x \circ g_2 = g_2^{-1} \cdot x$. Then we have

$$(g_1 \cdot x) \circ g_2 = g_2^{-1} \cdot (g_1 \cdot x) = g_1 \cdot (g_2^{-1} \cdot x) = g_1 \cdot (x \circ g_2).$$

Conversely, suppose that X is a left G_1-space as well as a right G_2-space satisfying $(g_1 \cdot x) \cdot g_2 = g_1 \cdot (x \cdot g_2)$. If we put $g_2 \circ x = x \cdot g_2^{-1}$, then we have

$$g_2 \circ (g_1 \cdot x) = (g_1 \cdot x) \cdot g_2^{-1} = g_1 \cdot (x \cdot g_2^{-1}) = g_1 \cdot (g_2 \circ x).$$

That is, both G_1- and G_2-actions commute.

Thus we have shown the following proposition.

Proposition 2.3 *The condition that two left G_1- and G_2-actions on X commute is equivalent to the condition that*

$$(g_1 \cdot x) \circ g_2 = g_1 \cdot (x \circ g_2) \qquad \textit{for} \quad g_i \in G_i \quad (i = 1, 2)$$

when we regard the left G_2-action as a right G_2-action as above. □

Whenever we say that X is a left G_1-space as well as right G_2-space, we assume the equality

$$(g_1 \cdot x)\cdot g_2 = g_1 \cdot (x \cdot g_2) \qquad \text{for} \quad g_i \in G_i,\ x \in X$$

hereafter. When we want to emphasize this, we say that both actions *commute.* Then X is called a *left G_1- right G_2-space* or a *G_1–G_2 bispace.*

Remark If X is only a left G_1-space, X can be regarded as a G_1–G_2 bispace by putting $G_2 = \{e\}$.

Example 2 If G is a commutative topological group, a (left) G-space can be regarded as a G–G bispace.

Example 3 Let H and K be subgroups of a topological group G. Then G is a H–K bispace with respect to the multiplication in G.

Let $G_1, \ldots, G_{n+1}$ be $n+1$ topological groups and X_i a G_i–G_{i+1} bispace for $i = 1, \ldots, n$. For $g_i \in G_i$ and $x_i \in X_i$, we put

$$(g_2, \ldots, g_n)\circ(x_1, \ldots, x_n) = (x_1 g_2^{-1}, g_2 x_2 g_3^{-1}, g_3 x_3 g_4^{-1}, \ldots, g_n x_n).$$

Then this defines a $G_2 \times \cdots \times G_n$-action on $X_1 \times \cdots \times X_n$. The orbit space of this action is denoted by

$$X_1 \underset{G_2}{\times} X_2 \underset{G_3}{\times} \cdots \underset{G_n}{\times} X_n$$

and is called a *twisted product* of $X_1, \ldots, X_n$. The orbit of $(x_1, \ldots, x_n)$ is denoted by $[x_1, \ldots, x_n]$.

Remark The construction above is similar to that of the tensor product of n bimodules.

Lemma 2.4 *The map*

$$\varphi: G_1 \times \left(X_1 \underset{G_2}{\times} X_2 \underset{G_3}{\times} \cdots \underset{G_n}{\times} X_n\right) \to X_1 \underset{G_2}{\times} X_2 \underset{G_3}{\times} \cdots \underset{G_n}{\times} X_n$$

defined by $\varphi(g_1, [x_1, \ldots, x_n]) = [g_1 x_1, x_2, \ldots, x_n]$ *gives a left* G_1*-action. Similarly the map*

$$\psi: \left(X_1 \underset{G_2}{\times} X_2 \underset{G_3}{\times} \cdots \underset{G_n}{\times} X_n\right) \times G_{n+1} \to X_1 \underset{G_2}{\times} X_2 \underset{G_3}{\times} \cdots \underset{G_n}{\times} X_n$$

defined by $\psi([x_1, \ldots, x_n], g_{n+1}) = [x_1, \ldots, x_{n-1}, x_n g_{n+1}]$ *gives a right* G_{n+1}*-action.*

Proof It is easy to see that φ and ψ are defined independently of the choice of representatives of orbits. Since the conditions (1) and (2) of actions evidently hold, we have only to prove the continuity of φ and ψ. For this, we consider the following commutative diagram:

$$\begin{array}{ccc} G_1\times(X_1\times X_2\times\cdots\times X_n) & \xrightarrow{\tilde{\varphi}} & X_1\times X_2\times\cdots\times X_n \\ \Big\downarrow{\scriptstyle \mathrm{id}\times\pi} & & \Big\downarrow{\scriptstyle\pi} \\ G_1\times\left(X_1\underset{G_2}{\times}X_2\underset{G_3}{\times}\cdots\underset{G_n}{\times}X_n\right) & \xrightarrow{\varphi} & X_1\underset{G_2}{\times}X_2\underset{G_3}{\times}\cdots\underset{G_n}{\times}X_n \end{array}$$

where $\tilde{\varphi}$ is the continuous map defined by $\tilde{\varphi}(g_1,(x_1,\ldots,x_n)) = (g_1x_1,x_2,\ldots,x_n)$. Since $\mathrm{id}\times\pi$ is open and surjective, it follows from Lemma 1.15 that φ is continuous. Similarly ψ is continuous. $\square$

Corollary 2.5 *The space $X_1\underset{G_2}{\times}X_2\underset{G_3}{\times}\cdots\underset{G_n}{\times}X_n$ is a G_1–G_{n+1} bispace.*

Proof Since the left G_1-action and the right G_{n+1}-action commute, we have Corollary 2.5. $\square$

Proposition 2.6 *For any i,j with $1\le i<j\le n$, the space*

$$X_1\underset{G_2}{\times}X_2\underset{G_3}{\times}\cdots\underset{G_i}{\times}\left(X_i\underset{G_{i+1}}{\times}X_{i+1}\times\cdots\underset{G_j}{\times}G_j\right)\underset{G_{j+1}}{\times}\cdots\underset{G_n}{\times}X_n$$

and the space $X_1\underset{G_2}{\times}X_2\underset{G_3}{\times}\cdots\underset{G_n}{\times}X_n$ are homeomorphic as G_1–G_{n+1} bispaces.

Proof Consider the following commutative diagram:

$$\begin{array}{ccc} X_1\times\cdots\times(X_i\times\cdots\times X_j)\times\cdots\times X_n & \xrightarrow{\mathrm{id}} & X_1\times\cdots\times X_n \\ \Big\downarrow{\scriptstyle\pi} & & \Big\downarrow{\scriptstyle\pi} \\ X_1\times\cdots\times\left(X_i\underset{G_{i+1}}{\times}\cdots\underset{G_j}{\times}X_j\right)\times\cdots\times X_n & \overset{\pi}{\searrow} & \\ \Big\downarrow{\scriptstyle\pi} & & \\ X_1\underset{G_2}{\times}X_2\underset{G_3}{\times}\cdots\underset{G_i}{\times}\left(X_i\underset{G_{i+1}}{\times}\cdots\underset{G_j}{\times}X_j\right)\times\cdots\underset{G_n}{\times}X_n & \xrightarrow{f} & X_1\underset{G_2}{\times}X_2\underset{G_3}{\times}\cdots\underset{G_n}{\times}X_n \end{array}$$

where f is defined by $f([x_1, \ldots, x_{i-1}, [x_i, \ldots, x_j], x_{j+1}, \ldots, x_n]) = [x_1, \ldots, x_{i-1}, x_i, \ldots, x_j, \ldots, x_n]$. Since all the orbit maps π are open and surjective, f is continuous. Obviously f is bijective and we see similarly that f^{-1} is also continuous. Moreover it is clear that the map f commutes with the G_1–G_{n+1}-action. □

Corollary 2.7 *Even if we put as many parentheses into* $X_1 \underset{G_2}{\times} X_2 \times \cdots \underset{G_n}{\times} X_n$ *as we want, it is naturally homeomorphic to* $X_1 \underset{G_2}{\times} X_2 \times \cdots \underset{G_n}{\times} X_n$ *as* G_1–G_{n+1} *bispaces.*

Proof This is deduced from Proposition 2.6 by induction. □

Example 4 Let us recall the induced G-actions in Section 1.9. Let G be a topological group and H a subgroup of G. Then we constructed the induced G-space $G \underset{H}{\times} X$ for any H-space X. This construction is nothing but that of the twisted product of G and X by regarding G as a G–H bispace and X as an H–$\{e\}$ bispace.

Let X and Y be G_1–G_2 bispaces. Then a map $f: X \to Y$ is called a G_1–G_2 *bimap*, if f satisfies

$$f((g_1 \cdot x) \cdot g_2) = (g_1 \cdot f(x)) \cdot g_2 \qquad \text{for} \quad g_i \in G_i, \quad x \in X.$$

Let X_i and Y_i be G_i–G_{i+1} bispaces and $f_i: X_i \to Y_i$ be G_i–G_{i+1} bimaps. Then f_i induce a map

$$f_1 \underset{G_2}{\times} f_2 \times \cdots \underset{G_n}{\times} f_n : X_1 \underset{G_2}{\times} X_2 \times \cdots \underset{G_n}{\times} X_n \to Y_1 \underset{G_2}{\times} Y_2 \times \cdots \underset{G_n}{\times} Y_n$$

by putting

$$f_1 \underset{G_2}{\times} f_2 \times \cdots \underset{G_n}{\times} f_n([x_1, \ldots, x_n]) = [f_1(x_1), \ldots, f_n(x_n)].$$

Clearly $f_1 \underset{G_2}{\times} f_2 \times \cdots \underset{G_n}{\times} f_n$ is a continuous G_1–G_{n+1} bimap.

Proposition 2.8 *If* f_i *are all open, then* $f_1 \underset{G_2}{\times} f_2 \times \cdots \underset{G_n}{\times} f_n$ *is also open.*

Proof Consider the following commutative diagram:

$$\begin{array}{ccc} X_1 \times X_2 \times \cdots \times X_n & \xrightarrow{f_1 \times \cdots \times f_n} & Y_1 \times Y_2 \times \cdots \times Y_n \\ \downarrow \pi & & \downarrow \pi \\ X_1 \underset{G_2}{\times} X_2 \times \cdots \underset{G_n}{\times} X_n & \xrightarrow{f_1 \underset{G_2}{\times} f_2 \times \cdots \underset{G_n}{\times} f_n} & Y_1 \underset{G_2}{\times} Y_2 \times \cdots \underset{G_n}{\times} Y_n \end{array}$$

By virtue of Lemma 1.13, it is shown by induction that $f_1 \times \cdots \times f_n$ is open. Since the vertical maps π are surjective, continuous and open, Proposition 2.8 is immediate from the commutative diagram above. □

2.3 Principal bundles and associated bundles

A fibre bundle (E, π, X, F, K) is called a *principal fibre bundle* (or *principal K-bundle* or briefly *principal bundle*) if $F = K$ and K operates on F by left translations. Then the total space is often written as P and this bundle is written as $\eta = (P, \pi, X, K)$ (or $\pi: P \to X$ or simply P).

Proposition 2.9 *For a principal K-bundle $\eta = (P, \pi, X, K)$, there is a canonical free right K-action on P such that each fibre $\pi^{-1}(x)$ is K-invariant. The projection $\pi: P \to X$ induces a homeomorphism $\bar{\pi}: P/K \to X$.*

Proof For each coordinate function $\varphi_\alpha: \pi^{-1}(U_\alpha) \to U_\alpha \times K$, we first define a right K-action on $U_\alpha \times K$ by

$$(x, k) \circ k' = (x, kk') \qquad \text{for} \quad x \in U_\alpha, \quad k, k' \in K.$$

This action induces a right K-action on $\pi^{-1}(U_\alpha)$ via φ_α. That is, we put

$$z \circ k' = \varphi_\alpha^{-1}(\varphi_\alpha(z) \circ k') \qquad \text{for} \quad z \in \pi^{-1}(U_\alpha), \quad k' \in K.$$

In the following we show that it is independent of the choice of a coordinate function φ_α. Let $\varphi_\beta: \pi^{-1}(U_\beta) \to U_\beta \times K$ be another coordinate function. For arbitrary $z \in \pi^{-1}(U_\alpha \cap U_\beta)$, there exists a unique $k \in K$ with $\varphi_\alpha(z) = (\pi(z), k)$ and we have

$$\begin{aligned}
\varphi_\beta \cdot \varphi_\alpha^{-1}(\varphi_\alpha(z) \circ k') &= \varphi_\beta \cdot \varphi_\alpha^{-1}((\pi(z), k) \circ k') = \varphi_\beta \cdot \varphi_\alpha^{-1}(\pi(z), kk') \\
&= (\pi(z), \theta_{\beta\alpha}(\pi(z))(kk')) = (\pi(z), (\theta_{\beta\alpha}(\pi(z))k)k') \\
&= (\pi(z), \theta_{\beta\alpha}(\pi(z))k) \circ k' = (\varphi_\beta \cdot \varphi_\alpha^{-1}(\pi(z), k)) \circ k' \\
&= \varphi_\beta(z) \circ k'.
\end{aligned}$$

Hence we have

$$\varphi_\alpha^{-1}(\varphi_\alpha(z) \circ k') = \varphi_\beta^{-1}(\varphi_\beta(z) \circ k'),$$

which means that the definition of $z \circ k'$ does not depend on the choice of a coordinate function φ_α.

Since the right K-action on $U_\alpha \times K$ is free, the K-action on P is also free. The projection $\pi: P \to X$ induces a bijective continuous map $\bar{\pi}: P/K \to X$. Since π is open by Proposition 2.1, $\bar{\pi}$ is also open by Lemma 1.18, and hence $\bar{\pi}$ is a homeomorphism. □

Hereafter we mean by the *right K-action on a principal K-bundle*, the K-action obtained in Proposition 2.9.

The notion of fibre bundles associated with principal bundles is important in general. When we construct and investigate these bundles, the fundamental knowledge of topological transformation groups is indispensable.

Let X be a right G-space and Y a left G-space. Then G operates on $X \times Y$ by $g \circ (x, y) = (xg^{-1}, gy)$ for $g \in G$, $x \in X$, $y \in Y$. As mentioned in the previous section, the orbit space $(X \times Y)/G$ of this action is denoted by $X \underset{G}{\times} Y$ and is called a twisted product of X and Y.

Theorem 2.10 *Let $\pi: P \to X$ be a principal K-bundle and let F be an effective left K-space. Then*

$$p: P \underset{K}{\times} F \to X$$

defined by $p([z, y]) = \pi(z)$ for $z \in P$, $y \in F$ is a fibre bundle with fibre F and structure group K.

Proof Let $\varphi: \pi^{-1}(U) \to U \times K$ be a coordinate function of P. Corresponding to φ, we consider the composition $\bar{\varphi}$ of the following maps:

$$\bar{\varphi}: p^{-1}(U) \xrightarrow{\varphi_1} \pi^{-1}(U) \underset{K}{\times} F \xrightarrow{\varphi_2} (U \times K) \underset{K}{\times} F$$

$$\xrightarrow{\varphi_3} U \times \left(K \underset{K}{\times} F \right) \xrightarrow{\varphi_4} U \times F.$$

Here φ_i $(i = 1, 2, 3, 4)$ are defined as follows. First note that $p^{-1}(U)$ coincides with $\pi^{-1}(U) \underset{K}{\times} F$ and φ_1 is the identity map. The second map φ_2 is given by $\varphi \underset{K}{\times} \mathrm{id}: \pi^{-1}(U) \underset{K}{\times} F \to (U \times K) \underset{K}{\times} F$ and the third map φ_3 is the canonical homeomorphism by virtue of Corollary 2.7. The last map φ_4 is induced by the K-homeomorphism of Lemma 1.87. Since all these maps are homeomorphisms, the composition is also a homeomorphism.

Next we check property (2) in Definition 2.1 of a coordinate bundle. Let $\psi: \pi^{-1}(V) \to V \times K$ be another coordinate function and $\theta: U \cap V \to K$ be the transition function for φ and ψ. That is, θ is the continuous map given by $\theta(x) = \psi_x \cdot \varphi_x^{-1}$. Let ψ_i $(i = 1, 2, 3, 4)$ be the maps defined in a similar manner to φ_i above. We identify $\{x\} \times F$ with F and note that $\varphi_1 = \mathrm{id}$ and $\psi_1 = \mathrm{id}$. Then we compute, for $x \in U \cap V$, $y \in F$,

$$\begin{aligned}
\bar{\psi}_x \cdot \bar{\varphi}_x^{-1}(y) &= \bar{\psi}_x \cdot \varphi_1^{-1} \cdot \varphi_2^{-1} \cdot \varphi_3^{-1} \cdot \varphi_4^{-1}(x, y) \\
&= \bar{\psi}_x \cdot \varphi_2^{-1} \cdot \varphi_3^{-1}(x, [e, y]) = \bar{\psi}_x \cdot \varphi_2^{-1}([(x, e), y]) \\
&= \bar{\psi}_x([\varphi_x^{-1}(e), y]) = \psi_4 \cdot \psi_3([(x, \psi_x \cdot \varphi_x^{-1}(e)), y]) \\
&= \psi_4(x, [\theta(x)e, y]) = \theta(x)y.
\end{aligned}$$

Hence we have $\bar{\psi}_x \cdot \bar{\varphi}_x^{-1} = \theta(x)$, that is, θ is the transition function for $\bar{\varphi}$ and $\bar{\psi}$ as well. □

Remark When the K-action on F is not necessarily effective in Theorem 2.10, $p\colon P \times_K F \to X$ is a fibre bundle with fibre F and structure group $K/\mathrm{Ker}\,\varphi$ (cf. Exercise 2 which will be mentioned later).

The fibre bundle ξ obtained in Theorem 2.10 is called the *bundle associated with the principal K-bundle P* (or simply the *associated bundle of P*). On the other hand, P is called the *principal K-bundle associated with ξ* (or simply *the associated principal bundle of ξ*).

Given a fibre bundle $\xi = (E, \pi, X, F, K)$, a principal K-bundle $\eta = (P, \pi, X, K)$ will be constructed from ξ as follows. Let $\{U_\alpha\}$ be coordinate neighbourhoods of ξ and $\theta_{\beta\alpha}$ the transition functions of ξ. Take the disjoint union $\bigcup_\alpha U_\alpha \times K$ over all α and, whenever $x \in U_\alpha \cap U_\beta$, identify $(x, k) \in U_\alpha \times K$ with $(x, \theta_{\beta\alpha}(x)k) \in U_\beta \times K$. We denote the identification space $(\cup_\alpha U_\alpha \times K)/\sim$ by P and define $\pi\colon P \to X$ by projection on the first factor of each product. Then one verifies easily that $\pi\colon P \to X$ is a principal K-bundle associated with ξ.

We now show the uniqueness of the associated principal K-bundle.

Theorem 2.11 *Two fibre bundles E and E′ having the same base space, fibre, and structure group are isomorphic if and only if their associated principal bundles are isomorphic.*

Proof The notion of isomorphisms is defined by the property of the function $\bar{\theta}_{\mu\alpha}$ in Section 2.1. If two fibre bundles E and E' are isomorphic, then one can easily construct an isomorphism between the principal bundles associated with E and E' by making use of the function $\bar{\theta}_{\mu\alpha}$. The converse procedure is similar. □

The following terminology will often be used later as well as in this section.

Definition 2.3 A *cross-section* for a surjective continuous map $f\colon X \to Y$ is a continuous map $s\colon Y \to X$ such that $f \cdot s$ is the identity on Y. For a subset U of Y, a *cross-section on U* is a cross-section for $f|f^{-1}(U)\colon f^{-1}(U) \to U$. When, for each point y of Y, there exist a neighbourhood U of y and a cross-section on U, we say that *f has a local cross-section.*

A cross-section is illustrated in Fig. 2.2.

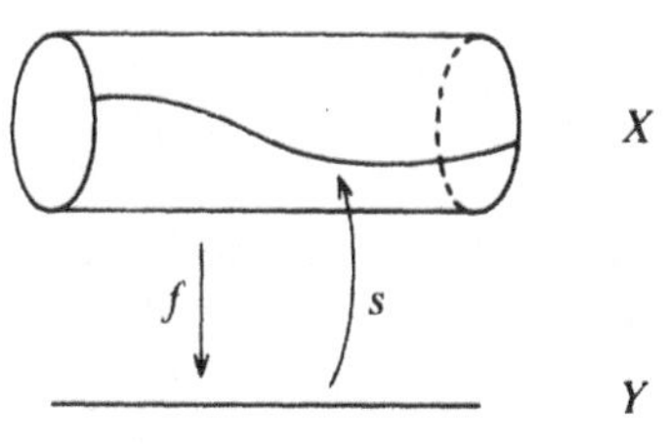

Fig. 2.2 A cross-section.

Theorem 2.12 *A principal K-bundle $\pi: P \to X$ is trivial if and only if there exists a cross-section for π.*

Proof If $\pi: P \to X$ is trivial, then π obviously has a cross-section. Conversely, suppose that there exists a cross-section $s: X \to P$ for π. Then define a map $f: X \times K \to P$ by $f(x, k) = s(x)k$ for $x \in X$, $k \in K$. Here $s(x)k$ denotes the right K-action on P obtained in Proposition 2.9. Recall that $\mathrm{id}: X \times K \to X \times K$ is the coordinate function for the product bundle $X \times K$. Let $\varphi_\alpha: \pi^{-1}(U_\alpha) \to U_\alpha \times K$ be a coordinate function for P. Then from the definition of the right K-action on P, we have, for $x \in U_\alpha$,

$$\varphi_{\alpha, x}(f(x, k)) = \varphi_{\alpha, x}(s(x)k) = \varphi_{\alpha, x}(s(x))k$$

and $\varphi_{\alpha, x}(s(x))$ is an element of K. Since $\varphi_{\alpha, x}(s(x)) = p'_\alpha \cdot \varphi_\alpha(s(x))$, $\varphi_{\alpha, x}(s(x))$ is continuous as a function of x. Hence f is a bundle map. Furthermore f is a bundle map covering $\mathrm{id}: X \to X$. That is, f is a bundle isomorphism. □

Combining Theorems 2.11 and 2.12, we have

Corollary 2.13 *A fibre bundle is trivial if and only if the associated principal bundle admits a cross-section.*

A covering with a principal bundle structure is said to be *regular*.

Example 1 Let $S^1 = \{z \in \mathbb{C} \mid |z| = 1\}$. For a positive integer n, we define a map $p: S^1 \to S^1$ by $p(z) = z^n$. Then p is a principal $\mathbb{Z}_n$-bundle, hence p is a regular covering.

We now consider the role of structure groups within fibre bundles. The prevailing point of view is that the smaller the structure group of a bundle, the simpler the bundle. However, a trivial bundle with a big structure group does exist. Accordingly, there might be readers feeling that the role of structure groups is not so clear. In the following, we hope to clarify this.

We begin by explaining the notion of extension and reduction of structure groups.

Given a homomorphism $f: K \to G$ of topological groups, K acts on G via f. Specifically we define it by

$$k \circ g = f(k)g \qquad \text{for } k \in K,\ g \in G.$$

Proposition 2.14 *Given a principal K-bundle P and a homomorphism $f: K \to G$ of topological groups, the associated bundle $P \times_K G$ becomes a principal G-bundle. If we define a map $f_1: P \to P \times_K G$ by $f_1(z) = [z, e]$, the following diagram commutes*

$$\begin{array}{ccc} P \times K & \longrightarrow & P \\ \Big\downarrow{\scriptstyle f_1 \times f} & & \Big\downarrow{\scriptstyle f_1} \\ (P \underset{K}{\times} G) \times G & \longrightarrow & P \underset{K}{\times} G \end{array}$$

where the two horizontal maps are those of the actions on respective principal bundles due to Proposition 2.9. If P and P' are equivalent as principal K-bundles, then $P \underset{K}{\times} G$ and $P' \underset{K}{\times} G$ are equivalent as principal G-bundles.

Proof It follows from Theorem 2.10 and the remark right after it that $P \underset{K}{\times} G$ is a fibre bundle with fibre G and structure group $f(K)$. Since the $f(K)$-space G can be regarded as a G-space by extension, a principal G-bundle structure can be induced on $P \underset{K}{\times} G$ naturally. The rest of the proof follows easily. □

Under the circumstances of Proposition 2.14, the principal G-bundle $\bar{P}$ which is equivalent to $P \underset{K}{\times} G$ is called an *extension of P by f*. Conversely, given a principal G-bundle $\bar{P}$, we say that $\bar{P}$ is *reducible* to a principal K-bundle if there is a principal K-bundle P with $\bar{P} = P \underset{K}{\times} G$, and we call P a *reduction* of $\bar{P}$. A *reduction* of a fibre bundle E with fibre F and structure group G is a reduction to a principal K-bundle P of the principal G-bundle $\bar{P}$ associated with E. Then it follows from the proof of Proposition 2.14 that two kinds of fibre bundle structures are induced on $P \underset{K}{\times} F$ with structure groups $f(K)$ and G, respectively, and it is easily verified that E is isomorphic to $P \underset{K}{\times} F$ as fibre bundles with structure group G.

Remark Usually extensions and reductions are considered in the case where K is a closed subgroup of G and f is the inclusion map. However, we shall deal with other cases in Sections 6.2 and 6.4 which are very important. Although an extension of a principal bundle is always possible, a reduction is not necessarily possible. For example, a bundle is trivial if and only if its structure group is reducible to the trivial group $\{e\}$. Even if there exists a reduction, it is not necessarily unique (see Exercise 3).

We are now ready to explain the meaning of structure groups as promised at the beginning. Fibre bundles are defined by designating structure groups, because the notion of transition functions and the notion of equivalence of bundles depend on structure groups. It often happens that inequivalent bundles with the same structure group become equivalent by extending their common structure group, that is, the notion of equivalence depends entirely on the designation of structure groups. For example, let K be a topological group and H a closed subgroup of K. If transition functions of two coordinate bundles E_1 and E_2 take values in H, then E_1 and E_2 can be regarded as coordinate bundles with structure group H (that is, reduction!). Then it often happens that E_1 and E_2 are not equivalent as bundles with structure group H, though they are equivalent as bundles with structure group K.

On the other hand there are bundles with a common structure group K such that their transition functions take values in K fully or take values only in the unit e. Accordingly the notion of reduction is introduced in order to see the essential part of the structure group.

These circumstances will be clarified by Theorem 2.29, Corollary 2.30 and the remark immediately after it, and Exercise 3.

2.4 Fibre products and induced bundles

Let X, Y, and Z be G-spaces and $f: X \to Z$ and $h: Y \to Z$ be G-maps. The *fibre product* $X \underset{Z}{\times} Y$ of X and Y over Z (or *pull-back of* Y *via* f) is defined to be the subspace.

$$X \underset{Z}{\times} Y = \{(x, y) \in X \times Y \mid f(x) = h(y)\}$$

of $X \times Y$. Observe that the fibre product is completely different from the twisted product in Section 2.2.

Since $X \underset{Z}{\times} Y$ is G-invariant under the diagonal G-action $(g, (x, y)) \mapsto (gx, gy)$ on $X \times Y$, $X \underset{Z}{\times} Y$ is a G-space.

By restricting the projections $X \times Y \to X$ and $X \times Y \to Y$ to the subspace $X \underset{Z}{\times} Y$, we have maps

$$h': X \underset{Z}{\times} Y \to X \qquad \text{and} \qquad f': X \underset{Z}{\times} Y \to Y,$$

which are clearly equivariant. These spaces and maps are collected into the following commutative diagram so that we may better see the situation:

$$\begin{array}{ccc} X \underset{Z}{\times} Y & \xrightarrow{\ f'\ } & Y \\ {\scriptstyle h'}\downarrow & & \downarrow{\scriptstyle h} \\ X & \xrightarrow{\ f\ } & Z \end{array}$$

This diagram is called a *pull-back diagram*.

The construction of fibre products is one of the methods of constructing new G-spaces out of old G-spaces and G-maps. Fibre products of topological spaces are dealt with broadly in topology without considering G-actions. The induced bundle which will appear later is one such example. On the other hand, there are structures in algebra similar to fibre products. For example, if X, Y, and Z are rings and f, h are homomorphisms of rings, then the fibre product $X \underset{Z}{\times} Y$ is constructed. More generally fibre products are considered in category theory.

The fibre product has the universal property in the following sense.

Proposition 2.15 (Universality). *Let X, Y, Z, and W be G-spaces and let $f: X \to Z$ and $h: Y \to Z$ be G-maps. If $\alpha: W \to X$ and $\beta: W \to Y$ are G-maps satisfying $f \cdot \alpha = h \cdot \beta$, then there is a unique G-map $\gamma: W \to X \underset{Z}{\times} Y$ such that the following diagram commutes.*

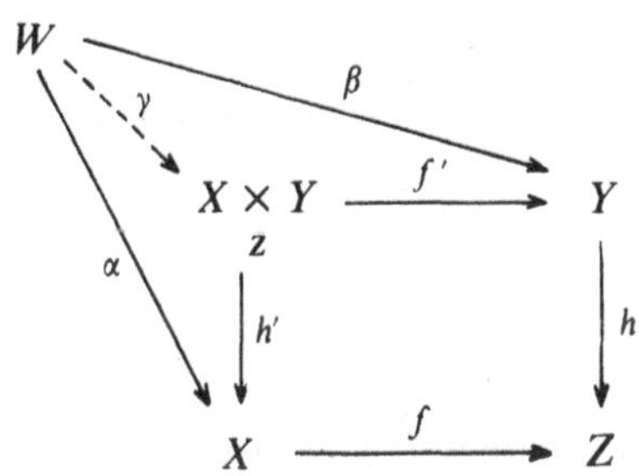

Proof If we put $\gamma(w) = (\alpha(w), \beta(w))$ for $w \in W$, then $\gamma(w)$ lies in $X \underset{Z}{\times} Y$ and the diagram commutes. The uniqueness follows from the commutativity of the triangles in the diagram above. □

Lemma 2.16 *In the pull-back diagram, we have:*

(1) *if f is surjective, then so is f';*

(2) *if f is injective, then so is f';*

(3) *if f is bijective, then so is f';*

(4) *if f is open, then so is f'.*

Similar results hold for h and h'.

Proof (1) It follows from the assumption that for any $y \in Y$, there exists $x \in X$ with $f(x) = h(y)$. Since (x, y) lies in $X \underset{Z}{\times} Y$ and satisfies $f'(x, y) = y$, f' is surjective.

(2) Let (x, y) and (x', y') be distinct elements in $X \underset{Z}{\times} Y$. That is, either $x \neq x'$ or $y \neq y'$ holds. If $x \neq x'$, we have by assumption

$$h(y) = f(x) \neq f(x') = h(y')$$

so that $y \neq y'$. Hence we have $y \neq y'$ in either case and

$$f'(x, y) = y \neq y' = f'(x', y'),$$

which implies that f' is injective.

(3) This is an immediate consequence of (1) and (2).

(4) Let O be a non-empty open set in $X \underset{Z}{\times} Y$. Given $(x, y) \in O$, we have only to find an open set V' of Y such that

$$f'(x, y) = y \in V' \subset f'(O).$$

It follows by definition of the topology of a product space and a subspace that there is an open set U of X and an open set V of Y satisfying

$$(x, y) \in X \underset{Z}{\times} Y \cap U \times V \subset O.$$

By assumption, $f(U)$ is open in Z, hence $h^{-1}(f(U))$ is an open set of Y containing y. Put

$$V' = V \cap h^{-1}(f(U)).$$

Then V' is an open set containing y. In the following we show the equality

$$f'\left(X \underset{Z}{\times} Y \cap U \times V'\right) = V'.$$

Obviously we have $f'(X \underset{Z}{\times} Y \cap U \times V') \subset V'$. Conversely, for $v' \in V'$, there exists $u \in U$ with $f(u) = h(v')$. Hence

$$(u, v') \in X \underset{Z}{\times} Y \cap U \times V'$$

and $f'(u, v') = v'$. That is, the reverse inclusion holds.

Summing up the results above, we have that V' is an open set of Y and that

$$y \in V' = f'\left(X \underset{Z}{\times} Y \cap U \times V'\right) \subset f'\left(X \underset{Z}{\times} Y \cap U \times V\right) \subset f'(O).$$

Thus we have shown that $f'(O)$ is an open set of Y. □

Theorem 2.17 *Given a fibre bundle $\xi = (E, \pi, X, F, K)$ and a continuous map $f\colon Y \to X$, we consider the pull-back diagram*

$$\begin{array}{ccc} Y \underset{X}{\times} E & \xrightarrow{\ f'\ } & E \\ \Big\downarrow{\scriptstyle \pi'} & & \Big\downarrow{\scriptstyle \pi} \\ Y & \xrightarrow{\ f\ } & X \end{array}$$

Then $\pi'\colon Y \underset{X}{\times} E \to Y$ is a fibre bundle with fibre F and structure group K, and f' is a bundle map.

Proof Let $\{\varphi_\alpha\colon \pi^{-1}(U_\alpha) \to U_\alpha \times F\}$ be the family of coordinate functions of ξ and $p'_\alpha\colon U_\alpha \times F \to F$ be the projections. We would like to define maps

$$\psi_\alpha\colon \pi'^{-1} f^{-1}(U_\alpha) \to f^{-1}(U_\alpha) \times F,$$

which will serve as coordinate functions of $Y \underset{X}{\times} E$. Since any $(y, z) \in \pi'^{-1} f^{-1}(U_\alpha)$ satisfies $f(y) = \pi(z)$, we can define ψ_α by

$$\psi_\alpha(y, z) = (y, p'_\alpha \cdot \varphi_\alpha(z)).$$

We now show that ψ_α is a homeomorphism.

Obviously ψ_α is continuous. Another map

$$\psi'_\alpha\colon f^{-1}(U_\alpha) \times F \to \pi'^{-1}(f^{-1}(U_\alpha))$$

is defined by

$$\psi'_\alpha(y, b) = (y, \varphi^{-1}_{\alpha, f(y)}(b)) \qquad \text{for } (y, b) \in f^{-1}(U_\alpha) \times F.$$

It is easy to see that $\psi'_\alpha \cdot \psi_\alpha = \psi_\alpha \cdot \psi'_\alpha = \mathrm{id}$, hence $\psi'_\alpha = \psi_\alpha^{-1}$. Since ψ'_α is continuous, ψ_α is a homeomorphism. Next we define

$$\psi_{\alpha, y}\colon \pi'^{-1}(y) = \{y\} \times \pi^{-1}(f(y)) \to F$$

by $\psi_{\alpha, y}(y, z) = \bar{p}'_\alpha \cdot \psi_\alpha(y, z)$ where $\bar{p}'_\alpha$ is the projection $\bar{p}'_\alpha\colon f^{-1}(U_\alpha) \times F \to F$.

If we put

$$\tau_{\beta\alpha}(y) = \psi_{\beta, y} \cdot \psi^{-1}_{\alpha, y}$$

for $y \in f^{-1}(U_\alpha) \cap f^{-1}(U_\beta)$, then we have

$$\tau_{\beta\alpha}(y) = \varphi_{\beta, f(y)} \cdot \varphi^{-1}_{\alpha, f(y)} = \theta_{\beta\alpha}(f(y)) \in K$$

where $\theta_{\beta\alpha}$ denotes the transition function for φ_α and φ_β. Hence $\tau_{\beta\alpha}\colon f^{-1}(U_\alpha) \cap f^{-1}(U_\beta) \to K$ is continuous. Thus we see that $\pi'\colon Y \underset{X}{\times} E \to Y$ is a fibre bundle with coordinate functions $\{\psi_\alpha\}$ and transition functions $\{\tau_{\beta\alpha}\}$.

Next we show that $f': Y \underset{X}{\times} E \to E$ is a bundle map. Let $f^{-1}(U_\alpha)$ (resp. U_β) be one of the coordinate neighbourhoods of the bundle $Y \underset{X}{\times} E$ (resp. E).

If we put

$$f_{\beta\alpha}(y) = \varphi_{\beta, f(y)} \cdot f' \cdot \psi_{\alpha, y}^{-1}$$

for $y \in f^{-1}(U_\alpha) \cap f^{-1}(U_\beta)$, then we have, for $b \in F$,

$$\begin{aligned} f_{\beta\alpha}(y)(b) &= \varphi_{\beta, f(y)} \cdot f'(y, \varphi_{\alpha, f(y)}^{-1}(b)) \\ &= \varphi_{\beta, f(y)} \cdot \varphi_{\alpha, f(y)}^{-1}(b) = \theta_{\beta\alpha}(f(y))(b). \end{aligned}$$

Hence $f_{\beta\alpha}(y) = \theta_{\beta\alpha}(f(y)) \in K$ and the map

$$f_{\beta\alpha}: f^{-1}(U_\alpha) \cap f^{-1}(U_\beta) \to K$$

is continuous. Thus f' is a bundle map. □

The fibre bundle $Y \underset{X}{\times} E$ obtained in Theorem 2.17 is called the *induced bundle* of E via f and is denoted by $f^*\xi$ or f^*E.

2.5 *G*-vector bundles

A fibre bundle with fibre $\mathbb{R}^n$ (an n-dimensional vector space over the real numbers) and structure group $GL(n, \mathbb{R})$ (the general linear group) is called a vector bundle. Here the action of $GL(n, \mathbb{R})$ on $\mathbb{R}^n$ is the ordinary one.

Although the above definition is transparent, the following definition is also easy to understand.

Definition 2.4 If topological spaces E and X and a continuous map $\pi: E \to X$ satisfy the following conditions (1) and (2), then (E, X, π) is called a *vector bundle*.

(1) For each $x \in X$, $\pi^{-1}(x)$ has the structure of a vector space over the real numbers $\mathbb{R}$.

(2) For each $x \in X$, there exist an open neighbourhood U of x and a homeomorphism $\varphi: \pi^{-1}(U) \to U \times \mathbb{R}^n$ such that $p \cdot \varphi = \pi$ and for each $y \in U$, $\varphi | \pi^{-1}(y): \pi^{-1}(y) \to \{y\} \times \mathbb{R}^n = \mathbb{R}^n$ is a linear isomorphism where $p: U \times \mathbb{R}^n \to U$ denotes the projection.

As in fibre bundles, we call E a total space, X a base space, $\pi^{-1}(x)$ a fibre over x, and $\varphi: \pi^{-1}(U) \to U \times \mathbb{R}^n$ a coordinate function. $\pi^{-1}(x)$ is often written as E_x.

Lemma 2.18 *This definition coincides with the above definition.*

Proof It is almost obvious that a vector bundle in the sense of the former definition is a vector bundle in the sense of the latter definition. Hence we will show that the latter definition implies the former. Let

$$\varphi_\alpha : \pi^{-1}(U_\alpha) \to U_\alpha \times \mathbb{R}^n, \qquad \varphi_\beta : \pi^{-1}(U_\beta) \to U_\beta \times \mathbb{R}^n$$

be two coordinate functions. For $x \in U_\alpha \cap U_\beta$, we put

$$\theta_{\beta\alpha}(x) = \varphi_{\beta,x} \cdot \varphi_{\alpha,x}^{-1} : \mathbb{R}^n \to \mathbb{R}^n$$

where $\varphi_{\alpha,x}$ and $\varphi_{\beta,x}$ are those maps defined as in Section 2.1. Since both $\varphi_{\alpha,x}$ and $\varphi_{\beta,x}$ are linear isomorphisms, $\theta_{\beta\alpha}(x)$ lies in $GL(n, \mathbb{R})$. By making use of the fact that the map

$$\varphi_\beta \cdot \varphi_\alpha^{-1} | (U_\alpha \cap U_\beta) \times \mathbb{R}^n : (U_\alpha \cap U_\beta) \times \mathbb{R}^n \to (U_\alpha \cap U_\beta) \times \mathbb{R}^n$$

is continuous, one easily verifies that the map

$$\theta_{\beta\alpha} : U_\alpha \cap U_\beta \to GL(n, \mathbb{R})$$

is continuous. Thus we see that E is a vector bundle in the former sense. □

Remark Lemma 2.18 asserts that a vector bundle is just a fibre bundle with fibre $\mathbb{R}^n$ which is associated with a principal $GL(n, \mathbb{R})$-bundle with respect to the natural $GL(n, \mathbb{R})$-action on $\mathbb{R}^n$.

Example 1 If we replace the fibre $I = [-1, 1]$ by $\mathbb{R}$ in Example 2 of Section 2.1, then it becomes a vector bundle.

Next we introduce the notion of G-vector bundles. A vector bundle with a G-action is not sufficient for the definition of a G-vector bundle. We demand that for each $g \in G$, g maps each fibre linearly onto the corresponding fibre.
We now give a strict definition.

Definition 2.5 Let G be a topological group and $\pi: E \to X$ be a vector bundle such that E and X are G-spaces and π is a G-map. If for each $g \in G$ and $x \in X$, the map

$$g : \pi^{-1}(x) \to \pi^{-1}(gx)$$

is a linear isomorphism, then $\pi: E \to X$ is called a *G-vector bundle.*

Remark When X is a point, the total space E is simply a G-representation space. (When G acts on a vector space V such that each $g \in G$ acts linearly, V together with a G-action is called a *G-representation space* in general.)

Let $\pi: E \to X$ and $\pi': E' \to X'$ be G-vector bundles. Suppose that two G-maps $\bar{f}: E \to E'$ and $f: X \to X'$ satisfy $\pi' \cdot \bar{f} = f \cdot \pi$, that is, the following diagram is commutative:

$$\begin{array}{ccc} E & \xrightarrow{\bar{f}} & E' \\ {\scriptstyle \pi}\downarrow & & \downarrow{\scriptstyle \pi'} \\ X & \xrightarrow{f} & X' \end{array}$$

If $\bar{f}|E_x\colon E_x \to E'_{f(x)}$ is linear for each $x \in X$, then $\bar{f}$ is called a *G-bundle homomorphism.*

If $\bar{f}|E_x\colon E_x \to E'_{f(x)}$ is a linear isomorphism for each $x \in X$, then $\bar{f}$ is called a *G-bundle map.*

If $X = X'$, $f = \mathrm{id}$, and $\bar{f}$ is a G-bundle map, then $\bar{f}$ is called a *G-bundle isomorphism.* Then we write $E \cong E'$.

The propriety of the term 'G-bundle isomorphism' will be clear from the following observation.

If $\bar{f}$ is a G-bundle isomorphism, $\bar{f}$ is a G-homeomorphism. Then by making use of the fact that the map $\beta\colon GL(n, \mathbb{R}) \to GL(n, \mathbb{R})$ defined by $\beta(A) = A^{-1}$ is continuous, we verify easily that $\bar{f}^{-1}$ is also a G-bundle isomorphism.

As in the case of fibre bundles, the notion of the induction of G-vector bundles is defined as follows. Let $\pi\colon E \to X$ be a G-vector bundle and $f\colon Y \to X$ be a G-map. The fibre product

$$f^*E = Y \underset{X}{\times} E = \{(y, v) \in Y \times E \mid f(y) = \pi(v)\}$$

is a G-invariant subspace of $Y \times E$ and we have G-maps $f'\colon f^*E \to E$ and $\pi'\colon f^*E \to Y$ by the restrictions of the natural projections. Then we have the following lemma whose proof is left to the reader, because if we have only to pay attention to the G-action, the proof is similar to that of Theorem 2.17.

Lemma 2.19 *$\pi'\colon f^*E \to Y$ is a G-vector bundle and $f'\colon f^*E \to E$ is a G-bundle map.*

Next we introduce the notion of the restriction of G-vector bundles and G-vector sub-bundles.

Let $\pi\colon E \to X$ be a G-vector bundle and $i\colon Y \to X$ be the inclusion map of a G-invariant subspace $Y \subset X$. If we put $E \mid Y = \pi^{-1}(Y)$, then $\pi\colon E \mid Y \to Y$ is a G-vector bundle. In fact, the local triviality is obtained by restricting that of E. If we define a map $\theta\colon E \mid Y \to i^*E$ by $\theta(v) = (y, v)$ for $v \in E_y (y \in Y)$, then θ is a natural G-bundle isomorphism. The bundle $E \mid Y$ so obtained is called the *restriction* of E to Y.

Let $\pi: E \to X$ be a G-vector bundle. If a G-invariant subspace E' of E and the restriction $\pi' = \pi | E': E' \to X$ satisfy the following conditions (1) and (2), then E' is called a *G-vector sub-bundle* of E.

(1) $E'_x = E' \cap E_x$ is a vector subspace of E_x for each $x \in X$.

(2) $\pi': E' \to X$ is a G-vector bundle with respect to the structure of the vector space on E'_x in (1).

Example 2 Given two G-vector bundles $\pi: E \to X$ and $\pi': E' \to X'$, the Cartesian product $\pi \times \pi': E \times E' \to X \times X'$ becomes a G-vector bundle as follows. Let $\varphi: \pi^{-1}(U) \to U \times \mathbb{R}^n$ and $\varphi': \pi'^{-1}(U') \to U' \times \mathbb{R}^{n'}$ be coordinate functions of E and E', respectively. Then the map

$$\begin{aligned}\varphi \times \varphi': (\pi \times \pi')^{-1}(U \times U') &= \pi^{-1}(U) \times \pi'^{-1}(U') \\ &\to U \times \mathbb{R}^n \times U' \times \mathbb{R}^{n'} = U \times U' \times \mathbb{R}^n \times \mathbb{R}^{n'}\end{aligned}$$

gives a coordinate function of $E \times E'$. Since $(\pi \times \pi')^{-1}(x, x') = \pi^{-1}(x) \times \pi'^{-1}(x')$ is the product of vector spaces, we put on it the structure of the direct sum of vector spaces. For each $g \in G$, the map

$$g: (\pi \times \pi')^{-1}(x, x') \to (\pi \times \pi')^{-1}(gx, gx')$$

is the direct sum of two linear isomorphisms, which is again a linear isomorphism. The G-vector bundle $\pi \times \pi': E \times E' \to X \times X'$ obtained in this way is called the *product* of E and E'.

Example 3 Consider two G-vector bundles $\pi: E \to X$ and $\pi': E' \to X$ over the same G-space X. Since the diagonal map $d: X \to X \times X$ defined by $d(x) = (x, x)$ is a G-map, we can consider the induced bundle $d^*(E \times E')$ of $E \times E'$ via d. We denote $d^*(E \times E')$ by $E \oplus E'$ which is called the *Whitney sum* of E and E'. Note that each fibre $(E \oplus E')_x$ is canonically isomorphic to the direct sum $E_x \oplus E'_x$. Moreover by identifying $\{(u, 0) \in E_x \times E'_x \mid x \in X\}$ (resp. $\{(0, v) \in E_x \times E'_x | x \in X\}$) with E (resp. E'), E and E' can be regarded as G-vector sub-bundles of $E \oplus E'$.

Lemma 2.20 *Let π: $E \to X$ be a G-vector bundle, and E_1 and E_2 be G-vector sub-bundles of E. If E_x is equal to the direct sum of the subspaces E_{1x} and E_{2x} for each $x \in X$, then E is isomorphic to the Whitney sum $E_1 \oplus E_2$.*

Proof An arbitrary element v of $(E_1 \oplus E_2)_x$ can be written uniquely as $v = (v_1, v_2)$ for $v_i \in E_{ix}$ and we define $f: E_1 \oplus E_2 \to E$ by $f(v) = v_1 + v_2$ where $+$ denotes the sum in E_x. Obviously f is continuous. Since $gv = (gv_1, gv_2)$ for $g \in G$, we have

$$f(gv) = gv_1 + gv_2 = g(v_1 + v_2) = gf(v),$$

that is, f is a G-map. By assumption, f is an isomorphism on each fibre. Hence f is a bundle isomorphism by definition. □

Theorem 2.21 *Let $\pi\colon E \to X$ be a G-vector bundle and $\pi'\colon E' \to X$ be a G-vector sub-bundle of E. For each $x \in X$, there exist an open neighbourhood U of x and a coordinate function $\varphi\colon \pi^{-1}(U) \to U \times \mathbb{R}^n$ of E such that $\varphi(\pi'^{-1}(U)) = U \times \mathbb{R}^m$ ($m \leq n$). Here $\mathbb{R}^m$ is regarded as a subspace of $\mathbb{R}^n$ by the following inclusion map:*

$$\mathbb{R}^m = \mathbb{R}^m \oplus \{0\} \subset \mathbb{R}^m \oplus \mathbb{R}^{n-m} \cong \mathbb{R}^n.$$

Proof Fix an arbitrary element x of X. There are two open neighbourhoods U_1 and U_2 of x over which there are coordinate functions of E and E', respectively. Hence by setting $U = U_1 \cap U_2$, we get coordinate functions

$$\varphi\colon \pi^{-1}(U) \to U \times \mathbb{R}^n, \qquad \psi\colon \pi'^{-1}(U) \to U \times \mathbb{R}^m.$$

By identifying $\{x\} \times \mathbb{R}^n$ with $\mathbb{R}^n$, $\varphi(E'_x)$ is an m-dimensional subspace of $\mathbb{R}^n$, which we denote by V. Since any two m-dimensional subspaces of $\mathbb{R}^n$ are transferred to each other by a linear isomorphism, there exists a matrix $A \in GL(n, \mathbb{R})$ such that $A(V) = \mathbb{R}^m$. Then the map

$$(\mathrm{id} \times A) \cdot \varphi\colon \pi^{-1}(U) \to U \times \mathbb{R}^n$$

is a coordinate function of E such that

$$(\mathrm{id} \times A) \cdot \varphi(E'_x) = \{x\} \times \mathbb{R}^m.$$

Therefore we may assume $\varphi(E'_x) = \{x\} \times \mathbb{R}^m$ without loss of generality. Although the equality $\varphi(E'_y) = \{y\} \times \mathbb{R}^m$ is not necessarily valid for every $y \in U$, there exists an open neighbourhood U' of x such that

$$\mathbb{R}^n = \varphi_y(E'_y) \oplus \mathbb{R}^{n-m} \qquad \text{for every } y \in U'.$$

This will be seen as follows. If

$$p\colon \mathbb{R}^n (= \mathbb{R}^m \oplus \mathbb{R}^{n-m}) \to \mathbb{R}^m$$

is the projection, we have $p \cdot \varphi(E'_x) = \mathbb{R}^m$. Choose a base of $\mathbb{R}^m$ so that $p \cdot \varphi_y \cdot \psi_y^{-1}$ is regarded as an element of $M(m, \mathbb{R})$ for $y \in U$ where $M(m, \mathbb{R})$ denotes the set of all $m \times m$ matrices over $\mathbb{R}$. Then the map $f\colon U \to M(m, \mathbb{R})$ defined by $f(y) = p \cdot \varphi_y \cdot \psi_y^{-1}$ is evidently continuous. Since $GL(m, \mathbb{R})$ is open in $M(m, \mathbb{R})$ and since $f(x) \in GL(m, \mathbb{R})$, the set U' defined by

$$U' = \{y \in U \mid p \cdot \varphi_y \cdot \psi_y^{-1} \in GL(m, \mathbb{R})\} = f^{-1}(GL(m, \mathbb{R}))$$

is open in U and contains x. Moreover U' coincides with $\{y \in U \mid p \cdot \varphi_y(E'_y) = \mathbb{R}^m\} = \{y \in U \mid \mathbb{R}^n = \varphi_y(E'_y) \oplus \mathbb{R}^{n-m}\}$, hence U' has the desired property. Hereafter we denote this U' by U for brevity.

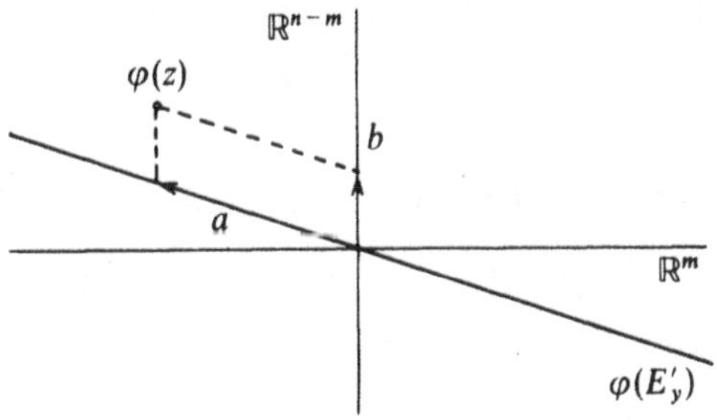

Fig. 2.3 The section $\{y\} \times \mathbb{R}^n$ through $y \in U$.

For arbitrary $y \in U$ and $z \in E_y$, we have a unique expression

$$\varphi(z) = a + b \qquad \text{for } a \in \varphi(E'_y),\ b \in \mathbb{R}^{n-m} = \{y\} \times \mathbb{R}^{n-m}$$

illustrated in Fig. 2.3.

Then we define a map

$$\Phi: \pi^{-1}(U) \to U \times (\mathbb{R}^m \oplus \mathbb{R}^{n-m}) = U \times \mathbb{R}^n$$

by $\Phi(z) = (y, \psi_y \cdot \varphi_y^{-1}(a) \oplus b)$.

Although it is obvious that the restriction of Φ to each fibre is a linear isomorphism, the homeomorphy of Φ is not so clear. To establish this, we first define a map

$$\Psi: U \times (\mathbb{R}^m \oplus \mathbb{R}^{n-m}) = U \times \mathbb{R}^n \to U \times \mathbb{R}^n$$

by

$$\Psi(y, v \oplus w) = (y, \varphi_y \cdot \psi_y^{-1}(v) + w)$$

for $y \in U$, $v \in \mathbb{R}^m$, and $w \in \mathbb{R}^{n-m}$. Evidently Ψ is continuous. Since $\Psi | \{y\} \times \mathbb{R}^n: \{y\} \times \mathbb{R}^n \to \{y\} \times \mathbb{R}^n$ is an isomorphism of vector spaces for every $y \in U$, the homeomorphy of Ψ will be shown in the same way as the proof that the inverse map of a bundle isomorphism is also a bundle isomorphism.

For arbitrary $y \in U$, $v \in \mathbb{R}^m$, $w \in \mathbb{R}^{n-m}$, we have

$$\begin{aligned}\Phi \cdot \varphi^{-1} \cdot \Psi(y, v \oplus w) &= \Phi \cdot \varphi^{-1}(y, \varphi_y \cdot \psi_y^{-1}(v) + w)\\ &= \Phi \cdot \varphi_y^{-1}(\varphi_y \cdot \psi_y^{-1}(v) + w)\\ &= (y, \psi_y \cdot \varphi_y^{-1} \cdot \varphi_y \cdot \psi_y^{-1}(v) \oplus w)\\ &= (y, v \oplus w),\end{aligned}$$

that is, $\Phi \cdot \varphi^{-1} \cdot \Psi = \mathrm{id}$.

Hence we have $\Phi = \Psi^{-1} \cdot \varphi | \pi^{-1}(U)$, which is a homeomorphism. By definition of Φ, we have

$$\Phi | \pi'^{-1}(U) = \psi | \pi'^{-1}(U).$$

Thus $\Phi\colon \pi^{-1}(U) \to U \times \mathbb{R}^n$ gives the desired coordinate function. □

Let $\pi\colon E \to X$ be a G-vector bundle and $\pi\colon E' \to X$ be a G-vector sub-bundle of E. Then we consider the relation $\sim$ in E defined by

$$z \sim z' \Leftrightarrow \pi(z) = \pi(z') \qquad \text{and} \qquad z - z' \in E'_{\pi(z)}$$

which is an equivalence relation. The quotient space $E/\sim$ is denoted by E/E'. Obviously there is a map $\bar{\pi}\colon E/E' \to X$ such that the following diagram

$$\begin{array}{ccc} E & \xrightarrow{\ p\ } & E/E' \\ & {\scriptstyle \pi}\searrow \quad \swarrow{\scriptstyle \bar{\pi}} & \\ & X & \end{array}$$

commutes, where p is the natural projection on the quotient space. By the definition of the quotient topology, $\bar{\pi}$ is continuous. Note that $(E/E')_x$ has the structure of the quotient vector space E_x/E'_x.

Theorem 2.22 *$\bar{\pi}\colon E/E' \to X$ is a G-vector bundle.*

Proof We show the local triviality of $\bar{\pi}\colon E/E' \to X$. Since there is a coordinate function $\varphi\colon \pi^{-1}(U) \to U \times \mathbb{R}^n$ with $\varphi(\pi'^{-1}(U)) = U \times \mathbb{R}^m$ by Theorem 2.21, there exists a unique map $\bar{\varphi}\colon \bar{\pi}^{-1}(U) \to U \times \mathbb{R}^{n-m}$ such that the following diagram

$$\begin{array}{ccc} \pi^{-1}(U) & \xrightarrow{\quad \varphi \quad} & U \times \mathbb{R}^n \\ {\scriptstyle p}\downarrow & & \downarrow{\scriptstyle p_1} \\ \bar{\pi}^{-1}(U) & \xrightarrow{\quad \bar{\varphi} \quad} & U \times (\mathbb{R}^n/\mathbb{R}^m) = U \times \mathbb{R}^{n-m} \end{array}$$

commutes, where p_1 denotes the natural projection. Then it is easy to see that $\bar{\varphi}$ is a homeomorphism and for each $y \in U$, $\bar{\varphi} | \bar{\pi}^{-1}(y)\colon \bar{\pi}^{-1}(y) \to \{y\} \times \mathbb{R}^{n-m} = \mathbb{R}^{n-m}$ is a linear isomorphism so that it gives a coordinate function of E/E'. Since the maps

$$g\colon E_x \to E_{gx}, \qquad g\colon E'_x \to E'_{gx}$$

of the G-action are linear isomorphisms, these maps induce a linear isomorphism

$$g: (E/E')_x = E_x/E'_x \to (E/E')_{gx} = E_{gx}/E'_{gx}.$$

Thus E/E' is a G-vector bundle. □

The G-vector bundle $\bar{\pi}$: $E/E' \to X$ so obtained is called the *quotient G-vector bundle* of E by E'.

Corollary 2.23 *Let π: $E \to X$ and π': $E' \to X$ be G-vector bundles over the same G-space X. Then E and E' are naturally G-vector sub-bundles of $E \oplus E'$ and satisfy*

$$E \cong (E \oplus E')/E'.$$

Proof Let i: $E \to E \oplus E'$ be the natural inclusion map. Since the composition of G-bundle homomorphisms

$$E \xrightarrow{i} E \oplus E' \xrightarrow{p} (E \oplus E')/E'$$

is also a G-bundle homomorphism and maps each fibre isomorphically onto the corresponding fibre, it follows by definition that the composition $p \cdot i$ is a G-bundle isomorphism. □

Definition 2.6 Let π: $E \to X$ be a G-vector bundle. Then the map s: $X \to E$ which assigns to each element x of X the zero vector of the fibre E_x is continuous and is called the *zero section.*

2.6 The classification of G-vector bundles over G/H

In this section we shall give the classification of G-vector bundles over a homogeneous space of G. Although this sort of G-vector bundle is special in comparison with general G-vector bundles, it is fundamental. The slice theorem which will be mentioned in Chapter 4 describes neighbourhoods of orbits and is expressed in terms of G-vector bundles over homogeneous spaces of G.

The purpose of this section is to see that G-vector bundles over G/H are determined completely by the H-action on the fibre over $eH \in GH$.

Let π: $E \to G/H$ be a G-vector bundle over G/H. Since the neutral coset eH is fixed by the H-action, the fibre $\pi^{-1}(eH)$ is H-invariant. Moreover, since E is a G-vector bundle, the map

$$h: \pi^{-1}(eH) \to \pi^{-1}(eH) \qquad \text{for each } h \in H$$

is a linear isomorphism. Thus we see that $\pi^{-1}(eH)$ is a representation space of H. For brevity, we denote $\pi^{-1}(eH)$ by V. Let H act on $G \times V$ by

$h \circ (g, v) = (gh^{-1}, hv)$ *for* $h \in H$, $g \in G$, $v \in V$ and let H act on G by $h \circ g = gh^{-1}$. Then the projection $G \times V \to G$ is an H-map and induces a map $q\colon G \times_H V \to G/H$ between their orbit spaces. Then we have the following theorem.

Theorem 2.24 *If we define a map* $f\colon G \times_H V \to E$ *by* $f([g, v]) = gv$, f *is a* G*-homeomorphism and the following diagram*

$$\begin{array}{ccc} G \times_H V & \xrightarrow{\ f\ } & E \\ & {\scriptstyle q}\searrow \quad \swarrow{\scriptstyle \pi} & \\ & G/H & \end{array}$$

commutes.

Proof It is easy to see that f is well defined and is a continuous G-map. To see that f is a homeomorphism, we consider a map $f'\colon E \to G \times_H V$ of the opposite direction as follows. For arbitrary $z \in E$, there exists an element g of G with $\pi(z) = gH$. If we set $v = g^{-1}z$, then we have

$$v = g^{-1}z \in \pi^{-1}(\pi(g^{-1}z)) = \pi^{-1}(g^{-1}\pi(z)) = \pi^{-1}(eH) = V.$$

We now define $f'\colon E \to G \times_H V$ by

$$f'(z) = [g, v] \in G \times_H V.$$

Since g is not unique in general, v is not unique either. Hence we need to show that the definition of f' does not depend on the choice of the representatives. Let $\pi(z) = g'H$ and $v' = g'^{-1}z$ be other such expressions. Put $h = g^{-1}g'$. Since $gH = g'H$, h lies in H and we have $hv' = g^{-1}g'g'^{-1}z = g^{-1}z = v$. Hence we have

$$[g', v'] = [g'h^{-1}, hv'] = [g, v],$$

that is, f' is well defined.

Next we show continuity of f'. For this, we consider the pull-back diagram of $\pi\colon E \to G/H$ via $p\colon G \to G/H$:

$$\begin{array}{ccc} G \times_{G/H} E & \xrightarrow{\ p'\ } & E \\ \Big\downarrow{\scriptstyle \pi'} & & \Big\downarrow{\scriptstyle \pi} \\ G & \xrightarrow{\ p\ } & G/H \end{array}$$

Since p is surjective and open, it follows from Lemma 2.16 that p' is also surjective and open. Recall the definition

$$G \underset{G/H}{\times} E = \{(g, z) \in G \times E \mid p(g) = \pi(z)\} \subset G \times E.$$

If we define $f_1\colon G \times E \to G \times E$ by $f_1(g, z) = (g, g^{-1}z)$, then f_1 is continuous and satisfies

$$f_1\,(G \underset{G/H}{\times} E) = G \times V.$$

Let $q_1\colon\ G \times V \to G \underset{H}{\times} V$ be the projection map and consider the following diagram:

$$\begin{array}{ccc} G \underset{G/H}{\times} E & \xrightarrow{\ f_1\ } & G \times V \\ {\scriptstyle p'}\big\downarrow & & \big\downarrow{\scriptstyle q_1} \\ E & \xrightarrow{\ f'\ } & G \underset{H}{\times} V \end{array}$$

For $(g, z) \in G \underset{G/H}{\times} E$, we have $p(g) = \pi(z)$ by definition and

$$q_1 \cdot f_1(g, z) = q_1(g, g^{-1}z) = [g, g^{-1}z],$$
$$f' \cdot p'(g, z) = f'(z) = [g, g^{-1}z],$$

that is, the diagram above commutes. Since both f_1 and q_1 are continuous and p' is surjective and open, it follows from Lemma 1.15 that f' is continuous.

For any $z \in E$, we choose $g \in G$ with $\pi(z) = gH$ and put $v = g^{-1}z$. Then we have

$$f \cdot f'(z) = f([g, v]) = gv = z,$$

that is, $f \cdot f' = \mathrm{id}$. For any $[g, v] \in G \underset{H}{\times} V$, we have

$$f' \cdot f([g, v]) = f'(gv) = [g, g^{-1}gv] = [g, v],$$

that is, $f' \cdot f = \mathrm{id}$.

Thus we have shown that f is a G-homeomorphism. Moreover the commutativity of the diagram in the theorem is easily seen and the proof of the theorem is completed. □

Theorem 2.25 *Let G be a topological group and H a closed subgroup of G. Then the projection $p\colon\ G \to G/H$ has a local cross-section if and only if $p\colon\ G \to G/H$ is a principal H-bundle.*

Proof If $p\colon G \to G/H$ is a principal H-bundle, then p evidently has a local cross-section. To show the converse, suppose that there exist an open neighbourhood U of $eH \in G/H$ and a continuous map $s\colon U \to G$ with $p \cdot s = \mathrm{id}$. Then we define a map

$$\varphi\colon p^{-1}(U) \to U \times H$$

by $\varphi(g) = (p(g), (s \cdot p(g))^{-1} g)$ and a map

$$\varphi'\colon U \times H \to p^{-1}(U)$$

by $\varphi'(u, h) = s(u)h$. In consequence,

$$\begin{aligned} \varphi' \cdot \varphi(g) &= \varphi'(p(g), (s \cdot p(g))^{-1} g) \\ &= (s \cdot p(g))(s \cdot p(g))^{-1} g = g, \\ \varphi \cdot \varphi'(u, h) &= \varphi(s(u)h) \\ &= (u, s(u)^{-1} s(u)h) = (u, h), \end{aligned}$$

that is, $\varphi' \cdot \varphi = \varphi \cdot \varphi' = \mathrm{id}$. Hence both φ and φ' are homeomorphisms and $\varphi' = \varphi^{-1}$.

Next, for any $g_0 \in G$, we define a map $s_{g_0}\colon g_0 U \to G$ by

$$s_{g_0}(x) = g_0 s(g_0^{-1} x) \qquad \text{for } x \in g_0 U$$

and a map $\varphi_{g_0}\colon p^{-1}(g_0 U) \to g_0 U \times H$ by

$$\varphi_{g_0}(g) = (p(g), (s_{g_0} \cdot p(g))^{-1} g) \qquad \text{for } g \in p^{-1}(g_0 U)$$

and a map $\varphi'_{g_0}\colon g_0 U \times H \to p^{-1}(g_0 U)$ by

$$\varphi'_{g_0}(x, h) = s_{g_0}(x)h = g_0 s(g_0^{-1} x)h \quad \text{for } x \in g_0 U, h \in H.$$

Then it can be shown, as in the case of φ and φ', that

$$\varphi_{g_0} \cdot \varphi'_{g_0} = \varphi'_{g_0} \cdot \varphi_{g_0} = \mathrm{id}.$$

Hence both φ_{g_0} and φ'_{g_0} are homeomorphisms and $\varphi'_{g_0} = \varphi_{g_0}^{-1}$.

Next we consider transition functions. For $g_0, g_1 \in G$ and $x \in g_0 U \cap g_1 U$, we have

$$\varphi_{g_1} \cdot \varphi'_{g_0}(x, h) = \varphi_{g_1}(s_{g_0}(x)h) = (x, (s_{g_1}(x))^{-1} s_{g_0}(x)h).$$

Observe that

$$s_{g_0}(x)H = p \cdot s_{g_0}(x) = x = p \cdot s_{g_1}(x) = s_{g_1}(x)H.$$

It turns out that $(s_{g_1}(x))^{-1} s_{g_0}(x) \in H$. Then we define a map

$$\theta_{g_1 g_0}\colon g_0 U \cap g_1 U \to H$$

by $\theta_{g_1 g_0}(x) = (s_{g_1}(x))^{-1} s_{g_0}(x)$ which is clearly continuous.

Thus we have shown that $p\colon G \to G/H$ is a principal fibre bundle with structure group H, coordinate functions $\varphi_{g_0}\colon p^{-1}(g_0 U) \to g_0 U \times H$, and transition functions $\theta_{g_1 g_0}\colon g_0 U \cap g_1 U \to H$. □

Remark It will be shown in Section 3.8 that $p: G \to G/H$ has a local cross-section when G is a Lie group and H is a closed subgroup of G.

Theorem 2.26 *If $p: G \to G/H$ has a local cross-section, then the converse of Theorem 2.24 holds. That is, for any given H-representation space V, $q: G \times_H V \to G/H$ is a G-vector bundle and $q^{-1}(eH)$ is isomorphic to V as H-representation spaces.*

Proof Although the H-action on V is not necessarily effective, we can consider a fibre bundle $q: G \times_H V \to G/H$ associated with the principal H-bundle $p: G \to G/H$ of Theorem 2.25 in a general sense as in Section 2.3. If we pass the value of each transition function from H into $GL(n, \mathbb{R})$ by H-representation (that is, 'extension' in Section 2.3), the bundle becomes a vector bundle. Then G acts on $G \times_H V$ by $(g, [g', v]) \mapsto [gg', v]$. It is immediate from the definition of the vector space structure on each fibre that the restriction of the action to each fibre

$$g: q^{-1}(g_0H) \to q^{-1}(gg_0H)$$

is linear. Thus we see that $q: G \times_H V \to G/H$ is a G-vector bundle. Since the H-action on $q^{-1}(eH)$ is given by

$$h \circ [e, v] = [h, v] = [e, hv],$$

it coincides with the H-action on V. □

Corollary 2.27 *If $p: G \to G/H$ has a local cross-section, then the assignment $V \mapsto G \times_H V$ gives a one-to-one correspondence between isomorphism classes of H-representation spaces and isomorphism classes of G-vector bundles over G/H.*

By combining Theorems 2.10 and 2.25, we have the following corollary.

Corollary 2.28 *If $p: G \to G/H$ has a local cross-section, then for any given effective H-space F, $G \times_H F \to G/H$ is a fibre bundle with fibre F.*

Theorem 2.29 *Let G be a topological group and H a closed subgroup of G. If $p: G \to G/H$ has a local cross-section, then $G \times_H G \to G/H$ is a principal G-bundle and the G-homeomorphism $f: G \times_H G \to (G/H) \times G$ in the proof of Proposition 1.98 gives an isomorphism of principal G-bundles. Here $(G/H) \times G$ denotes the trivial principal G-bundle over G/H.*

Proof It follows from Proposition 2.14 that $G \times_H G \to G/H$ is a principal G-bundle. It is easy to see that the map $f_{\mu\alpha}: g_0 U \to G$ with respect to the coordinate functions φ_{g_0} and id, respectively, in the definition of a bundle map is given by $f_{\mu\alpha}(x) = g_0 s(g_0^{-1}x)$ if we employ the notation in the proof of Theorem 2.25. Hence f gives an isomorphism of principal G-bundles. □

The following corollary will be of great value when we determine the stable tangent vector bundles of homogeneous spaces in Chapter 3.

Corollary 2.30 *Let G and H be as in Theorem 2.29 and F be a G-space. The G-homeomorphism $f\colon G \times_H F \to (G/H) \times F$ in the proof of Proposition 1.98 gives an isomorphism of fibre bundles with fibre F and base space G/H. In particular, if F is a G-representation space, then f gives an isomorphism of G-vector bundles.*

Remark The bundles appearing in Theorem 2.29 and Corollary 2.30 can be regarded as fibre bundles with structure group H (reducible!). However they are not isomorphic as bundles with structure group H (cf. Exercise 3).

2.7 Invariant integration on a compact topological group

In this section we summarize integration on compact groups. This will be needed in the next section. The reader may skip the present section, returning to it when necessary.

Let G be a topological group and F be the set of real valued continuous functions defined on G. That is, we put

$$F = \{f\colon G \to \mathbb{R} \text{ continuous}\}$$

For $h \in G$, define three maps

$$R_h\colon F \to F, \qquad L_h\colon F \to F, \qquad B\colon F \to F$$

by $(R_h f)(g) = f(gh)$, $(L_h f)(g) = f(hg)$, and $(Bf)(g) = f(g^{-1})$ for $g \in G$.

The *invariant integral* (or the *Haar integral*) on a compact topological group G is a map

$$I\colon F \to \mathbb{R}$$

satisfying the following conditions.

(1) $I(f_1 + f_2) = I(f_1) + I(f_2)$ for $f_i \in F$.

(2) $I(cf) = cI(f)$ for $c \in \mathbb{R}$, $f \in F$.

(3) If $f \in F$ satisfies $f(g) \geq 0$ for all $g \in G$, then $I(f) \geq 0$.

(4) If $f(g) = 1$ for all $g \in G$, then $I(f) = 1$.

(5) $I(R_h f) = I(L_h f) = I(f)$ for all $f \in F$, $h \in G$.

(6) $I(Bf) = I(f)$ for $f \in F$.

Among the above conditions, (1)–(3) are the usual properties of an integral. Condition (4) is a normalization and conditions (5) and (6) mean that the integral is invariant under left and right translations and conjugations.

The following theorem is fundamental.

Theorem 2.31 *There exists a unique invariant integral on every compact topological group.*

Proof We will not give the proof of this theorem here, but the reader is referred to Pontrjagin.[111] We will, however, briefly indicate the meaning of the theorem and the idea of its proof.

Let G be a finite group. Denote by $|G|$ the order of G. For a function $f\colon G \to \mathbb{R}$, we put

$$I(f) = \frac{1}{|G|} \sum_{g \in G} f(g).$$

Then it is easily verified that the map $I\colon F \to \mathbb{R}$ satisfies conditions (1)–(6) and hence I is the invariant integral. Thus the invariant integral is, as it were, a functional which assigns to each function its mean value. The proof in the case of a compact topological group uses essentially the same idea. □

We shall often use the following notation for $I(f)$:

$$I(f) = \int f(g)\,dg \qquad \text{or} \qquad \int_G f(g)\,dg,$$

so that conditions (5) and (6) can be written

(5) $\int f(gh)\,dg = \int f(hg)\,dg = \int f(g)\,dg$

(6) $\int f(g^{-1})\,dg = \int f(g)\,dg.$

While the former notation is well formed and easy to understand for the purpose of definition, the latter is more conveniently suited to computation.

In the following we prepare some well-known results which follow from Theorem 2.31.

Corollary 2.32 *If $f \in F$ satisfies $f(g) \geq 0$ for all $g \in G$ and is not identically* 0, *then we have*

$$I(f) > 0.$$

Proof Put $\mathbb{R}^+ = \{x \in \mathbb{R} \mid x > 0\}$. Then $U = f^{-1}(\mathbb{R}^+)$ is a non-empty open set of G. Since $\bigcup_{h \in G} hU$ is an open covering of G, we can find elements $h_1, \ldots, h_n$ in G such that

$$G = \bigcup_{i=1}^{n} h_i U$$

by compactness of G. Note that each function $L_{h_i^{-1}}(f)$ is positive on $h_i U$, hence the function

$$F = \sum_{i=1}^{n} L_{h_i^{-1}}(f)$$

is everywhere positive. Since G is compact, there exists a positive constant c such that $F(g) \geq c$ for all $g \in G$. By making use of the properties of the invariant integral, we compute

$$nI(f) - c = I\left(\sum_{i=1}^{n} L_{h_i^{-1}}(f) - c\right) = I(F - c) \geq 0.$$

Thus we have $I(f) \geq c/n > 0$. □

Theorem 2.33 *Let G be a compact topological group, A a topological space, and $f: G \times A \to \mathbb{R}$ a continuous function. Then the function $F: A \to \mathbb{R}$ defined by*

$$F(a) = \int f(g, a)\,\mathrm{d}g$$

is continuous.

Proof Let any positive number ε and any $a \in A$ be given. It follows from the compactness of G and the continuity of f that there is a neighbourhood U of a such that

$$|f(g, b) - f(g, a)| < \varepsilon$$

for all $b \in U$ and $g \in G$. Then

$$\begin{aligned}|F(b) - F(a)| &= \left|\int f(g, b)\,\mathrm{d}g - \int f(g, a)\,\mathrm{d}g\right| \\ &= \left|\int (f(g, b) - f(g, a))\,\mathrm{d}g\right| \\ &\leq \int |f(g, b) - f(g, a)|\,\mathrm{d}g < \varepsilon.\end{aligned}$$

Here the first inequality is easily proved from conditions (1)–(3) of the invariant integral. □

Theorem 2.34 *Let G be a compact topological group and $f: G \times \mathbb{R} \to \mathbb{R}$ be a continuous function. If $f(g, t)$ is differentiable in t with a derivative $\mathrm{d}f(g, t)/\mathrm{d}t$ which is continuous on $G \times \mathbb{R}$, then the function*

$$F(t) = \int f(g, t)\,\mathrm{d}g$$

is also differentiable and the equality

$$\frac{\mathrm{d}}{\mathrm{d}t}F(t) = \int \frac{\mathrm{d}}{\mathrm{d}t} f(g, t)\,\mathrm{d}g$$

holds.

Proof It follows from the mean value theorem that for a real number $s \neq 0$, there exists a real number θ with $0 < \theta < 1$ such that

$$\frac{f(g, t+s) - f(g, t)}{s} = \frac{\mathrm{d}}{\mathrm{d}t} f(g, t + \theta s).$$

Since G is compact and $\mathrm{d}f(g, t)/\mathrm{d}t$ is continuous, it follows that, for a fixed real number t and arbitrary $\varepsilon > 0$, there exists a $\delta > 0$ such that

$$\left| \frac{\mathrm{d}}{\mathrm{d}t} f(g, t+s) - \frac{\mathrm{d}}{\mathrm{d}t} f(g, t) \right| < \varepsilon$$

for all $g \in G$ and $|s| < \delta$. Therefore, for s with $0 \neq |s| < \delta$, we have

$$\left| \frac{f(g, t+s) - f(g, t)}{s} - \frac{\mathrm{d}}{\mathrm{d}t} f(g, t) \right| < \varepsilon$$

for all $g \in G$. Accordingly we have

$$\left| \int_G \left(\frac{f(g, t+s) - f(g, t)}{s} \right) \mathrm{d}g - \int_G \left(\frac{\mathrm{d}}{\mathrm{d}t} f(g, t) \right) \mathrm{d}g \right| < \varepsilon,$$

that is,

$$\left| \frac{F(t+s) - F(t)}{s} - \int_G \left(\frac{\mathrm{d}}{\mathrm{d}t} f(g, t) \right) \mathrm{d}g \right| < \varepsilon.$$

Since this inequality holds for arbitrary $\varepsilon > 0$ and for sufficiently small $s \neq 0$, this means that the limit

$$\lim_{s \to 0} \frac{F(t+s) - F(t)}{s}$$

exists and that it coincides with

$$\int_G \left(\frac{\mathrm{d}}{\mathrm{d}t} f(g, t) \right) \mathrm{d}g. \qquad \square$$

Corollary 2.35 *Let G be a compact topological group and $f\colon G \times \mathbb{R}^n \to \mathbb{R}$ be a continuous function. If $f(g, t_1, \ldots, t_n)$ is a function of class C^r on $\mathbb{R}^n$ for fixed $g \in G$ and the sth order partial derivatives are all continuous on $G \times \mathbb{R}^n$ for all s* *with* $0 \leq s \leq r$, *then the function* $F\colon \mathbb{R}^m \to \mathbb{R}$ *defined by*

$$F(t_1, \ldots, t_n) = \int_G f(g, t_1, \ldots, t_n)\,dg$$

is also C^r. *This holds for* $r = \infty$ *as well.*

Proof This is easily shown by an inductive use of Theorem 2.34. □

Although, so far, we have considered the invariant integral for functions $f\colon G \to \mathbb{R}$, it can be extended to an integral on vector valued continuous functions $f\colon G \to \mathbb{R}^n$, because we have only to integrate componentwise. That is, when we write $f(g)$ as

$$f(g) = (f_1(g), \ldots, f_n(g)) \qquad \text{for } g \in G,$$

we define it by

$$\int_G f(g)\,dg = \left(\int_G f_1(g)\,dg, \ldots, \int_G f_n(g)\,dg\right).$$

Theorem 2.36 *Let* $T\colon \mathbb{R}^n \to \mathbb{R}^k$ *be a linear map. For a continuous function* $f\colon G \to \mathbb{R}^n$, *we have*

$$\int_G T(f(g))\,dg = T\left(\int_G f(g)\,dg\right).$$

Proof This follows immediately from the linearity conditions (1) and (2) of the invariant integral. □

The decomposition theorem of the normal G-vector bundle of the fixed point set will play an important role in the study of G-manifolds in Chapters 4 and 6, including the localization theorem, and the following observation will be employed in the proof of the decomposition theorem of G-vector bundles over trivial G-spaces in general.

Let V be a representation space of a compact topological group G. For fixed $v \in V$, we define a map

$$f_v\colon G \to V$$

by

$$f_v(g) = gv \qquad \text{for } g \in G.$$

Since f_v is a vector valued continuous function defined on G, it follows from the preceding observation that we can consider the invariant integral

$$\int_G f_v(g)\,dg.$$

Then we define a map

$$\mu\colon V \to V$$

by

$$\mu(v) = \int_G f_v(g)\,dg = \int_G gv\,dg.$$

Theorem 2.37
(1) *μ is linear.*
(2) *μ is a projection operator, that is, $\mu^2 = \mu$.*
(3) *$\mu(V) = V^G$ (the fixed point set).*

Proof (1) Since the G-action on V is linear, the linearity of μ follows from the linearity conditions (1) and (2) of the invariant integral.

(2) Each element g of G defines a linear map $T(g)\colon V \to V$ given by $v \mapsto gv$. By making use of Theorem 2.36, we have, for fixed $g' \in G$,

$$\begin{aligned} g'\int_G gv\,dg &= T(g')\int_G T(g)v\,dg = \int_G T(g')T(g)v\,dg \\ &= \int_G T(g'g)v\,dg = \int_G (L_{g'}T)(g)v\,dg \\ &= \int_G T(g)v\,dg = \int_G gv\,dg. \end{aligned}$$

Hence we have, for any $v \subset V$,

$$\begin{aligned} \mu(\mu(v)) &= \int_G g'\left(\int_G gv\,dg\right)dg' = \int_G\left(\int_G gv\,dg\right)dg' \\ &= \int_G gv\,dg = \mu(v). \end{aligned}$$

(3) As shown in the proof of (2), the equality

$$g'\int_G gv\,dg = \int_G gv\,dg$$

holds for arbitrary $g' \in G$. Hence $\mu(v) \in V^G$, that is, $\mu(V) \subset V^G$. Conversely, for any $v \in V^G$, we have

$$\mu(v) = \int_G gv\,dg = \int_G v\,dg = v,$$

that is, $\mu \mid V^G = \mathrm{id}$. In particular, we have $\mu(V) \supset V^G$. □

Theorem 2.38 *Let X be a trivial G-space and $\pi: E \to X$ be a G-vector bundle. Then by considering the above map $\mu: V \to V$ on each fibre, we have a continuous map $\mu: E \to E$ with $\mu^2 = \mu$ and $\mu(E) = E^G$.*

Proof The above map $\mu_x: E_x \to E_x$ on each fibre E_x induces

$$\mu = \bigcup_{x \in X} \mu_x: E \to E.$$

The continuity of μ will be shown by making use of coordinate functions and Theorem 2.33. Since μ preserves each fibre and since $\mu_x^2 = \mu_x$ and $\mu_x(E_x) = E_x^G$ on each fibre E_x, μ has the desired property. □

Remark Since it is known in general that the image of a projection operator on a vector bundle is always a vector sub-bundle, E^G is a vector sub-bundle of E.

2.8 *G*-invariant metrics on *G*-vector bundles and orthogonal complements

A metric on a vector bundle $\pi: E \to X$ is a function which assigns to each point $x \in X$ an inner product (positive definite, symmetric, bilinear form) $\langle \cdot, \cdot \rangle_x$ on the fibre E_x in a continuous way. Here continuity of the metric is defined strictly as follows. Define a map $f: E \to \mathbb{R}$ by

$$f(v) = \langle v, v \rangle_{\pi(v)}.$$

When f is continuous, we say that the metric is continuous.

We now take G-actions into consideration. Let $\pi: E \to X$ be a G-vector bundle. A metric $\langle \cdot, \cdot \rangle$ on E is said to be *G-invariant* if

$$\langle gv, gw \rangle_{gx} = \langle v, w \rangle_x$$

for all $g \in G$, all $v, w \in E_x$, and all $x \in X$.

Theorem 2.39 *Let G be a compact topological group and $\pi: E \to X$ be a G-vector bundle over a paracompact G-space X. Then there exists a G-invariant metric on E.*

Proof First we show that there is a metric without considering G-actions. To see this, let

$$\varphi_\alpha: \pi^{-1}(U_\alpha) \to U_\alpha \times \mathbb{R}^n$$

be coordinate functions. Over each U_α we can consider an inner product $(\cdot, \cdot)_\alpha$ naturally. That is, for $v, w \in E_x$, $x \in U_\alpha$, we put

$$(v, w)_\alpha = (\varphi_\alpha^2(v), \varphi_\alpha^2(w)).$$

Here $\varphi_\alpha^2(v)$ denotes the $\mathbb{R}^n$ factor of $\varphi_\alpha(v)$, and $(\cdot,\cdot)$ of the right-hand side denotes the usual inner product in $\mathbb{R}^n$. Since, however, this depends on α, we cannot define a metric on E as it is. Therefore we piece them together by making use of a partition of unity as follows. Since X is paracompact, there is a locally finite open refinement $\{V_\mu\}$ of the covering $\{U_\alpha\}$, and a partition of unity $\{\lambda_\mu\}$ subordinate to $\{V_\mu\}$. For each V_μ, we choose U_α such that $V_\mu \subset U_\alpha$, which we denote by $U_{\alpha(\mu)}$. We now put

$$(v, w)_x = \sum_\mu \lambda_\mu(x)(v, w)_{\alpha(\mu)}$$

for $v, w \in E_x$ and $x \in X$. Since the indices μ with $\lambda_\mu(x) \neq 0$ are finite for each point x, the summation $\sum_\mu$ makes sense. Since each $(\cdot,\cdot)_{\alpha(\mu)}$ is a positive definite inner product, the $(\cdot,\cdot)_x$ defined above is also a positive definite inner product. On the other hand, continuity of $(\cdot,\cdot)_x$ concerning x follows from the definition of the partition of unity. Thus we obtain a metric $(\cdot,\cdot)$ on E.

Next we take the G-action into consideration. Corresponding to the metric $(\cdot,\cdot)$, we define a new metric $\langle\cdot,\cdot\rangle$ on E by putting

$$\langle v, w\rangle_x = \int_G (gv, gw)_{gx}\, dg,$$

where the right-hand side denotes the invariant integral. It follows from the properties of the invariant integral and Corollary 2.32 that $\langle\cdot,\cdot\rangle_x$ is a positive definite inner product. Since the integrand is continuous in x and g, it follows from Theorem 2.33 that $\langle\cdot,\cdot\rangle_x$ is continuous. For $h \in G$, we have

$$\begin{aligned}\langle hv, hw\rangle_{hx} &= \int_G (ghv, ghw)_{ghx}\, dg \\ &= \int_G (gv, gw)_{gx}\, dg \\ &= \langle v, w\rangle_x\end{aligned}$$

so that $\langle\cdot,\cdot\rangle$ is G-invariant. □

Let V be a vector space and W a subspace of V. If we choose a positive definite inner product in V, there exists the orthogonal complement $W^\perp$ of W in V so that we have a direct sum decomposition

$$V \cong W \oplus W^\perp.$$

This is extended to the case of a vector bundle and its vector sub-bundle. Furthermore, if G is a compact topological group, this is extended to the case of a G-vector bundle and its G-vector sub-bundle. Since the former is a special case of the latter, we show the latter.

Theorem 2.40 *Let G be a compact topological group and X be a paracompact G-space. Given a G-vector bundle $\pi: E \to X$ and its G-vector sub-bundle E', there exists a G-vector sub-bundle $E'^{\perp}$ of E such that*

$$E \cong E' \oplus E'^{\perp} \quad \textit{(Whitney sum).}$$

Proof Fix a G-invariant metric $\langle \cdot, \cdot \rangle$ on E. Then we put

$$E'^{\perp} = \bigcup_{x \in X} E_x'^{\perp}$$

where $E_x'^{\perp}$ denotes the orthogonal complement of E'_x in E_x with respect to the inner product $\langle \cdot, \cdot \rangle_x$. In the following we show that $E'^{\perp}$ is a G-vector sub-bundle of E and satisfies $E \cong E' \oplus E'^{\perp}$.

We consider the vector

$$e_i = (0, \ldots, 0, \overset{i}{1}, 0, \ldots, 0)$$

of $\mathbb{R}^n$ whose ith component is 1 and other components are all zero. Then the set $\{e_i \mid i = 1, \ldots, n\}$ is a basis of $\mathbb{R}^n$. Choose a coordinate function $\varphi: \pi^{-1}(U) \to U \times \mathbb{R}^n$ satisfying the conditions of Theorem 2.21. If we define $v_i: U \to \pi^{-1}(U)$ by $v_i(x) = \varphi^{-1}(x, e_i)$ for $x \in U$, then $\{v_1(x), \ldots, v_n(x)\}$ is a basis of E_x and v_i is a cross-section over U. Applying the Gram–Schmidt process to $v_1(x), \ldots, v_n(x)$ with respect to the inner product $\langle \cdot, \cdot \rangle_x$ we obtain a normal orthogonal basis $u_1(x), \ldots, u_n(x)$ for E_x. Since $\langle \cdot, \cdot \rangle$ is continuous, $u_i(x)$ is continuous and is a cross-section over U. By the property of the φ chosen above and the method of the Gram–Schmidt process, we see that for $x \in U$,

$$E'_x = \mathrm{Sp}\{v_1(x), \ldots, v_m(x)\} = \mathrm{Sp}\{u_1(x), \ldots, u_m(x)\},$$
$$E_x'^{\perp} = \mathrm{Sp}\{u_{m+1}(x), \ldots, u_n(x)\}$$

where $\mathrm{Sp}(S)$ denotes the linear subspace spanned by S and m denotes the fibre dimension of the bundle E'.

Next we define a new map

$$\psi: \pi^{-1}(U) \to U \times \mathbb{R}^n$$

as follows. When we write $v \in E_x$ as

$$v = \sum a_i u_i(x) \qquad a_i \in \mathbb{R},$$

we put

$$\psi(v) = (x, (a_1, \ldots, a_n)).$$

Obviously ψ is a coordinate function for E and satisfies

$$\psi(E' \mid U) = U \times \mathbb{R}^m$$
$$\psi(E'^{\perp} \mid U) = U \times \mathbb{R}^{n-m}$$

where we identify $\mathbb{R}^{n-m}$ with $\{0\} \times \mathbb{R}^{n-m}$ in $\mathbb{R}^m \times \mathbb{R}^{n-m} = \mathbb{R}^n$. Thus $E'^{\perp}$ is a vector sub-bundle of E with these coordinate functions

$$\psi \mid (E'^{\perp} \mid U) \colon E'^{\perp} \mid U \to U \times \mathbb{R}^{n-m}.$$

Moreover since the metric is G-invariant, $E'^{\perp}$ is a G-invariant subspace of E and the map

$$g \colon E_x'^{\perp} \to E_{gx}'^{\perp}$$

is a linear isomorphism. Hence $E'^{\perp}$ is a G-vector sub-bundle of E.

Since $E_x = E'_x \oplus E_x'^{\perp}$ for each $x \in X$, it follows from Lemma 2.20 that

$$E \cong E' \oplus E'^{\perp}. \qquad \square$$

The bundle $E'^{\perp}$ obtained in Theorem 2.40 is called the *orthogonal complement* of E' in E.

By combining Corollary 2.23 and Theorem 2.40, we have the following corollary.

Corollary 2.41 *Let G be a compact topological group and X be a paracompact G-space. Given a G-vector bundle $\pi \colon E \to X$ and a G-vector sub-bundle E' of E, we have*

$$E'^{\perp} \cong E/E'.$$

Since a G-vector bundle over one point is simply a representation space of G, the following corollary is an immediate consequence of Theorem 2.40.

Corollary 2.42 *Every representation of a compact topological group is completely reducible; that is, it is equivalent to the direct sum of irreducible representations.*

Given a G-vector bundle $\pi \colon E \to X$ with G-invariant metric $\langle \cdot, \cdot \rangle$, we put

$$D(E) = \{v \in E \mid \langle v, v \rangle_{\pi(v)} \leq 1\},$$

$$S(E) = \{v \in E \mid \langle v, v \rangle_{\pi(v)} = 1\}.$$

These are called the *unit disc bundle* and the *unit sphere bundle*, respectively. Both are G-invariant subspaces of E.

Theorem 2.43 *A real (resp. complex) representation of a compact topological group G is equivalent to an orthogonal (resp. unitary) representation of G.*

Proof It is well known that a real (resp. complex) representation V of a topological group is equivalent to an orthogonal (resp. unitary) representation if and only if there is a positive definite inner product (resp. positive definite Hermitian inner product) in V which is invariant under G. In the real case, Theorem 2.39 guarantees the existence of such an inner product. A similar proof shows that a positive definite Hermitian inner product exists for a complex G-vector bundle over a paracompact G-space. □

In fact, Theorem 2.43 is a special case of the following theorem.

Theorem 2.44 *Let G be a compact topological group and $\pi: E \to X$ be a real (resp. complex) G-vector bundle over a paracompact G-space. Then there is a reduction of the structure group of E from the general linear group over $\mathbb{R}$ (resp. $\mathbb{C}$) to the orthogonal (resp. unitary) group.*

Proof Theorem 2.39 guarantees the existence of a G-invariant metric on E, which gives rise to a reduction of the structure group of E from the general linear group over $\mathbb{R}$ to the orthogonal group. The proof for the complex case is similar. Compare Steenrod.[120] □

Exercises for Chapter 2

1. Define a map $p: \mathbb{R} \to S^1$ by $p(t) = \exp 2\pi i t$. Show that $p: \mathbb{R} \to S^1$ is a regular covering.

2. Prove the statement of Example 1 in Section 2.3.

3. Consider the bundle $S^1 \times_{\mathbb{Z}_n} S^1 \to S^1$ with fibre S^1 associated with the principal $\mathbb{Z}_n$-bundle $p: S^1 \to S^1$ of Example 1 in Section 2.3 where $\mathbb{Z}_n$ acts on S^1 in a natural way. Show that $S^1 \times_{\mathbb{Z}_n} S^1 \to S^1$ is not trivial as a fibre bundle with structure group $\mathbb{Z}_n$, but is trivial as a principal S^1-bundle.

4. Given a principal G-bundle $P \to X$ and a closed subgroup H of G, show that $P/H \to X$ becomes a fibre bundle with fibre G/H associated with $P \to X$. Show that if H is a closed normal subgroup of G, then $P/H \to X$ becomes a principal G/H-bundle.

(5) Let $\mathbb{Z}_2 = \{\pm 1\}$ act on $\mathbb{R}^2$ by $(-1)\cdot(a_1, a_2) = (-a_1, -a_2)$. Show that the projection $p: \mathbb{R}^2 \to \mathbb{R}^2/\mathbb{Z}_2$ does not have a local cross-section.

3 Manifolds and Lie groups

The aim of this chapter is to provide an introduction to the theory of differentiable manifolds and Lie groups.

The Euclidean space $\mathbb{R}^n$ is the most important space in mathematics in general; for example, differential calculus is performed extensively on Euclidean space. It is natural enough that we are then led to study those spaces which are locally like $\mathbb{R}^n$, and more precisely those spaces in which each point has a neighbourhood homeomorphic to $\mathbb{R}^n$. We call such spaces manifolds.

In other words, a manifold is a topological space constructed by piecing together copies of Euclidean space. The most familiar examples of manifolds are curves and surfaces like the sphere or torus; the surface of the earth is constructed by gluing two copies of Euclidean space (planes) together.

Calculus on Euclidean spaces extends to calculus on manifolds. Furthermore, it has come to be recognized in many areas of mathematics that the study of local and global structures of manifolds is a major subject, having its own point of view and methods. Thus it is no exaggeration to say that, in general, manifolds are basic concepts in mathematics. A Lie group is, roughly speaking, a manifold which is also a group in such a way that the group structure is compatible with the manifold structure. Naturally, Lie groups are suitable groups which serve as transformation groups on manifolds. Moreover, homogeneous spaces of Lie groups provide us with abundant examples of manifolds and clearly homogeneous spaces are closely related to transformation groups.

3.1 Manifolds and examples

In short, manifolds are topological spaces obtained by gluing copies of the Euclidean space $\mathbb{R}^n$ together.

There are several ways of piecing them together: through the use of homeomorphisms; diffeomorphisms; and analytic homeomorphisms, in which cases we call the resulting manifolds topological manifolds; differentiable (or smooth) manifolds; and analytic manifolds, respectively. Since we deal only with differentiable manifolds in this book, we mean differentiable manifolds whenever we use the term manifolds hereafter.

We now give a precise definition of manifolds. In order to simplify the terminology, we introduce the notions of smooth map and diffeomorphism between open sets of Euclidean spaces. Although we will make use of these terms in a more general setting later, we need them now for the definition of manifolds.

Let U and V be open sets of $\mathbb{R}^n$ and $\mathbb{R}^m$, respectively. A map $f\colon U \to V$ can be written as

$$f(x) = (f_1(x), \ldots, f_m(x)) \qquad \text{for } x \in U.$$

Then each $f_i(x)$ is called a *coordinate function* of f. We say that f is *smooth* or *of class* C^∞ if each coordinate function of f possesses continuous partial derivatives of all orders. If f is a homeomorphism and both f and f^{-1} are smooth, we say that f is a *diffeomorphism.*

Definition 3.1 Let M be a topological space and $\mathbb{R}^n$ be n-dimensional Euclidean space. A pair (U, φ) consisting of an open set U of M and a homeomorphism φ of U onto an open set of $\mathbb{R}^n$ is called a *chart* (or *coordinate neighbourhood*) of M. When we write $\varphi(x)$ as

$$\varphi(x) = (\varphi_1(x), \ldots, \varphi_n(x)) \qquad \text{for } x \in U,$$

$\varphi_i(x)$ is called the ith *coordinate function.* If a point x of M lies in U, then we say that (U, φ) is a *chart at x.*

A collection $\{(U_\alpha, \varphi_\alpha)\}$ of charts is called an *atlas* (or *system of coordinate neighbourhoods*) on M if $\{U_\alpha\}$ forms an open covering of M. The atlas is *smooth* if for each pair of indices α, β with $U_\alpha \cap U_\beta \neq \varnothing$, the map

$$\varphi_\beta \cdot \varphi_\alpha^{-1}\colon \varphi_\alpha(U_\alpha \cap U_\beta) \to \varphi_\beta(U_\alpha \cap U_\beta)$$

is smooth (that is, of class C^∞). These definitions are illustrated in Fig. 3.1. Unless otherwise stated we will use *atlas* to mean smooth atlas hereafter.

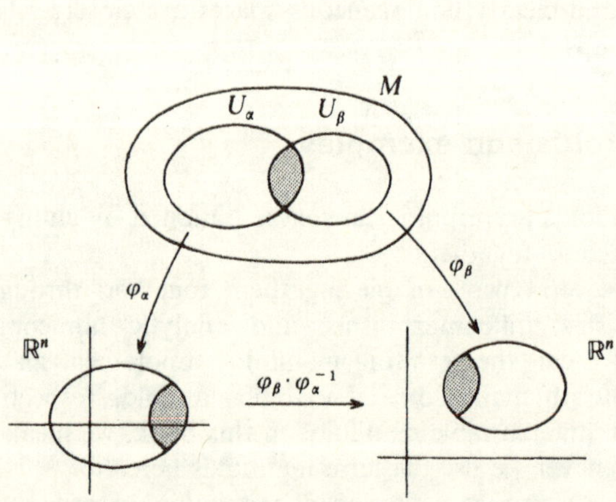

Fig. 3.1 A smooth atlas.

Remark Since $(\varphi_\beta \cdot \varphi_\alpha^{-1})^{-1} = \varphi_\alpha \cdot \varphi_\beta^{-1}$, $(\varphi_\beta \cdot \varphi_\alpha^{-1})^{-1}$ is also smooth by definition, and hence the map $\varphi_\beta \cdot \varphi_\alpha^{-1}$ is in fact a diffeomorphism.

Definition 3.2 We say that two atlases $\{(U_\alpha, \varphi_\alpha)\}$ and $\{(V_\beta, \psi_\beta)\}$ on M are *equivalent* if the union $\{(U_\alpha, \varphi_\alpha)\} \cup \{(V_\beta, \psi_\beta)\}$ is smooth. That is, for each pair of indices α, β with $U_\alpha \cap V_\beta \neq \varnothing$, the map

$$\psi_\beta \cdot \varphi_\alpha^{-1}: \varphi_\alpha(U_\alpha \cap V_\beta) \to \psi_\beta(U_\alpha \cap V_\beta)$$

is a diffeomorphism. An equivalence class of atlases on M is called a *differential* or *smooth structure* on M. The smooth structure represented by an atlas $\{(U_\alpha, \varphi_\alpha)\}$ is denoted by $[(U_\alpha, \varphi_\alpha)]$.

A pair $(M, [(U_\alpha, \varphi_\alpha)])$ consisting of a Hausdorff space M satisfying the second countability axiom and a smooth structure $[(U_\alpha, \varphi_\alpha)]$ on M is called a *differentiable* or *smooth manifold* of dimension n (or n-dimensional *manifold*, or simply *manifold*).

Remark The countability condition is intended to avoid some pathological cases. Since the expression $(M, [(U_\alpha, \varphi_\alpha)])$ is too long, we often denote it briefly by M. In this case, we have a tacit understanding that a smooth structure $[(U_\alpha, \varphi_\alpha)]$ is appointed on M.

The notion that a map from a manifold to another manifold is smooth is introduced naturally as follows.

A continuous map $f: M \to N$ between two smooth manifolds $(M, [(U_\alpha, \varphi_\alpha)])$ and $(N, [(V_\beta, \psi_\beta)])$ is said to be *smooth* or C^∞ (or *differentiable*) if for each pair of indices α, β with $U_\alpha \cap f^{-1}(V_\beta) \neq \varnothing$, the map

$$\psi_\beta \cdot f \cdot \varphi_\alpha^{-1}: \varphi_\alpha(U_\alpha \cap f^{-1}(V_\beta)) \to \psi_\beta(V_\beta)$$

is smooth in the previous sense. Figure 3.2 illustrates this definition.

A smooth map $f: M \to N$ between smooth manifolds is a *diffeomorphism* if it is a homeomorphism and the inverse map f^{-1} is also smooth. M and N are *diffeomorphic* if there exists a diffeomorphism $f: M \to N$.

This extends the concept of smooth map and diffeomorphism, previously defined for open subsets of $\mathbb{R}^n$ only, to arbitrary smooth manifolds.

We now give two methods of manufacturing new manifolds out of old.

Open submanifolds Let $(M, [(U_\alpha, \varphi_\alpha)])$ be a smooth manifold and O be an open subset of M. Then it is easily seen that $\{(O \cap U_\alpha, \varphi_\alpha | O \cap U_\alpha)\}$ is a smooth atlas on O. Hence O is a smooth manifold with smooth structure $[(O \cap U_\alpha, \varphi_\alpha | O \cap U_\alpha)]$, which is called an *open submanifold* of M.

Product manifolds. Let $(M, [(U_\alpha, \varphi_\alpha)])$ and $(N, [(V_\beta, \psi_\beta)])$ be two smooth manifolds. Then it is easy to see that $\{(U_\alpha \times V_\beta, \varphi_\alpha \times \psi_\beta)\}$ is a smooth atlas on the product space $M \times N$. Hence $M \times N$ is a smooth manifold with smooth

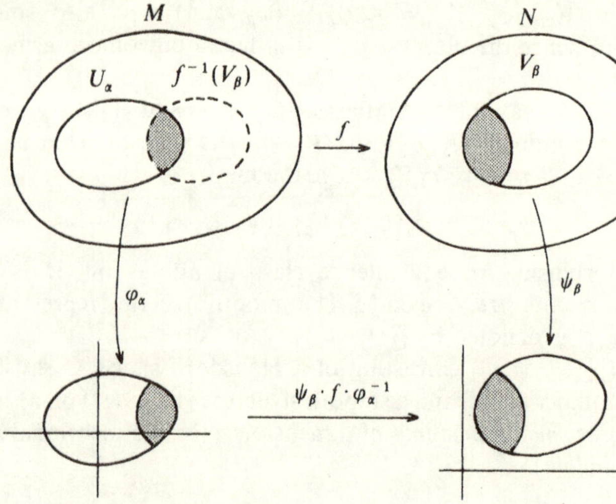

Fig. 3.2 A smooth map between two smooth manifolds.

structure $[(U_\alpha \times V_\beta, \varphi_\alpha \times \psi_\beta)]$, which is called the *product manifold of M and N.*

Proposition 3.1

(1) *The projections $\pi_1 : M \times N \to M$ and $\pi_2 : M \times N \to N$ are smooth.*

(2) *Let W be another smooth manifold. Then a map $f: W \to M \times N$ is smooth if and only if $\pi_1 \cdot f$ and $\pi_2 \cdot f$ are smooth.*

Proof (1) is obvious by definition.

(2) Since each chart of $M \times N$ has the form $\varphi_\alpha \times \psi_\beta$ by definition, (2) follows immediately from the following observation. A map $f: \mathbb{R}^a \to \mathbb{R}^b \times \mathbb{R}^c$ between Euclidean spaces is smooth if and only if the compositions $\pi_1 \cdot f$ and $\pi_2 \cdot f$ are smooth where $\pi_1 : \mathbb{R}^b \times \mathbb{R}^c \to \mathbb{R}^b$ and $\pi_2 : \mathbb{R}^b \times \mathbb{R}^c \to \mathbb{R}^c$ are the projection maps. □

Definition 3.3 Let M be an m-dimensional manifold, N be a subset of M and n be an integer with $0 \leq n \leq m$. If for each $a \in N$, there exists a chart (U, φ) of M at a such that

$$N \cap U = \{x \in U \mid \varphi_{n+1}(x) = \cdots = \varphi_m(x) = 0\},$$

then we call N a (regular) *submanifold* of M. Here $\varphi_i(x)$ denotes the ith coordinate function of $\varphi(x)$. A submanifold is sketched in Fig. 3.3.

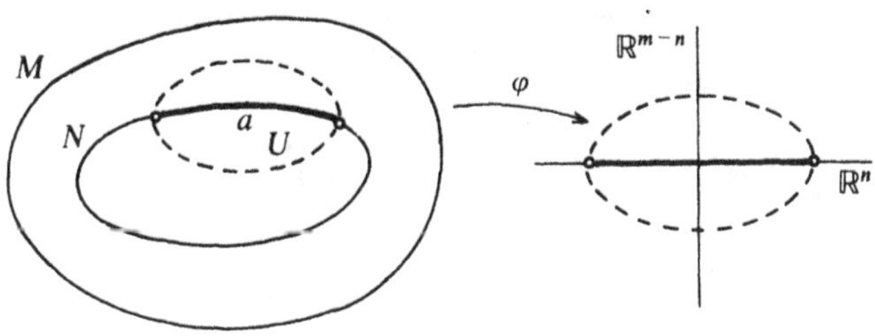

Fig. 3.3 A submanifold

Theorem 3.2 *A submanifold N of a smooth manifold M is a smooth manifold.*

Proof Corresponding to each chart $(U_\alpha, \varphi_\alpha)$ which appeared in the definition of submanifold, we define a map $\overline{\varphi_\alpha}: N \cap U_\alpha \to \mathbb{R}^n$ by

$$\overline{\varphi_\alpha}(x) = (\varphi_{\alpha 1}(x), \ldots, \varphi_{\alpha n}(x)) \qquad \text{for } x \in N \cap U_\alpha$$

where $\varphi_{\alpha i}(x)$ denotes the ith coordinate function of $\varphi_\alpha(x)$. Then N is a smooth manifold with atlas $\{(N \cap U_\alpha, \overline{\varphi_\alpha})\}$. The proof of this can be left safely to the reader as an exercise. □

We now describe a lemma which will be needed later.

Lemma 3.3 *Let M_i be a smooth manifold and N_i be a submanifold of M_i where $i = 1, 2$. If a smooth map $f: M_1 \to M_2$ satisfies $f(N_1) \subset N_2$, then the restriction $f | N_1: N_1 \to N_2$ is smooth.*

Proof This follows immediately from the definition of submanifolds. □

We now give some fundamental examples of manifolds.

Example 1 *The Euclidean space $\mathbb{R}^n$.* Put $M = U = \mathbb{R}^n$ and $\varphi = \mathrm{id}: U \to \mathbb{R}^n$. Then the atlas $\{(U, \varphi)\}$ consisting only of a single chart (U, φ) determines a smooth structure on M.

Example 2 *The n-sphere S^n.* The n-sphere S^n is given by the set

$$S^n = \{x = (x_1, \ldots, x_{n+1}) \in \mathbb{R}^{n+1} \mid x_1^2 + \cdots + x_{n+1}^2 = 1\}.$$

Take an arbitrary real number a with $0 < a \leq 1$. Then the sets

$$U_+ = \{x = (x_1, \ldots, x_{n+1}) \in S^n \mid x_{n+1} < a\},$$
$$U_- = \{x = (x_1, \ldots, x_{n+1}) \in S^n \mid x_{n+1} > -a\}$$

are open and cover S^n. We now define maps $\varphi_+: U_+ \to \mathbb{R}^n$ and $\varphi_-: U_- \to \mathbb{R}^n$

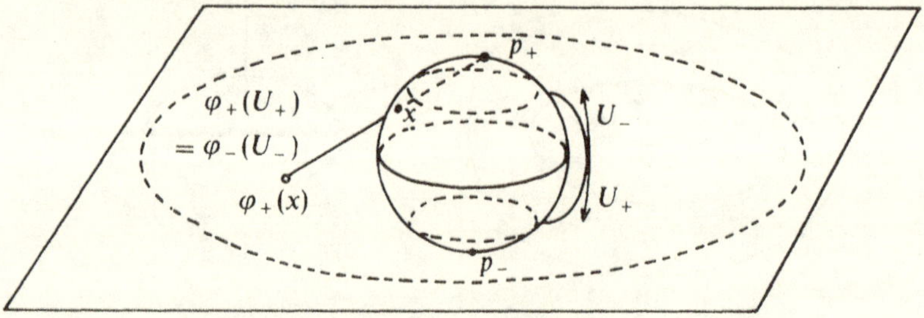

Fig. 3.4 Stereographic projections from S^n.

by the stereographic projections from the north pole $p_+ = (0, \ldots, 0, 1)$ and the south pole $p_- = (0, \ldots, 0, -1)$, respectively.

Referring to Fig. 3.4, we easily compute the following:

$$\varphi_+(x_1, \ldots, x_{n+1}) = \left(\frac{x_1}{1 - x_{n+1}}, \ldots, \frac{x_n}{1 - x_{n+1}}\right) \quad \text{for} \quad x_{n+1} < a,$$

$$\varphi_-(x_1, \ldots, x_{n+1}) = \left(\frac{x_1}{1 + x_{n+1}}, \ldots, \frac{x_n}{1 + x_{n+1}}\right) \quad \text{for} \quad x_{n+1} > -a.$$

Then it is easily verified that if $a < 1$, then both images $\varphi_+(U_+)$ and $\varphi_-(U_-)$ coincide with the open n-disc of radius $\sqrt{(1+a)/(1-a)}$:

$$\mathring{D} = \left\{ y = (y_1, \ldots, y_n) \in \mathbb{R}^n \,\middle|\, \sum_{i=1}^{n} y_i^2 < \frac{1+a}{1-a} \right\}$$

and we have for $y \in \mathring{D}$,

$$(\varphi_+)^{-1}(y) = \left(\frac{2y_1}{1 + \|y\|^2}, \ldots, \frac{2y_n}{1 + \|y\|^2}, \frac{-1 + \|y\|^2}{1 + \|y\|^2}\right),$$

$$(\varphi_-)^{-1}(y) = \left(\frac{2y_1}{1 + \|y\|^2}, \ldots, \frac{2y_n}{1 + \|y\|^2}, \frac{1 - \|y\|^2}{1 + \|y\|^2}\right)$$

where $\|y\| = \sqrt{y_1^2 + \cdots + y_n^2}$. Moreover we have

$$\varphi_+(U_+ \cap U_-) = \varphi_-(U_+ \cap U_-) = \left\{ y \in \mathbb{R}^n \,\middle|\, \frac{1-a}{1+a} < \|y\|^2 < \frac{1+a}{1-a} \right\}$$

and

$$\varphi_- \cdot \varphi_+^{-1}(y) = \frac{y}{\|y\|^2}, \qquad \varphi_+ \cdot \varphi_-^{-1}(y) = \frac{y}{\|y\|^2}$$

for $y \in \varphi_+(U_+ \cap U_-)$. Therefore the maps

$$\varphi_- \cdot \varphi_+^{-1} : \varphi_+(U_+ \cap U_-) \to \varphi_-(U_+ \cap U_-)$$

$$\varphi_+ \cdot \varphi_-^{-1} : \varphi_-(U_+ \cap U_-) \to \varphi_+(U_+ \cap U_-)$$

are smooth. Thus we see that S^n is a manifold with atlas $\{(U_+, \varphi_+), (U_-, \varphi_-)\}$.

The proof is similar when $a = 1$, the only difference being that

$$\varphi_+(U_+) = \varphi_-(U_-) = \mathbb{R}^n,$$

$$\varphi_+(U_+ \cap U_-) = \varphi_-(U_+ \cap U_-) = \mathbb{R}^n - \{0\}$$

in this case.

Example 3 *Projective space* ΛP^n. Denote by Λ either the real numbers $\mathbb{R}$ or the complex numbers $\mathbb{C}$. Put $\Lambda^* = \Lambda - \{0\}$ which is a topological group with respect to multiplication. Denote by $\Lambda^{n+1} - \{0\}$ the set of all $(n+1)$-tuples $x = (x_0, \ldots, x_n)$ of elements of Λ except $0 = (0, \ldots, 0)$. Then Λ^* acts freely on $\Lambda^{n+1} - \{0\}$ by

$$(\lambda, x) \mapsto \lambda x = (\lambda x_0, \ldots, \lambda x_n)$$

for $\lambda \in \Lambda^*$, $x = (x_0, \ldots, x_n) \in \Lambda^{n+1} - \{0\}$. Its orbit space $(\Lambda^{n+1} - \{0\})/\Lambda^*$ is denoted by ΛP^n, that is, if $\Lambda = \mathbb{R}$ (resp. $\mathbb{C}$), then it is denoted by $\mathbb{R}P^n$ (resp. $\mathbb{C}P^n$) and is called the *real* (resp. *complex*) *projective space* of dimension n.

We now give an atlas on ΛP^n as follows. Let $\pi: \Lambda^{n+1} - \{0\} \to \Lambda P^n$ be the projection and denote the image $\pi(x_0, \ldots, x_n)$ of $(x_0, \ldots, x_n) \in \Lambda^{n+1} - \{0\}$ by $[x_0, \ldots, x_n]$. Put

$$U_i = \{[x_0, \ldots, x_n] \in \Lambda P^n \mid x_i \neq 0\} \qquad \text{for} \quad i = 0, 1, \ldots, n$$

and define maps $\varphi_i: U_i \to \Lambda^n$ by

$$\varphi_i([x_0, \ldots, x_n]) = \left(\frac{x_0}{x_i}, \ldots, \frac{\hat{x}_i}{x_i}, \ldots, \frac{x_n}{x_i}\right)$$

where the caret ^ means that the term with ^ is deleted. It follows from the definition of the quotient topology that the φ_i are continuous. On the other hand, define maps $\psi_i: \Lambda^n \to U_i$ by

$$\psi_i(y_1, \ldots, y_n) = [y_1, \ldots, y_i, 1, y_{i+1}, \ldots, y_n].$$

Then the ψ_i are also continuous and satisfy $\psi_i \cdot \varphi_i = \varphi_i \cdot \psi_i = \mathrm{id}$. Hence the φ_i are homeomorphisms and we have $\varphi_i^{-1} = \psi_i$.

Suppose now that $i < j$. Then the map $\varphi_j \cdot \varphi_i^{-1}: \varphi_i(U_i \cap U_j) \to \varphi_j(U_i \cap U_j)$ is given by

$$\begin{aligned} \varphi_j \cdot \varphi_i^{-1}(y_1, \ldots, y_n) &= \varphi_j \cdot \psi_i(y_1, \ldots, y_n) \\ &= \left(\frac{y_1}{y_j}, \ldots, \frac{y_i}{y_j}, \frac{1}{y_j}, \ldots, \frac{\hat{y}_j}{y_j}, \ldots, \frac{y_n}{y_j}\right) \end{aligned}$$

and hence is smooth. The proof for the case $i > j$ is similar.

Since $\Lambda^{n+1} - \{0\}$ satisfies the second countability axiom, it follows from Corollary 1.56 that ΛP^n satisfies the second countability axiom as well.

To see that ΛP^n is Hausdorff, consider the groups

$$Z_2 = \{r \in \mathbb{R} \mid |r| = 1\} = \{\pm 1\}$$
$$S^1 = \{z \in \mathbb{C} \mid |z| = 1\}$$

which are subgroups of $\mathbb{R}^*$ $(=\mathbb{R} - \{0\})$ and $\mathbb{C}^*$ $(=\mathbb{C} - \{0\})$, respectively. Hence Z_2 and S^1 act freely on $\mathbb{R}^{n+1} - \{0\}$ and $\mathbb{C}^{n+1} - \{0\}$, respectively, by the restriction. Evidently the unit spheres $S^n \subset \mathbb{R}^{n+1} - \{0\}$ and $S^{2n+1} \subset \mathbb{C}^{n+1} - \{0\}$ are invariant under these actions. Then it is easily verified that $\mathbb{R}P^n$ (resp. $\mathbb{C}P^n$) is canonically homeomorphic to the orbit space S^n/Z_2 (resp. S^{2n+1}/S^1). Hence it follows from Proposition 1.58 that both $\mathbb{R}P^n$ and $\mathbb{C}P^n$ are Hausdorff. Thus ΛP^n is endowed with a manifold structure.

Finally we shall show that the projection $\pi : \Lambda^{n+1} - \{0\} \to \Lambda P^n$ has a smooth local cross-section, a result which will be needed later. Let U_i be as above. Define a map $s_i : U_i \to \Lambda^{n+1} - \{0\}$ by

$$s_i([x_0, \ldots, x_n]) = \left(\frac{x_0}{x_i}, \ldots, \frac{x_n}{x_i}\right).$$

Note that the ith term is not deleted this time. Evidently s_i does not depend on the choice of the representatives and satisfies $\pi \cdot s_i = \mathrm{id}$. Since $s_i \cdot \psi_i : \Lambda^n \to \Lambda^{n+1} - \{0\}$ is given by

$$s_i \cdot \psi_i(y_1, \ldots, y_n) = (y_1, \ldots, y_i, 1, y_{i+1}, \ldots, y_n),$$

we see that s_i is smooth. Thus we have shown that s_i is a smooth local cross-section. Needless to say, the projection π is smooth as well.

Example 4 By the following natural inclusions, the former are submanifolds of the latter. For $n < m$, $\mathbb{R}^n \subset \mathbb{R}^m$, $S^n \subset \mathbb{R}^{n+1}$, $S^n \subset S^m$, $\mathbb{R}P^n \subset \mathbb{R}P^m$, and $\mathbb{C}P^n \subset \mathbb{C}P^m$.

3.2 Lie groups and their actions

As defined previously, a topological group is a topological space which is also a group in such a way that the group structure is compatible with the topological structure.

Similarly, a Lie group is a manifold which is also a group in such a way that the group structure is compatible with the smooth structure. The following is a precise definition of a Lie group.

Definition 3.4 G is a *Lie group* provided that:

(1) G is a manifold;

(2) G is a group;

(3) the group operation is smooth, that is, the maps $\alpha: G \times G \to G$ and $\beta: G \to G$ defined by $\alpha(g, h) = gh$ and $\beta(g) = g^{-1}$ for g, $h \in G$ are both smooth. Here $G \times G$ is endowed with the structure of the product manifold.

A map $f: G_1 \to G_2$ between two Lie groups is called a *homomorphism of Lie groups* provided that:

(1) f is smooth;

(2) f is a homomorphism of groups.

Example 1 *Euclidean space* $\mathbb{R}^n$. The space $\mathbb{R}^n$ is a manifold, as mentioned previously, and at the same time an abelian group with group operation given by componentwise addition. Moreover the group operations $(x, y) \mapsto x + y$ and $x \mapsto -x$ are evidently smooth. Hence $\mathbb{R}^n$ is a Lie group.

Example 2 *The torus* T^n. Let Z^n be the set of points in $\mathbb{R}^n$ all of whose coordinates are integers. Then Z^n is a normal subgroup of $\mathbb{R}^n$. The quotient group $\mathbb{R}^n/Z^n$ is denoted by T^n. Since the projection $\pi: \mathbb{R}^n \to T^n$ is a (regular) covering, we have charts given by the restriction of π to small open sets. The atlas consisting of these charts is evidently smooth so that T^n is a manifold. Since the group operations of $\mathbb{R}^n$ are smooth, those of T^n are also smooth.

Example 3 *General linear groups* $GL(n, \mathbb{R})$, $GL(n, \mathbb{C})$. Let us first recall some notation. Let $M(n, \mathbb{R})$ be the set of $n \times n$ real matrices. We identify $M(n, \mathbb{R})$ with the n^2-dimensional Euclidean space $\mathbb{R}^{n^2}$, so that $M(n, \mathbb{R})$ is a manifold. Evidently the determinant map $\det: M(n, \mathbb{R}) \to \mathbb{R}$ is continuous. Since $\mathbb{R}^*$ $(= \mathbb{R} - \{0\})$ is open in $\mathbb{R}$ and since

$$GL(n, \mathbb{R}) = \det^{-1}(\mathbb{R}^*),$$

$GL(n, \mathbb{R})$ is an open set of $M(n, \mathbb{R})$. Accordingly $GL(n, \mathbb{R})$ is an open submanifold of $M(n, \mathbb{R})$ as we mentioned previously. Moreover $GL(n, \mathbb{R})$ is a group with respect to matrix multiplication. The product AB of elements A and B in $GL(n, \mathbb{R})$ has entries which are polynomials in the entries of A and B, and these entries are exactly the expressions in local coordinates of the product map, which is thus smooth. The entries of the inverse A^{-1} of A are rational functions of the entries of A with non-vanishing denominators, and are hence smooth. Thus $GL(n, \mathbb{R})$ is a Lie group. Similarly it can be shown that $GL(n, \mathbb{C})$ is a Lie group.

Example 4 The following groups can be shown to be Lie groups directly or by the theorem that a closed subgroup of a Lie group is also a Lie group, which we will obtain in Section 3.7.

$$O(n) = \{A \in GL(n, \mathbb{R}) \mid {}^tAA = I\},$$
$$SL(n, \mathbb{R}) = \{A \in GL(n, \mathbb{R}) \mid \det A = 1\},$$
$$SO(n) = O(n) \cap SL(n, \mathbb{R}),$$
$$U(n) = \{A \in GL(n, \mathbb{C}) \mid A{}^t\bar{A} = I\},$$
$$SL(n, \mathbb{C}) = \{A \in GL(n, \mathbb{C}) \mid \det A = 1\},$$
$$SU(n) = U(n) \cap SL(n, \mathbb{C}).$$

Here ${}^tA$ denotes the transpose of A, ${}^t\bar{A}$ denotes the conjugate transpose of A, and I denotes the identity matrix.

We now consider Lie group actions on manifolds. Thus far we have dealt with transformation groups with respect to topological groups and topological spaces; in other words the topological category. Hereafter we shall be concerned with the differential category. This category holds many possibilities for developing an expanded theory and might be central to the study of transformation groups.

Definition 3.5 Let G be a Lie group and M be a manifold. If the map $\varphi: G \times M \to M$ of a G-action on M is smooth, the action is called a *smooth G-action*. Here $G \times M$ is endowed with the structure of the product manifold. Then the triple (G, M, φ) is called a *smooth* (or *differentiable*) *transformation group* and M is called a *smooth* (or *differentiable*) *G-manifold* or simply *G-manifold*. A smooth G-map $f: M \to N$ between G-manifolds is a *G-diffeomorphism* if it is a homeomorphism and the inverse f^{-1} is also smooth. M and N are *G-diffeomorphic* if there exists a G-diffeomorphism $f: M \to N$.

A fibre bundle is called a *smooth* fibre bundle if the base space is a manifold, the structure group is a Lie group, the fibre is a K-manifold, and the transition functions are smooth.

Example 5 A Lie group G is a G-manifold. In fact, the multiplication $\alpha: G \times G \to G$ gives a smooth G-action on G.

Example 6 Linear algebra provides us with a simple but important example. Let G be one of $GL(n, \mathbb{R})$, $O(n)$, $SL(n, \mathbb{R})$, or $SO(n)$, and let M be $\mathbb{R}^n$. Then G acts on M in the natural manner. Since the action is given by polynomials in the entries of $A \in G$ and $x \in M$, it is smooth.

Example 7 Let G be one of $GL(n, \mathbb{C})$, $U(n)$, $SL(n, \mathbb{C})$, or $SU(n)$, and let M be $\mathbb{C}^n$. Then G acts on M in the natural manner. As in Example 6, the action is smooth.

Example 8 Let G be an arbitrary Lie group, G' be one of the G in Example 6 and $f: G \to G'$ be a homomorphism of Lie groups. Then G acts on $M = \mathbb{R}^n$ via f as follows. We define a map $\varphi: G \times M \to M$ by

$$\varphi(g, x) = \varphi'(f(g), x) \qquad \text{for } g \in G,\ x \in M.$$

Here φ' indicates the action of Example 6. Since φ can be decomposed as

$$G \times M \xrightarrow{f \times \mathrm{id}} G' \times M \xrightarrow{\varphi'} M,$$

it is smooth. It is easy to check that φ gives a G-action. A similar statement holds for an arbitrary Lie group G, one of the G' in Example 7, and $M = \mathbb{C}^n$.

The G-action obtained in this way is nothing but a representation of G.

Example 9 Let G be either $O(n)$ or $SO(n)$ and let M be the $(n-1)$-sphere S^{n-1}. Since S^{n-1} is invariant under the natural G-action $\varphi: G \times \mathbb{R}^n \to \mathbb{R}^n$, S^{n-1} is a G-space. Moreover we show in the following that S^{n-1} is a G-manifold. To see this, consider the following commutative diagram:

$$\begin{array}{ccc} G \times \mathbb{R}^n & \xrightarrow{\varphi} & \mathbb{R}^n \\ \cup & & \cup \\ G \times S^{n-1} & \xrightarrow{\varphi|} & S^{n-1} \end{array}$$

Since S^{n-1} is a submanifold of $\mathbb{R}^n$, $G \times S^{n-1}$ is a submanifold of $G \times \mathbb{R}^n$. It follows from Lemma 3.3 that the restriction $\varphi|$ is smooth, hence S^{n-1} is a smooth G-manifold.

Let G be either $U(n)$ or $SU(n)$ and M be $S^{2n-1}(\subset \mathbb{C}^n)$. Then a similar statement holds for G and M.

Example 10 Let G be one of the groups in Example 6 and M be the real projective space $\mathbb{R}P^{n-1}$ of dimension $n-1$. Note that $\mathbb{R}^n - \{0\}$ is invariant under the G-action on $\mathbb{R}^n$ in Example 6. Moreover the action commutes with the $\mathbb{R}^*$ action on $\mathbb{R}^n - \{0\}$ which appeared in the definition of $\mathbb{R}P^{n-1}$, that is,

$$g \cdot (ax) = ag \cdot x \qquad \text{for } g \in G,\ a \in \mathbb{R}^*,\ x \in \mathbb{R}^n - \{0\}.$$

Therefore the G-action on $\mathbb{R}^n - \{0\}$ induces the G-action on $\mathbb{R}P^{n-1}$. We now show that the action is smooth. For this, consider the following commutative diagram

$$\begin{array}{ccc} G \times (\mathbb{R}^n - \{0\}) & \xrightarrow{\varphi} & \mathbb{R}^n - \{0\} \\ \Big\downarrow{\scriptstyle \mathrm{id} \times \pi} & & \Big\downarrow{\scriptstyle \pi} \\ G \times \mathbb{R}P^{n-1} & \xrightarrow{\bar{\varphi}} & \mathbb{R}P^{n-1} \end{array}$$

where φ is the action induced by that of Example 6 and $\bar{\varphi}$ is the action obtained from φ by the above observation. Since π has a smooth local cross-section as we showed in Example 3 of Section 3.1, $\mathrm{id} \times \pi$ has also a smooth local cross-section. Accordingly it follows from the commutativity of the above diagram that $\bar{\varphi}$ is locally smooth. Since the concept of smoothness is local, we can conclude that $\bar{\varphi}$ is smooth globally.

Let G be one of the groups in Example 7 and M be the complex projective space $\mathbb{C}P^{n-1}$ of dimension $n-1$. Then we have similarly that G acts naturally on M.

Example 11 When a Lie group G acts on $\mathbb{R}^n$ in the manner of Example 8, it induces a G-action on $\mathbb{R}P^{n-1}$ in the manner of Example 10. Similarly we have a G-action on $\mathbb{C}P^{n-1}$.

In these cases, we say that G acts on $\mathbb{R}P^{n-1}$ or $\mathbb{C}P^{n-1}$ *linearly*.

Example 12 Let G be a Lie group and H be a closed subgroup of G. Then the homogeneous space G/H becomes a manifold as follows (cf. Section 3.8). Let $\pi: G \to G/H$ be the projection. There exists a unique differential structure on G/H such that:

(1) π is smooth;

(2) there exists a smooth local cross-section around eH.

Then the G-action on G/H by left translation is smooth.

Given a smooth transformation group (G, M, φ), the map $\varphi_g: M \to M$ of Lemma 1.29 is now a diffeomorphism for any $g \in G$.

Example 13 A smooth $\mathbb{R}$-action on a manifold M is called a *smooth dynamical system* or a *1-parameter group of diffeomorphisms* of M.

3.3 Tangent spaces

The concepts of tangent spaces, tangent bundles, and normal bundles which we are going to discuss will provide clues to the study of manifolds and transformation groups. In particular, the slice theorem, which is basic to the study of G-manifolds, is described in terms of these structures.

An example of a two-dimensional manifold lying in the three-dimensional Euclidean space $\mathbb{R}^3$ is illustrated in Fig. 3.5. In this case we can consider a plane tangent to M at p. In this way, if a manifold lies in a Euclidean space, 'a plane tangent to the manifold' makes sense.

A space tangent to a one-dimensional manifold lying in $\mathbb{R}^2$ is simply a tangent line.

A natural question arises: is it possible to define a space tangent to a manifold which does not lie in a Euclidean space? It is possible and there are

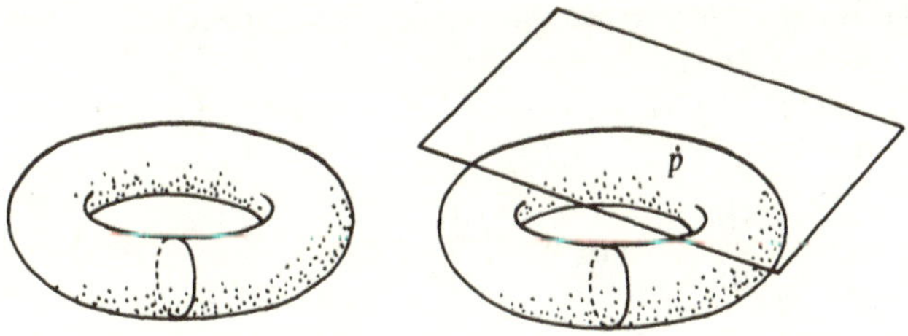

Fig. 3.5 A two-dimensional manifold and a tangent plane.

several definitions, but the methods employed in almost all books on manifolds seem to be difficult to understand for a beginner. Accordingly, we will try to introduce these ideas in a manner which encourages geometric intuition and imagery.

First of all, we define tangent space on the simplest manifold $\mathbb{R}^n$ and then we transfer it to a general manifold through charts. Notice that we will make use of $\mathbb{R}^n$ as both manifold and vector space over the real numbers $\mathbb{R}$ below.

Let p be a fixed point of $\mathbb{R}^n$. We first define a smooth curve in $\mathbb{R}^n$ through p as follows. Since the open interval $(-\varepsilon, \varepsilon)$ is an open set of $\mathbb{R}$ for a positive number ε, it is a manifold. Then a smooth map

$$l:(-\varepsilon, \varepsilon) \to \mathbb{R}^n, \qquad l(0) = p$$

is called a *smooth curve* (or *path*) through p in $\mathbb{R}^n$.

Let $l_1:(-\varepsilon_1, \varepsilon_1) \to \mathbb{R}^n$ and $l_2:(-\varepsilon_2, \varepsilon_2) \to \mathbb{R}^n$ be two smooth curves through p. If there exists $\varepsilon > 0$ with $\varepsilon_1 > \varepsilon$, $\varepsilon_2 > \varepsilon$ such that the restrictions $l_1|(-\varepsilon, \varepsilon)$ and $l_2|(-\varepsilon, \varepsilon)$ agree with each other, then they will be said to have the same *germ*. This is an equivalence relation among the smooth curves through p. The equivalence class of all smooth curves through p that have the same germ as l will be called the germ of l.

The set L of all germs of smooth curves through p is given the structure of a vector space as follows. Hereafter, for simplicity, we use the same symbol l for the germ of l.

First we translate curves through p to curves through the origin by parallel displacement. We consider the ordinary operations of addition and scalar multiplication there, and then transform them back to curves through p. That is to say, for $l_i \in L$, $a \in \mathbb{R}$, we define $l_1 + l_2 \in L$ and $al_1 \in L$ by

$$(l_1 + l_2)(t) = (l_1(t) - p) + (l_2(t) - p) + p = l_1(t) + l_2(t) - p,$$

$$(al_1)(t) = a(l_1(t) - p) + p = al_1(t) + (1 - a)p.$$

Here $l_1 + l_2$ is a smooth curve defined on $(-\varepsilon, \varepsilon)$ where l_i is defined on $(-\varepsilon_i, \varepsilon_i)$ and $\varepsilon = \min\{\varepsilon_1, \varepsilon_2\}$. The addition and scalar multiplication in the centres and the right-hand sides of the above equalities mean those of the vector space $\mathbb{R}^n$. It is easy to see that L is a vector space with respect to these operations. Note that $\dim L = \infty$ if $n \geq 1$.

For $l \in L$, the *velocity vector* $v(l)$ of l at p is defined as follows. When we write $l(t)$ as $l(t) = (x_1(t), \ldots, x_n(t))$, we put

$$v(l) = \frac{\mathrm{d}l}{\mathrm{d}t}(0) = l'(0) = \left(\frac{\mathrm{d}x_1}{\mathrm{d}t}(0), \ldots, \frac{\mathrm{d}x_n}{\mathrm{d}t}(0)\right) \in \mathbb{R}^n$$

where $(\mathrm{d}x_i/\mathrm{d}t)(0)$ denotes the ordinary derivative of the smooth function $x_i(t)$ at 0. Figure 3.6 shows an example of a velocity vector.

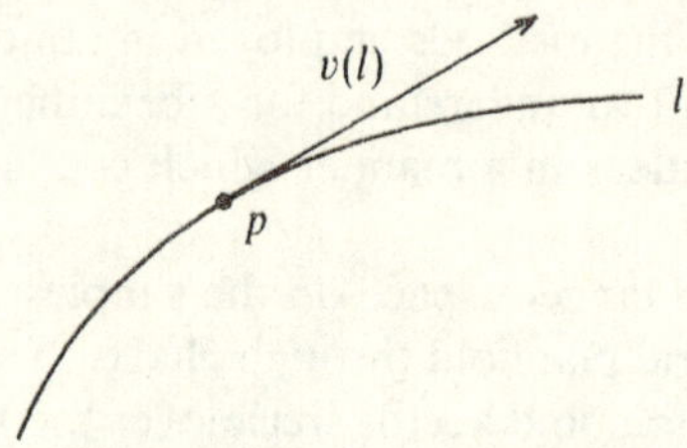

Fig. 3.6 The velocity vector.

We now consider a relation $\sim$ on L as follows: for $l_1, l_2 \in L$,

$$l_1 \sim l_2 \Leftrightarrow v(l_1) = v(l_2).$$

Then $\sim$ is evidently an equivalence relation. Let $[l]$ denote the equivalence class of l ($\in L$). Then we say that $[l]$ is the *tangent vector* to l at p or $[l]$ is a *tangent vector* to $\mathbb{R}^n$ at p.

We shall presently show that the set of all tangent vectors to $\mathbb{R}^n$ at p may be given the structure of a vector space.

The following lemma is immediate from differential calculus.

Lemma 3.4 *If $l_i, k_i \in L$ satisfy $l_1 \sim l_2$ and $k_1 \sim k_2$, then $l_1 + k_1 \sim l_2 + k_2$ and $al_1 \sim al_2$ for $a \in \mathbb{R}$.*

Let $L/\sim$ denote the classification of L with respect to the equivalence relation $\sim$ above. It follows from Lemma 3.4 that the vector space operations in L induce those in $L/\sim$. That is, if we define addition and scalar multiplication in $L/\sim$ by

$$[l] + [k] = [l + k]$$
$$a[l] = [al],$$

then these do not depend on the choice of representatives.

Proposition 3.5 $L/\sim$ *becomes a vector space with respect to these operations and is isomorphic to* $\mathbb{R}^n$.

Proof Obviously $L/\sim$ becomes a vector space. We shall show the isomorphism. Define a map $f: L/\sim \to \mathbb{R}^n$ by

$$f([l]) = v(l).$$

It follows from the definition of the relation $\sim$ that f is well defined and injective. Since differentiation is a linear operator, we have

$$\begin{aligned} f([l]+[k]) &= f([l+k]) = v(l+k) = v(l) + v(k) \\ &= f([l]) + f([k]), \\ f(a[l]) &= f([al]) = v(al) = av(l) \\ &= af([l]), \end{aligned}$$

that is, f is a linear map.

Given an arbitrary vector $v = (v_1, \ldots, v_n) \in \mathbb{R}^n$, we put

$$l(t) = (v_1 t, \ldots, v_n t) + p \qquad \text{for } t \in \mathbb{R}.$$

Then $l \in L$ and $f([l]) = v$, that is, f is surjective. Thus f gives an isomorphism of vector spaces. □

The vector space $L/\sim$ obtained in Proposition 3.5 will be called the *tangent space* to $\mathbb{R}^n$ at p.

We now proceed to define the tangent space of a general manifold. Let M be a smooth manifold and let p be a point of M. Take a chart $(U_\alpha, \varphi_\alpha)$ at p. As in $\mathbb{R}^n$, a smooth curve through p in U_α will mean a smooth map

$$l: (-\varepsilon, \varepsilon) \to U_\alpha$$

defined on some interval $(-\varepsilon, \varepsilon)$ of real numbers, with $l(0) = p$.

The notion of germ is introduced as before and we denote by L_α the set of all germs of smooth curves through p in U_α. Then for $l \in L_\alpha$, the map

$$\varphi_\alpha \cdot l: (-\varepsilon, \varepsilon) \to \mathbb{R}^n$$

is a smooth curve through $\varphi_\alpha(p)$ in $\mathbb{R}^n$, that is, $\varphi_\alpha \cdot l \in L$ where L denotes the set of all germs of smooth curves through $\varphi_\alpha(p)$ in $\mathbb{R}^n$.

By making use of the equivalence relation $\sim$ defined previously, we introduce a new relation $\underset{\alpha}{\sim}$ on L_α as follows: for $l_1, l_2 \in L_\alpha$,

$$l_1 \underset{\alpha}{\sim} l_2 \Leftrightarrow \varphi_\alpha \cdot l_1 \sim \varphi_\alpha \cdot l_2,$$

which is obviously an equivalence relation. Let $L_\alpha / \underset{\alpha}{\sim}$ denote the classification of L_α with respect to the equivalence relation $\underset{\alpha}{\sim}$ and let $[l]_\alpha$ denote the equivalence class of $l \in L_\alpha$.

We now show that the equivalence relation does not depend on the choice of charts. Let (U_β, φ_β) be another chart at p. First observe that L_α and L_β coincide with each other as sets, because we are considering germs of smooth curves through p.

Lemma 3.6 *For $l_1, l_2 \in L_\alpha = L_\beta$, we have*

$$l_1 \underset{\alpha}{\sim} l_2 \Leftrightarrow l_1 \underset{\beta}{\sim} l_2.$$

That is, we have $L_\alpha / \underset{\alpha}{\sim} = L_\beta / \underset{\beta}{\sim}$.

Proof The reader should bear in mind that the domain and range of the map $\varphi_\beta \cdot \varphi_\alpha^{-1}$ are subsets of $\mathbb{R}^n$, so that it has the explicit form

$$(y_1, \ldots, y_n) = \varphi_\beta \cdot \varphi_\alpha^{-1}(x_1, \ldots, x_n)$$

where $(x_1, \ldots, x_n)$ ranges over the points of the set $\varphi_\alpha(U_\alpha \cap U_\beta)$. Each y_i is a smooth function of $x_1, \ldots, x_n$. Denote by $\varphi_{\alpha i}(q)$ the ith component of $\varphi_\alpha(q)$.

Suppose now that $l_1 \underset{\alpha}{\sim} l_2$. Using the formula for differentiation of composite functions, we compute

$$\begin{aligned} v(\varphi_\beta \cdot l_1) &= v((\varphi_\beta \cdot \varphi_\alpha^{-1}) \cdot (\varphi_\alpha \cdot l_1)) \\ &= \left(\sum_i \frac{\partial y_1}{\partial x_i}(\varphi_\alpha(p)) \frac{\mathrm{d}\varphi_{\alpha i} \cdot l_1}{\mathrm{d}t}(0), \ldots, \sum_i \frac{\partial y_n}{\partial x_i}(\varphi_\alpha(p)) \frac{\mathrm{d}\varphi_{\alpha i} \cdot l_1}{\mathrm{d}t}(0) \right) \\ &= \left(\sum_i \frac{\partial y_1}{\partial x_i}(\varphi_\alpha(p)) \frac{\mathrm{d}\varphi_{\alpha i} \cdot l_2}{\mathrm{d}t}(0), \ldots, \sum_i \frac{\partial y_n}{\partial x_i}(\varphi_\alpha(p)) \frac{\mathrm{d}\varphi_{\alpha i} \cdot l_2}{\mathrm{d}t}(0) \right) \\ &= v((\varphi_\beta \cdot \varphi_\alpha^{-1}) \cdot (\varphi_\alpha \cdot l_2)) = v(\varphi_\beta \cdot l_2). \end{aligned}$$

Hence we have $\varphi_\beta \cdot l_1 \sim \varphi_\beta \cdot l_2$, that is, $l_1 \underset{\beta}{\sim} l_2$. Similarly it can be shown that $l_1 \underset{\beta}{\sim} l_2 \Rightarrow l_1 \underset{\alpha}{\sim} l_2$.

Thus the equivalence class $[l]_\alpha$ of l does not depend on the choice of a chart $(U_\alpha, \varphi_\alpha)$ at p. Accordingly the classification of L_α with respect to the equivalence relation $\underset{\alpha}{\sim}$ coincides with that of L_β $(=L_\alpha)$ with respect to the equivalence relation $\underset{\beta}{\sim}$, that is, $L_\alpha / \underset{\alpha}{\sim} = L_\beta / \underset{\beta}{\sim}$. □

Owing to Lemma 3.6, we can delete α of $[l]_\alpha$ so that we denote $[l]_\alpha$ by $[l]$ (or $(\mathrm{d}l/\mathrm{d}t)(0)$ or $l'(0)$).

By making use of a chart $(U_\alpha, \varphi_\alpha)$ at p, we introduce vector space operations in $L_\alpha / \underset{\alpha}{\sim}$ by

$$[l] \underset{\alpha}{+} [k] = [\varphi_\alpha^{-1}(\varphi_\alpha \cdot l + \varphi_\alpha \cdot k)]$$

$$a \underset{\alpha}{\cdot} [l] = [\varphi_\alpha^{-1}(a\varphi_\alpha \cdot l)] \qquad \text{for } a \in \mathbb{R}.$$

Here we need to explain the meaning of the right-hand sides of the above equations. If we choose a sufficiently small number $\varepsilon > 0$, then we have

$$(\varphi_\alpha \cdot l + \varphi_\alpha \cdot k)(t) \in \varphi_\alpha(U_\alpha), \ (a\varphi_\alpha \cdot l)(t) \in \varphi_\alpha(U_\alpha)$$

for $t \in (-\varepsilon, \varepsilon)$. Hence the insides of [ ] of the right-hand sides make sense and represent elements of L_α.

Lemma 3.7 *The operations do not depend on the choice of representatives, that is,*

$$l_1 \underset{\alpha}{\sim} l_2, k_1 \underset{\alpha}{\sim} k_2 \Rightarrow \varphi_\alpha^{-1}(\varphi_\alpha \cdot l_1 + \varphi_\alpha \cdot k_1) \underset{\alpha}{\sim} \varphi_\alpha^{-1}(\varphi_\alpha \cdot l_2 + \varphi_\alpha \cdot k_2)$$

$$\varphi_\alpha^{-1}(a\varphi_\alpha \cdot l_1) \underset{\alpha}{\sim} \varphi_\alpha^{-1}(a\varphi_\alpha \cdot l_2) \qquad \textit{for } a \in \mathbb{R}.$$

Proof It follows from the assumption that $\varphi_\alpha \cdot l_1 \sim \varphi_\alpha \cdot l_2$ and $\varphi_\alpha \cdot k_1 \sim \varphi_\alpha \cdot k_2$. By the linearity of differentiation, we have

$$\varphi_\alpha \cdot l_1 + \varphi_\alpha \cdot k_1 \sim \varphi_\alpha \cdot l_2 + \varphi_\alpha \cdot k_2 \qquad \text{and} \qquad a\varphi_\alpha \cdot l_1 \sim a\varphi_\alpha \cdot l_2,$$

which are the desired relations. □

The following corollary can be proved in the same manner as Proposition 3.5.

Corollary 3.8 $L_\alpha / \underset{\alpha}{\sim}$ *becomes a vector space with respect to these operations and the map* $\mathrm{d}\varphi_\alpha : L_\alpha / \underset{\alpha}{\sim} \to \mathbb{R}^n$ *defined by* $\mathrm{d}\varphi_\alpha([l]) = v(\varphi_\alpha \cdot l)$ *is an isomorphism of vector spaces.*

We now show that the vector space operations do not depend on the choice of a chart $(U_\alpha, \varphi_\alpha)$ at p.

Lemma 3.9 *Let* $(U_\alpha, \varphi_\alpha)$, (U_β, φ_β) *and* $(U_\gamma, \varphi_\gamma)$ *be charts at p. Given smooth curves l and k through p in* $U_\alpha \cap U_\beta \cap U_\gamma$, *we have*

$$\varphi_\alpha^{-1}(\varphi_\alpha \cdot l + \varphi_\alpha \cdot k) \underset{\gamma}{\sim} \varphi_\beta^{-1}(\varphi_\beta \cdot l + \varphi_\beta \cdot k),$$

$$\varphi_\alpha^{-1}(a\varphi_\alpha \cdot l) \underset{\gamma}{\sim} \varphi_\beta^{-1}(a\varphi_\beta \cdot l), \qquad \text{for } a \in \mathbb{R}.$$

That is, $[l] \underset{\alpha}{+} [k] = [l] \underset{\beta}{+} [k]$ *and* $a \underset{\alpha}{\cdot} [l] = a \underset{\beta}{\cdot} [l]$.

Proof By virtue of Lemma 3.6, it suffices to show the lemma in the case where $\gamma = \beta$. That is, we need only show that

$$\varphi_\alpha^{-1}(\varphi_\alpha \cdot l + \varphi_\alpha \cdot k) \underset{\beta}{\sim} \varphi_\beta^{-1}(\varphi_\beta \cdot l + \varphi_\beta \cdot k),$$

$$\varphi_\alpha^{-1}(a\varphi_\alpha \cdot l) \underset{\beta}{\sim} \varphi_\beta^{-1}(a\varphi_\beta \cdot l), \qquad \text{for } a \in \mathbb{R}.$$

By definition, these relations are equivalent to

$$\varphi_\beta \cdot \varphi_\alpha^{-1}(\varphi_\alpha \cdot l + \varphi_\alpha \cdot k) \sim \varphi_\beta \cdot l + \varphi_\beta \cdot k,$$
$$\varphi_\beta \cdot \varphi_\alpha^{-1}(a\varphi_\alpha \cdot l) \sim a\varphi_\beta \cdot l.$$

We now employ the notation

$$(y_1, \ldots, y_n) = \varphi_\beta \cdot \varphi_\alpha^{-1}(x_1, \ldots, x_n)$$

introduced in the proof of Lemma 3.6. Then, by making use of the linearity of differentiation, we compute:

$$\begin{aligned}
&v(\varphi_\beta \cdot \varphi_\alpha^{-1}(\varphi_\alpha \cdot l + \varphi_\alpha \cdot k)) \\
&= \Bigg(\sum_i \frac{\partial y_1}{\partial x_i}(\varphi_\alpha(p)) \frac{\mathrm{d}(\varphi_{\alpha i} \cdot l + \varphi_{\alpha i} \cdot k - \varphi_{\alpha i}(p))}{\mathrm{d}t}(0), \ldots, \\
&\quad \sum_i \frac{\partial y_n}{\partial x_i}(\varphi_\alpha(p)) \frac{\mathrm{d}(\varphi_{\alpha i} \cdot l + \varphi_{\alpha i} \cdot k - \varphi_{\alpha i}(p))}{\mathrm{d}t}(0) \Bigg) \\
&= \Bigg(\sum_i \frac{\partial y_1}{\partial x_i}(\varphi_\alpha(p)) \frac{\mathrm{d}\varphi_{\alpha i} \cdot l}{\mathrm{d}t}(0), \ldots, \sum_i \frac{\partial y_n}{\partial x_i}(\varphi_\alpha(p)) \frac{\mathrm{d}\varphi_{\alpha i} \cdot l}{\mathrm{d}t}(0) \Bigg) \\
&\quad + \Bigg(\sum_i \frac{\partial y_1}{\partial x_i}(\varphi_\alpha(p)) \frac{\mathrm{d}\varphi_{\alpha i} \cdot k}{\mathrm{d}t}(0), \ldots, \sum_i \frac{\partial y_n}{\partial x_i}(\varphi_\alpha(p)) \frac{\mathrm{d}\varphi_{\alpha i} \cdot k}{\mathrm{d}t}(0) \Bigg) \\
&= v((\varphi_\beta \cdot \varphi_\alpha^{-1}) \cdot (\varphi_\alpha \cdot l)) + v((\varphi_\beta \cdot \varphi_\alpha^{-1}) \cdot (\varphi_\alpha \cdot k)) \\
&= v(\varphi_\beta \cdot l) + v(\varphi_\beta \cdot k) = v(\varphi_\beta \cdot l + \varphi_\beta \cdot k).
\end{aligned}$$

Hence we have one of the desired relations:

$$\varphi_\beta \cdot \varphi_\alpha^{-1}(\varphi_\alpha \cdot l + \varphi_\alpha \cdot k) \sim \varphi_\beta \cdot l + \varphi_\beta \cdot k.$$

The proof of the other relation is similar. □

Putting together Lemmas 3.6, 3.7, and 3.9, and Corollary 3.8, we have the following theorem.

Theorem 3.10 *The set $L_\alpha / \underset{\alpha}{\sim}$ and the vector space structure induced on it via the chart $(U_\alpha, \varphi_\alpha)$ do not depend on the choice of a chart $(U_\alpha, \varphi_\alpha)$ at p.*

Definition 3.6 By virtue of Theorem 3.10, we obtain a unique vector space $L_\alpha / \underset{\alpha}{\sim}$ that does not depend on the choice of a chart $(U_\alpha, \varphi_\alpha)$ at p. This common vector space is called the *tangent space* of M at p and is denoted by $T_p(M)$. An element of the tangent space $T_p(M)$ is called a *tangent vector* of M at p.

Given a smooth curve l through p in M, we say also that $[l]$ is the *tangent vector to l at p*. More generally, given a smooth curve $l:(a, b) \to M$, the tangent vector to l at $l(t)$ ($t \in (a, b)$) is defined as follows. For fixed $t \in (a, b)$, we introduce a new curve

$$k:(a-t, b-t) \to M$$

by $k(s) = l(s+t)$. Then k is a smooth curve through $l(t)$. Its equivalence class $[k]$ is called the *tangent vector to l at $l(t)$ ($t \in (a, b)$)* and is denoted by $\mathrm{d}l(t)/\mathrm{d}t$ or $l'(t)$.

When we assign to a tangent vector of M at p a vector of $\mathbb{R}^n$ by the map $\mathrm{d}\varphi_\alpha$, the corresponding element depends on the choice of a chart $(U_\alpha, \varphi_\alpha)$. Although we have described this relationship already in the proof of Lemma 3.6, we re-examine it in the following. That is, we would like to investigate the relation between the following two vectors:

$$\mathrm{d}\varphi_\alpha([l]) = v(\varphi_\alpha \cdot l) = \left(\frac{\mathrm{d}\varphi_{\alpha 1} \cdot l}{\mathrm{d}t}(0), \ldots, \frac{\mathrm{d}\varphi_{\alpha n} \cdot l}{\mathrm{d}t}(0)\right)$$

$$\mathrm{d}\varphi_\beta([l]) = v(\varphi_\beta \cdot l) = \left(\frac{\mathrm{d}\varphi_{\beta 1} \cdot l}{\mathrm{d}t}(0), \ldots, \frac{\mathrm{d}\varphi_{\beta n} \cdot l}{\mathrm{d}t}(0)\right).$$

For this we introduce some terminology.

As before, we express $\varphi_\beta \cdot \varphi_\alpha^{-1}$ as

$$(y_1, \ldots, y_n) = \varphi_\beta \cdot \varphi_\alpha^{-1}(x_1, \ldots, x_n)$$

for $(x_1, \ldots, x_n) \in \varphi_\alpha(U_\alpha \cap U_\beta)$. Then each y_i is a smooth function of $x_1, \ldots, x_n$. Since $\varphi_\beta \cdot \varphi_\alpha^{-1}: \varphi_\alpha(U_\alpha \cap U_\beta) \to \varphi_\beta(U_\alpha \cap U_\beta)$ is a diffeomorphism, we have

$$(x_1, \ldots, x_n) = \varphi_\alpha \cdot \varphi_\beta^{-1}(y_1, \ldots, y_n)$$

for $(y_1, \ldots, y_n) \in \varphi_\beta(U_\alpha \cap U_\beta)$. Hence each x_i is also a smooth function of $y_1, \ldots, y_n$.

Then the $n \times n$ matrix

$$\begin{bmatrix} \dfrac{\partial y_1}{\partial x_1}(\varphi_\alpha(p)) & \cdots & \dfrac{\partial y_1}{\partial x_n}(\varphi_\alpha(p)) \\ \vdots & & \vdots \\ \dfrac{\partial y_n}{\partial x_1}(\varphi_\alpha(p)) & \cdots & \dfrac{\partial y_n}{\partial x_n}(\varphi_\alpha(p)) \end{bmatrix}$$

is called the *Jacobian matrix* of $\varphi_\beta \cdot \varphi_\alpha^{-1}$ and is denoted by $(\varphi_\beta \cdot \varphi_\alpha^{-1})'$. Since

$$(\varphi_\alpha \cdot \varphi_\beta^{-1})(\varphi_\beta \cdot \varphi_\alpha^{-1}) = \mathrm{id}$$

on $\varphi_\alpha(U_\alpha \cap U_\beta)$, it follows from the formulae for differentiation of composite functions that

$$(\varphi_\alpha \cdot \varphi_\beta^{-1})' \cdot (\varphi_\beta \cdot \varphi_\alpha^{-1})' = I \qquad (I: \text{unit matrix}).$$

In particular, the Jacobian matrix $(\varphi_\beta \cdot \varphi_\alpha^{-1})'$ is non-singular. As shown in the proof of Lemma 3.6, the corresponding vectors in $\mathbb{R}^n$ of a tangent vector of M via different charts are related in terms of the Jacobian matrix as follows.

Corollary 3.11 *If we write elements of* $\mathbb{R}^n$ *as column vectors* ($n \times 1$ *matrices*), *then we have*

$$(\varphi_\beta \cdot \varphi_\alpha^{-1})' \mathrm{d}\varphi_\alpha([l]) = \mathrm{d}\varphi_\beta([l])$$

where the left-hand side denotes a matrix product.

Corresponding to a chart $(U_\alpha, \varphi_\alpha)$ of M, we consider the copy $\mathbb{R}^n_\alpha$ of $\mathbb{R}^n$. Let p be a fixed point of M. Take the disjoint union of the copies $\mathbb{R}^n_\alpha$ over all α with $p \in U_\alpha$ and identify $v \in \mathbb{R}^n_\alpha$ with $(\varphi_\beta \cdot \varphi_\alpha^{-1})' v \in \mathbb{R}^n_\beta$. We denote the identification space by

$$\left\{ \bigcup_{p \in U_\alpha} \mathbb{R}^n_\alpha \right\} \Big/ \sim .$$

Here p is fixed and the union is taken over all the charts $(U_\alpha, \varphi_\alpha)$ at p.

Corollary 3.12 *The tangent space* $T_p(M)$ *of* M *at* p *and the space* $\{\bigcup_{p \in U_\alpha} \mathbb{R}^n_\alpha\}/\sim$ *can be identified naturally.*

Remark Because of Corollary 3.12, the space $\{\bigcup_{p \in U_\alpha} \mathbb{R}^n_\alpha\}/\sim$ is also called the tangent space of M at p and is denoted by $T_p(M)$ as well. This observation will guide us to the concept of tangent bundles which we will discuss in the next section.

The following lemma will be needed in Section 3.6 and its proof is left to the reader as an exercise.

Lemma 3.13 *Let* l *be a smooth curve through* p *in* M *and* a *be a number. Then the smooth curve* l_a *defined by* $l_a(t) = l(at)$ *satisfies*

$$[l_a] = a[l]$$

in $T_p(M)$.

3.4 Tangent *G*-vector bundles

In the previous section, we defined the tangent space of a manifold M at a point of M. In this section, we topologize the union of all the tangent spaces of M and show that it can be given the structure of a vector bundle. Moreover, we will see that if M is a G-manifold, then this vector bundle becomes a G-vector bundle.

First we consider a manifold without G-actions. Let $\{(U_\alpha, \varphi_\alpha)\}$ be an atlas of M. Take the disjoint union

$$\bigcup_\alpha U_\alpha \times \mathbb{R}^n_\alpha$$

over all α and, whenever $p \in U_\alpha \cap U_\beta$, identify $(p, v) \in U_\alpha \times \mathbb{R}^n_\alpha$ with $(p, (\varphi_\beta \circ \varphi_\alpha^{-1})' v) \in U_\beta \times \mathbb{R}^n_\beta$. Here $(\varphi_\beta \cdot \varphi_\alpha^{-1})'$ indicates the Jacobian matrix which appeared in the preceding section. We denote the identification space by $T(M)$ and define $\pi: T(M) \to M$ by projection on the first factor of each product. Then we will show that $\pi: T(M) \to M$ is a vector bundle. Hereafter we abbreviate $\mathbb{R}^n_\alpha$ to $\mathbb{R}^n$ for brevity. It follows from the method of topologizing $T(M)$ that $U_\alpha \times \mathbb{R}^n$ is included naturally in $T(M)$ as the open subset $\pi^{-1}(U_\alpha)$. We denote this natural homeomorphism by

$$\bar{\varphi}_\alpha : \pi^{-1}(U_\alpha) \to U_\alpha \times \mathbb{R}^n.$$

If we put

$$\theta_{\beta\alpha}(p) = \bar{\varphi}_{\beta,p} \cdot \bar{\varphi}_{\alpha,p}^{-1} \qquad \text{for } p \in U_\alpha \cap U_\beta$$

as in Section 2.5, then we have

$$\theta_{\beta\alpha}(p) = (\varphi_\beta \cdot \varphi_\alpha^{-1})'(\varphi_\alpha(p)) = \left(\frac{\partial y_i}{\partial x_j}(\varphi_\alpha(p))\right) \in GL(n, \mathbb{R})$$

and the map

$$\theta_{\beta\alpha} : U_\alpha \cap U_\beta \to GL(n, \mathbb{R})$$

is smooth. Thus $T(M)$ becomes a vector bundle with coordinate functions $\bar{\varphi}_\alpha$ and transition functions $\theta_{\beta\alpha}$. Each fibre $\pi^{-1}(p)$ is just the tangent space $T_p(M)$ obtained in Corollary 3.12. Hence we have

$$T(M) = \bigcup_{p \in M} T_p(M)$$

as sets.

Definition 3.7 The vector bundle $T(M)$ obtained in this way is called the *tangent bundle* of M.

Example 1 *Tangent bundle $T(S^n)$ of the n-sphere S^n.* We make use of the same notation as in Example 2 of Section 3.1. Using $\varphi_- \cdot \varphi_+^{-1}(y) = y/\|y\|^2$, we compute its Jacobian matrix as follows. Since

$$\frac{\partial}{\partial y_j}\left(\frac{y_i}{\|y\|^2}\right) = \begin{cases} \dfrac{-2y_i y_j}{\|y\|^4} & \text{for } i \neq j \\[2ex] \dfrac{\|y\|^2 - 2y_i^2}{\|y\|^2} & \text{for } i = j, \end{cases}$$

the Jacobian matrix is given by

$$(\varphi_- \cdot \varphi_+^{-1})'(y) = \begin{bmatrix} \dfrac{\|y\|^2 - 2y_1^2}{\|y\|^4} & & \overset{j}{\vdots} & \\ & \ddots & \vdots & \\ & & \ddots & \\ i \cdots & \cdots & \dfrac{-2y_i y_j}{\|y\|^4} & \dfrac{\|y\|^2 - 2y_n^2}{\|y\|^4} \end{bmatrix}$$

Hence the transition function $\theta_{-+}: U_+ \cap U_- \to GL(n, \mathbb{R})$ is given as follows. For $x \in U_+ \cap U_-$, $\theta_{-+}(x)$ is the matrix obtained by substituting $x_i/(1 - x_{n+1})$ for y_i in the above matrix. If we define a map

$$f: (U_+ \cap U_-) \times \mathbb{R}^n \to (U_+ \cap U_-) \times \mathbb{R}^n$$

by $f(x, v) = (x, \theta_{-+}(x)v)$, f is a diffeomorphism. The total space $T(S^n)$ is the identification space by the map f:

$$T(S^n) = U_+ \times \mathbb{R}^n \bigcup_f U_- \times \mathbb{R}^n$$

and the projection $\pi: T(S^n) \to S^n$ is induced by the projections $U_\pm \times \mathbb{R}^n \to U_\pm$. The coordinate functions are given by the natural homeomorphisms:

$$\bar{\varphi}_\pm : \pi^{-1}(U_\pm) \to U_\pm \times \mathbb{R}^n$$

and the transition functions are given by θ_{-+} and θ_{+-} whose definition is similar to θ_{-+}.

Incidentally the principal $O(n)$-bundle associated with the tangent bundle of S^n is given by the quotient map $\pi: O(n+1) \to O(n+1)/O(n) = S^n$ so that we have

$$T(S^n) = O(n+1) \underset{O(n)}{\times} \mathbb{R}^n$$

where $O(n)$ acts on $\mathbb{R}^n$ in the canonical way. For this, see Theorem 3.15, Proposition 3.46, and Corollary 3.47 which will be mentioned later.

A vector bundle is called a *smooth vector bundle* if the base space is a manifold and the transition functions are smooth. We have shown that tangent bundles are smooth vector bundles. Moreover it is evident that $T(M)$ is a $2n$-manifold whose charts are given by the compositions of the following maps

$$\pi^{-1}(U_\alpha) \xrightarrow{\bar{\varphi}_\alpha} U_\alpha \times \mathbb{R}^n \xrightarrow{\varphi_\alpha \times \mathrm{id}} \mathbb{R}^n \times \mathbb{R}^n.$$

Given a smooth map $f: M \to N$ between two manifolds M^m and N^n, we construct a bundle homomorphism

$$f_*: T(M) \to T(N)$$

as follows. Corresponding to the two methods of defining tangent spaces, we define it in two ways. Each has strong and weak points.

The first is a method which is easy to understand intuitively. Since $T(M)$ consists of tangent vectors, we designate their destinations. Given a smooth curve l through p in M, we define a curve $f \cdot l$ by $(f \cdot l)(t) = f \cdot l(t)$ which is a smooth curve through $f(p)$ in N. The assignment $[l] \mapsto [f \cdot l]$ defines f_* (see Fig. 3.7). The proof that it does not depend on the choice of representatives is easy and is left to the reader.

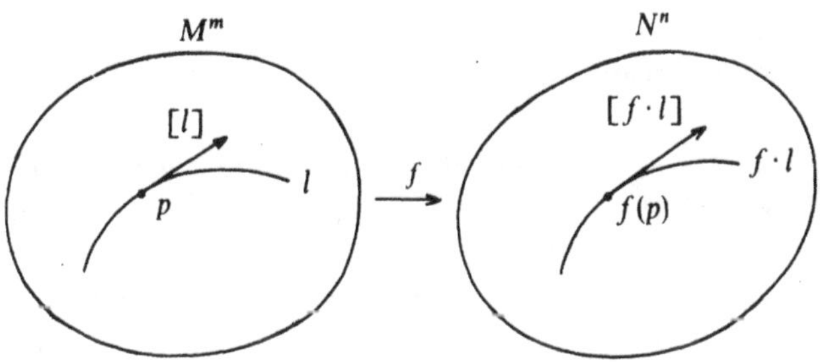

Fig. 3.7 A bundle homomorphism.

The second is a method which enables us to see that it is a bundle homomorphism. Let $(U_\alpha, \varphi_\alpha)$ and (V_β, ψ_β) be charts at p and $f(p)$ of M and N, respectively. For $v \in \mathbb{R}^m$, we set

$$f_*(p, v) = (f(p), (\psi_\beta \cdot f \cdot \varphi_\alpha^{-1})' v).$$

It follows from the latter definition of a tangent space that f_* is well defined.

The proof that these two definitions agree with each other and the proof of the following lemma are safely entrusted to the reader.

Lemma 3.14 (1) *For* $\mathrm{id}: M \to M$, *we have*

$$\mathrm{id}_* = \mathrm{id}: T(M) \to T(M).$$

(2) *For* $M_1 \xrightarrow{f_1} M_2 \xrightarrow{f_2} M_3$, *we have*

$$(f_2 \cdot f_1)_* = f_{2*} \cdot f_{1*}: T(M_1) \to T(M_3).$$

In particular, if $f: M \to N$ *is a diffeomorphism, then* $f_*: T(M) \to T(N)$ *is a bundle isomorphism.*

That is, the correspondence $f \mapsto f_*$ is functorial. f_* is called the *differential* of f and is often written as df.

Next we take G-actions into consideration. Let M be a G-manifold. Since the G-action on M is smooth, the map $g: M \to M$ is a diffeomorphism for each $g \in G$. Hence a bundle isomorphism

$$g_*: T(M) \to T(M)$$

is induced by Lemma 3.14. If we define a map

$$\Phi: G \times T(M) \to T(M)$$

by $\Phi(g, v) = g_* v$, Φ is smooth because of the method of imposing the smooth structure on $T(M)$. Furthermore Lemma 3.14 implies that Φ is a G-action on $T(M)$. Thus we see that $T(M)$ is a G-manifold. Since

$$g_*: T_p(M) \to T_{gp}(M)$$

is a linear isomorphism, $T(M)$ is a G-vector bundle over M.

Definition 3.8 The G-vector bundle obtained in this way is called the *tangent G-vector bundle* of M.

When a Lie group G acts on a smooth vector bundle E smoothly and E is a G-vector bundle with respect to this action, we say that E is a *smooth G-vector bundle*. The tangent G-vector bundle of a G-manifold is an example of a smooth G-vector bundle.

Let M be a G-manifold. Fix an arbitrary point p of M. If we restrict the G-action on $T(M)$ to the isotropy group G_p, then the tangent space $T_p(M)$ at p is G_p-invariant.

Since G_p acts linearly on $T_p(M)$, $T_p(M)$ is a representation space of G_p. This representation is called the *isotropy representation* at p of the G-manifold M.

The construction of the homogeneous space G/H of a Lie group G by a closed subgroup H of G provides us with abundant examples of manifolds as will be seen later. In addition to that, these are even G-manifolds. It is easy to see that the classification of G-vector bundles over G/H in Section 2.6, is valid also in smooth category as it is. Moreover it will be shown in Section 3.8 that the projection $\pi: G \to G/H$ has a smooth local cross-section. Since the tangent bundle $T(G/H)$ of G/H is a G-vector bundle, it follows from Corollary 2.27 that $T(G/H)$ is determined by the isotropy representation of H at eH. Summing up the above results, we have

Theorem 3.15 *The tangent space $T_{eH}(G/H)$ of G/H at eH is the representation space of the isotropy representation of H and the bundle*

$$G \times_H T_{eH}(G/H) \to G/H$$

associated with the principal H-bundle $G \to G/H$ is isomorphic to the tangent G-vector bundle $T(G/H)$ as G-vector bundles. In particular, the tangent G-vector bundle $T(G)$ of G is trivial:

$$T(G) \cong G \times T_e(G).$$

Remark Characteristic classes of homogeneous spaces G/H are also investigated by this method.[13]

3.5 Vector fields and 1-parameter groups

Let M be a manifold and $\pi: T(M) \to M$ be the tangent bundle of M. Then a smooth cross-section $X: M \to T(M)$ of π is called a *smooth vector field* on M. Unless otherwise stated we will use *vector field* to mean smooth vector field hereafter.

Given a vector field X on a manifold M, and a point x of M, we shall say that a curve $l:(a, b) \to M$ is an *integral curve* of X through x if

$$\frac{\mathrm{d}l(t)}{\mathrm{d}t} = X(l(t)), \qquad l(0) = x.$$

Here $\mathrm{d}l(t)/\mathrm{d}t$ denotes the tangent vector to l at $l(t)$ defined in Section 3.3.

Given a vector field X on M and a point x of M, we consider whether there exists an integral curve of X through x.

First we deal with the case where M is an open set of $\mathbb{R}^n$. In this case, the tangent bundle of M is trivial. Then we denote the composition of a cross-section $X: M \to T(M) = M \times \mathbb{R}^n$ with the projection $M \times \mathbb{R}^n \to \mathbb{R}^n$ by X, again for brevity. If we write the above equation in terms of coordinates, we have a system of ordinary differential equations:

$$l(t) = (l_1(t), \ldots, l_n(t)),$$

$$X(l(t)) = (X_1(l_1(t), \ldots, l_n(t)), \ldots, X_n(l_1(t), \ldots, l_n(t))),$$

$$\frac{\mathrm{d}l_i(t)}{\mathrm{d}t} = X_i(l_1(t), \ldots, l_n(t)), \qquad i = 1, \ldots, n, \tag{3.1}$$

for unknown functions $l_1(t), \ldots, l_n(t)$ of one variable t.

Then we have the following existence and uniqueness theorem from the theory of differential equations.

Since $X_i(x)$ is smooth, it satisfies the Lipshitz condition locally. Thus for each point p of M, there exist an open neighbourhood U of p and a number $\varepsilon_1 > 0$ such that for each $x = (x_1, \ldots, x_n) \in U$, (3.1) has a unique set of solutions $l_1(t), \ldots, l_n(t)$, defined for $t \in (-\varepsilon_1, \varepsilon_1)$, which satisfy the initial conditions $l_i(0) = x_i$. We denote the solution with the initial condition x by

$l(t, x)$. The map

$$l:(-\varepsilon_1, \varepsilon_1) \times U \to M$$

obtained in this way is smooth, since the solution depends smoothly on the initial condition.

The smooth curve $l(t, x)$ obtained by fixing x is an integral curve of X through x by definition.

Lemma 3.16 *In the above circumstances, there exist an open neighbourhood V of p and a number $\varepsilon > 0$ such that*

$$l(t_0 + t, x) = l(t, l(t_0, x))$$

for any $x = (x_1, \ldots, x_n) \in V$ and $|t_0|, |t|, |t_0 + t| < \varepsilon$.

Proof Since l is continuous in t and x, there exist an open neighbourhood V of p and a number ε with $0 < \varepsilon \le \varepsilon_1$ such that

$$l(t, x) \in U \qquad \text{for} \quad |t| < \varepsilon, x \in V.$$

Hence $l(t_0, x) \in U$ for $|t_0| < \varepsilon$, $x \in V$ so that the right-hand side of the equality in the lemma exists uniquely.

We now fix $x \in V$ and t_0, and regard t as a variable. Then both $l(t_0 + t, x)$ and $l(t, l(t_0, x))$ are curves through $l(t_0, x)$ at $t = 0$ and satisfy

$$\frac{\mathrm{d}l(t_0 + t, x)}{\mathrm{d}t} = X(l(t_0 + t, x)),$$

$$\frac{\mathrm{d}l(t, l(t_0, x))}{\mathrm{d}t} = X(l(t, l(t_0, x))).$$

Hence we have

$$l(t_0 + t, x) = l(t, l(t_0, x))$$

by the uniqueness of the solution of ordinary differential equations. □

Next we will deal with a general manifold M. Let $X: M \to T(M)$ be a vector field on M. Take a chart $\varphi: U \to \mathbb{R}^n$ and define a map $Y: \varphi(U) \to T(\varphi(U))$ by

$$Y(\varphi(x)) = \varphi_* X(x) = \mathrm{d}\varphi(X(x)).$$

Then Y is a smooth vector field on the open set $\varphi(U)$ of $\mathbb{R}^n$. As observed above, for any point x of U, there exists a unique integral curve l of Y through $\varphi(x)$ in $\varphi(U)$ locally. Then the curve $\varphi^{-1} \cdot l$ is an integral curve of X through x in U. This is easily seen from the definition of the tangent vector to a smooth curve. Thus we have the existence of an integral curve of X through x in U locally.

Next we show uniqueness. Let l_1 and l_2 be two integral curves of X through x in U. Then $\varphi \cdot l_1$ and $\varphi \cdot l_2$ are integral curves of Y through $\varphi(x)$ by definition. It follows from the uniqueness of the solution of ordinary differential equations that these two curves coincide locally. This means also that the integral curve constructed above by making use of a chart does not depend on the chart selected. Thus we have the following lemma.

Lemma 3.17 *Let X be a smooth vector field on a manifold M. For each point p of M, there exist open neighbourhoods $V \subset U$ of p, a number $\varepsilon > 0$, and a unique smooth map*

$$l:(-\varepsilon, \varepsilon) \times U \to M$$

such that

$$\frac{dl(t, x)}{dt} = X(l(t, x)), \qquad l(0, x) = x$$

and

$$l(t_0 + t, x) = l(t, l(t_0, x))$$

for $|t_0|, |t|, |t_0 + t| < \varepsilon$ and $x \in V$. □

Generally $l(t, x)$ does not necessarily exist globally. However, we show in the following that if $l(t, x)$ exists for arbitrary t, then it is unique.

Lemma 3.18 *Let X be a smooth vector field on a manifold M and x be a point of M. If an integral curve $l(t, x)$ of X through x exists, then it is unique.*

Proof Let $l_1(t, x)$ and $l_2(t, x)$ be two integral curves of X through x. It follows from Lemma 3.17 that both curves coincide in a sufficiently small neighbourhood of 0. We now put

$$\varepsilon_0 = \sup \{\varepsilon \mid l_1(t, x) = l_2(t, x) \quad \text{for} \quad |t| < \varepsilon\}.$$

If ε_0 is finite, we have

$$l_1(t, x) = l_2(t, x) \qquad \text{for} \quad |t| = \varepsilon_0,$$

since l_1 and l_2 are continuous. Moreover an integral curve of X through $l_1(\varepsilon_0, x)$ is unique in a neighbourhood of 0. Hence it follows from Lemma 3.17 that l_1 and l_2 coincide up to ε which is slightly larger than ε_0. Similarly we can show that both curves coincide down to $-\varepsilon'$ which is a bit smaller than $-\varepsilon_0$. But this contradicts the definition of ε_0, that is, ε_0 cannot be finite. □

Given a 1-parameter group $\varphi: \mathbb{R} \times M \to M$ of diffeomorphisms of M (see Example 13 in Section 3.2), we define a map $\varphi_t: M \to M$ by $\varphi_t(x) = \varphi(t, x)$ for $t \in \mathbb{R}$, which is a diffeomorphism. If we fix x and regard t as a variable, then $\varphi_t(x)$ is a smooth curve through x. We now define a smooth vector field

$X: M \to T(M)$ on M by

$$X(x) = \left.\frac{d\varphi_t(x)}{dt}\right|_{t=0}.$$

This vector field X is said to *generate* the group φ.

Let us consider the converse.

Theorem 3.19 *A smooth vector field X on M which vanishes outside of a compact set $K \subset M$ generates a unique 1-parameter group of diffeomorphisms of M.*

Proof In view of Lemma 3.17, for each point p of M, there exist open neighbourhoods $V \subset U$ of p, a number $\varepsilon > 0$, and a unique integral curve $l(t, x)$ of X through x for $|t| < \varepsilon$ and $x \in U$ which is smooth as a function of both variables. Moreover it satisfies the second equation in Lemma 3.17 for $|t| < \varepsilon$ and $x \in V$. Since K is compact, K is covered by a finite number of such open sets V_i. Let ε_i be the numbers corresponding to ε in Lemma 3.17 and put

$$\varepsilon_0 = \min\{\varepsilon_i\}.$$

Setting $l(t, x) = x$ for $x \notin K$, we have a unique integral curve $l(t, x)$ for $|t| < \varepsilon_0$ and for all $x \in M$. Furthermore, it satisfies

$$l(t_1 + t_2, x) = l(t_1, l(t_2, x))$$

for $|t_1|, |t_2|, |t_1 + t_2| < \varepsilon_0$. We now define a map $l_t: M \to M$ by $l_t(x) = l(t, x)$ for $|t| < \varepsilon_0$, which satisfies

$$l_{t_1} \cdot l_{t_2} = l_{t_1 + t_2} = l_{t_2} \cdot l_{t_1} \tag{3.2}$$

for $|t_1|, |t_2|, |t_1 + t_2| < \varepsilon_0$.

Given an arbitrary number t, choose a positive integer a with $|t|/a < \varepsilon_0$, and set

$$l(t, x) = l_{t/a} \cdots l_{t/a}(x) = (l_{t/a})^a(x)$$

where the transformation $l_{t/a}$ is iterated a times.

We now show that this is well defined. Although $l(t, x)$ is already defined for $|t| < \varepsilon_0$, it coincides with the previous one on account of Lemma 3.17 so that there is no contradiction in terms.

Let b be another positive integer with $|t|/b < \varepsilon_0$. Then we have

$$(l_{t/a})^a(x) = ((l_{t/ab})^b)^a(x) = ((l_{t/ab})^a)^b(x) = (l_{t/b})^b(x)$$

by Lemma 3.17, so the definition of $l(t, x)$ does not depend on the choice of a.

Next we show that $l(t, x)$ is the desired 1-parameter group of diffeomorphisms of M. For this, we first verify the formula

$$l(t_1 + t_2, x) = l(t_1, l(t_2, x))$$

for arbitrary $t_1, t_2 \in \mathbb{R}$. Given any $t_1, t_2 \in \mathbb{R}$, choose a positive integer a such that

$$\frac{|t_1|}{a}, \frac{|t_2|}{a}, \frac{|t_1 + t_2|}{a} < \varepsilon_0.$$

Then it follows from (3.2) that

$$\begin{aligned} l(t_1 + l_2, x) &= (l_{(t_1+t_2)/a})^a(x) \\ &= (l_{t_1/a} \cdot l_{t_2/a})^a(x) = (l_{t_1/a})^a \cdot (l_{t_2/a})^a(x) \\ &= l(t_1, l(t_2, x)). \end{aligned}$$

To see that $l(t, x)$ is smooth in both variables t and x, we need only show it locally. Given any $t_0 \in \mathbb{R}$, choose a positive integer a such that $|t_0|/a < \varepsilon_0$. Since $l(t, x) = (l_{t/a})^a(x)$, we see that $l(t, x)$ is smooth around (t_0, x).

Moreover, it follows by definition that

$$\left.\frac{dl(t, x)}{dt}\right|_{t=0} = X(x).$$

Thus we see that $l(t, x)$ is a 1-parameter group of diffeomorphisms generated by X, that is, existence is shown.

Next we shall show uniqueness.

Let $\varphi: \mathbb{R} \times M \to M$ be an arbitrary 1-parameter group of diffeomorphisms generated by X. Define a map $\varphi_t: M \to M$ by $\varphi_t(x) = \varphi(t, x)$ for $t \in \mathbb{R}$, $x \in M$. We now fix x and regard t as a variable so that $\varphi_t(x)$ is a smooth curve through x. Then we shall show that $\varphi_t(x)$ is an integral curve of X through x. Since φ is generated by X, we have

$$\left.\frac{d\varphi_t(x)}{dt}\right|_{t=0} = X(x)$$

by definition. Moreover, since φ is a 1-parameter group of diffeomorphisms by assumption, it satisfies, for fixed $t_0 \in \mathbb{R}$,

$$\left.\frac{d\varphi_{t+t_0}(x)}{dt}\right|_{t=0} = \left.\frac{d\varphi_t(\varphi_{t_0}(x))}{dt}\right|_{t=0} = X(\varphi_{t_0}(x)).$$

That is, we have

$$\frac{d\varphi_t(x)}{dt} = X(\varphi_t(x))$$

for any $t \in \mathbb{R}$. This means that, for each fixed x, the curve $\varphi_t(x)$ is an integral curve of X through x. Therefore it follows from the uniqueness of an integral curve in Lemma 3.18 that such a $\varphi_t(x)$ is unique. □

Corollary 3.20 *A smooth vector field on a compact manifold M generates a unique 1-parameter group of diffeomorphisms of M.*

Remark Even if a curve $f: \mathbb{R} \to M = \mathbb{R}$ is smooth and bijective, the vector field of tangent vectors to f is not necessarily smooth. For instance, let $x = f(t) = t^3$. Then we have

$$\frac{\mathrm{d}f}{\mathrm{d}t} = 3t^2 = 3x^{2/3}$$

and hence it is not a smooth vector field with respect to the natural smooth structure on M. Needless to say, f is an integral curve of the above vector field through 0. On the other hand, the constant curve $\bar{f}(t) \equiv 0$ is also an integral curve of the vector field through 0. Hence uniqueness does not hold. This is due to the lack of smoothness.

Corollary 3.21 *Given a smooth vector field X on M and a 1-parameter group $\varphi: \mathbb{R} \times M \to M$ of diffeomorphisms generated by X, the following holds*

$$M^{\mathbb{R}} = \{x \in M \mid X(x) = 0\}$$

where $M^{\mathbb{R}}$ denotes the fixed point set of the $\mathbb{R}$-action φ.

Proof If $X(x) = 0$, then the constant curve $l(t, x) \equiv x$ is an integral curve of X through x. As we saw in the proof of Theorem 3.19, for fixed $x \in M$, $\varphi(t, x)$ is an integral curve of X through x. It follows from the uniqueness of an integral curve in Lemma 3.18 that

$$\varphi(t, x) = l(t, x) \equiv x,$$

that is, $x \in M^{\mathbb{R}}$. The converse is obvious. □

Remark The example in the above remark shows that although the tangent vector to f at 0 is 0, f moves the point 0. This is also due to the lack of smoothness.

Example 1 As a vector field on $M = \mathbb{R}$, we consider $X(x) = x$. That is, we assign to each point x of M the vector x (see Fig. 3.8).

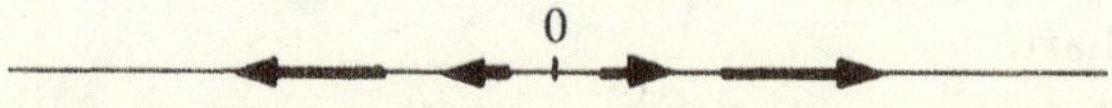

Fig. 3.8 A vector field on $\mathbb{R}$.

An integral curve $x(t)$ of X will be obtained by solving the following differential equation

$$\frac{\mathrm{d}x(t)}{\mathrm{d}t} = x(t), \qquad x(0) = x_0.$$

This will be obtained by

$$\int \frac{\mathrm{d}x(t)}{x(t)} = \int \mathrm{d}t,$$

which is solved by elementary calculus as follows:

$$\log|x(t)| = t + c \qquad (c\text{: constant}).$$

Hence we have

$$|x(t)| = \exp(t + c).$$

If the initial condition $x(0) = x_0$ is non-zero, then we have

$$x(t) = x_0 \exp t.$$

If $x_0 = 0$, then the constant function $x(0) \equiv 0$ is evidently a solution, which is also included in the above expression. Thus the integral curve of X through x_0 is given by

$$x(t) = x_0 \exp t$$

in general so that the 1-parameter group $\varphi: \mathbb{R} \times \mathbb{R} \to \mathbb{R}$ generated by X is given by

$$\varphi(t, x) = x \exp t.$$

Example 2 As a vector field on $M = \mathbb{R}$, we consider $X(x) = x^2$. An integral curve $x(t)$ of X will be obtained by solving the following differential equation

$$\frac{\mathrm{d}x(t)}{\mathrm{d}t} = x(t)^2, \qquad x(0) = x_0.$$

The solution will be obtained by

$$\int \frac{\mathrm{d}x(t)}{x(t)^2} = \int \mathrm{d}t$$

which is solved as

$$-\frac{1}{x(t)} = t + c \qquad (c\text{: constant}).$$

Hence we have

$$x(t) = \frac{-1}{t + c}.$$

If the initial condition $x(0) = x_0$ is non-zero, then we have

$$x(t) = \frac{x_0}{-x_0 t + 1}.$$

If $x_0 = 0$, then the constant function $x(t) \equiv 0$ is evidently a solution, which is also included in the above expression. Thus the integral curve of X through x_0 is given by

$$x(t) = \frac{x_0}{-x_0 t + 1}$$

in general. The graph of this function in the case $x_0 > 0$ is shown in Fig. 3.9.

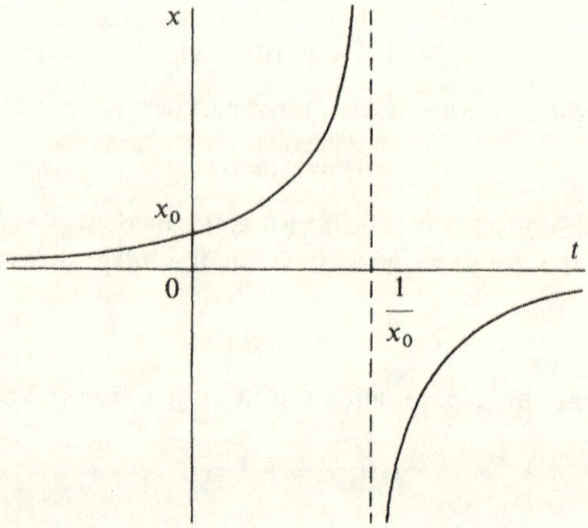

Fig. 3.9 The function $x(t) = x_0/(-x_0 t + 1)$.

Hence the domain of this function is $(-\infty, 1/x_0)$ and t cannot move into $[1/x_0, \infty)$. While t starts from 0 and approaches $1/x_0$, $x(t)$ moves from x_0 to ∞. While t moves from 0 to $-\infty$, $x(t)$ moves from x_0 to 0. A similar phenomenon occurs in the case $x_0 < 0$. Thus this vector field is an example which does not generate a 1-parameter group.

3.6 Left invariant vector fields on Lie groups and the exponential map

On a manifold there is in general no natural way to construct a vector field. On a Lie group, however, there is a natural method to construct vector fields by using the group structure. In this section, we shall show that an integral curve of such a vector field through the identity element e is a subgroup of G and shall define the exponential map by making use of it.

Let G be a Lie group and $T_e(G)$ be the tangent space of G at the identity element e. Then for each $v \in T_e(G)$, we construct a smooth vector field X_v on G by

$$X_v(g) = L_{g*}(v) \qquad \text{for } g \in G$$

where $L_{g*}\colon T(G) \to T(G)$ denotes the differential of the left translation $L_g\colon G \to G$ defined in Section 1.2. Note that L_{g*} is often written as g_* for simplicity.

Here we must show that X_v, so determined, is smooth. As we mentioned in Section 3.4, the tangent G-vector bundle of a G-manifold is a smooth G-vector bundle. In particular, $T(G)$ is a smooth G-vector bundle. Hence the map of the G-action

$$\varphi\colon G \times T(G) \to T(G)$$

is smooth. Since $G \times T_e(G)$ is a smooth submanifold of $G \times T(G)$, the restriction

$$\varphi|\colon G \times T_e(G) \to T(G)$$

is also smooth by Lemma 3.3. Note that $\varphi(g, v) = g_* v = X_v(g)$ for $g \in G$ and $v \in T_e(G)$ so that $X_v(g)$ depends smoothly on both variables $g \in G$ and $v \in T_e(G)$. In particular, for fixed $v \in T_e(G)$, X_v is a smooth cross-section. Then the vector field X_v satisfies the relation

$$\begin{aligned} L_{g*} X_v(h) &= L_{g*} \cdot L_{h*}(v) = (L_g \cdot L_h)_*(v) \\ &= L_{gh*}(v) = X_v(gh) = X_v(L_g(h)), \end{aligned}$$

that is, the following diagram is commutative for each $g \in G$:

$$\begin{array}{ccc} T(G) & \xrightarrow{\;L_{g*}\;} & T(G) \\ {\scriptstyle X_v}\big\downarrow & & \big\downarrow{\scriptstyle X_v} \\ G & \xrightarrow{\;L_g\;} & G \end{array}$$

The vector field X_v obtained in this way is called the *left invariant vector field defined by* v. Generally any vector field X on G which satisfies the relation $L_{g*} \cdot X = X \cdot L_g$ for all $g \in G$ is said to be *left invariant.*

Let $\mathscr{G}$ denote the set of left invariant vector fields on G. Given X, $Y \in \mathscr{G}$ and $a \in \mathbb{R}$, we define new vector fields $X + Y$ and aX by

$$(X + Y)(g) = X(g) + Y(g),$$

$$(aX)(g) = aX(g)$$

where the operations on the right-hand sides are those in the tangent space at g. Then the following lemma is easily proved.

Lemma 3.22 *The set $\mathscr{G}$ becomes a vector space under the above operations and the correspondence $v \mapsto X_v$ gives rise to an isomorphism $T_e(G) \to \mathscr{G}$ of vector spaces.*

Remark Actually the vector space $\mathscr{G}$ is equipped with a structure called a Lie algebra, which we do not discuss in this book.

Given a left invariant vector field X on a Lie group G, we consider an integral curve

$$l(t, e), \qquad l(0, e) = e$$

of X through e.

Lemma 3.23 *There exists* $\varepsilon > 0$ *such that*

$$l(t_0 + t, e) = l(t_0, e) \cdot l(t, e) = l(t, e) \cdot l(t_0, e)$$

for $|t_0|, |t|, |t_0 + t| < \varepsilon$, *where the products indicate those in* G.

Proof In the following we fix t_0 and regard t as a variable. As we showed in Section 3.5, there exist $\varepsilon > 0$ and an integral curve $l(t, e)$ of X through e for $|t| < \varepsilon$. Then

$$\frac{\mathrm{d}l(t + t_0, e)}{\mathrm{d}t} = X(l(t + t_0, e)) \tag{3.3}$$

for $|t + t_0| < \varepsilon$.

On the other hand, since $l(t_0, e) \cdot l(t, e)$ is a smooth curve through $l(t_0, e)$, we denote it by $k(t, l(t_0, e))$, that is, we set

$$k(t, l(t_0, e)) = l(t_0, e) \cdot l(t, e).$$

Then we have

$$\begin{aligned} \frac{\mathrm{d}k(t, l(t_0, e))}{\mathrm{d}t} &= \frac{\mathrm{d}l(t_0, e) \cdot l(t, e)}{\mathrm{d}t} \qquad (3.4) \\ &= l(t_0, e)_* \frac{\mathrm{d}l(t, e)}{\mathrm{d}t} = l(t_0, e)_* X(l(t, e)) \\ &= X(l(t_0, e) \cdot l(t, e)) = X(k(t, l(t_0, e))). \end{aligned}$$

Here the fourth equality follows from the assumption that X is left invariant.

It follows from (3.3) and (3.4) that both $l(t + t_0, e)$ and $k(t, l(t_0, e))$ satisfy the same differential equation and the same initial condition

$$l(t_0, e) = k(0, l(t_0, e)).$$

Hence by the uniqueness of the solution of differential equations we have

$$l(t + t_0, e) = k(t, l(t_0, e))$$

for $|t_0|, |t|, |t + t_0| < \varepsilon$. Thus we have

$$l(t + t_0, e) = l(t_0, e) \cdot l(t, e)$$

for $|t_0|, |t|, |t+t_0| < \varepsilon$. By exchanging the roles of t_0 and t, we also have

$$l(t+t_0, e) = l(t, e)\cdot l(t_0, e). \qquad \square$$

Definition 3.9 A 1-*parameter subgroup* of a Lie group G is a homomorphism

$$\rho\colon \mathbb{R} \to G$$

of Lie groups, where $\mathbb{R}$ is the Lie group $\mathbb{R}^1$ of Example 1 in Section 3.2.

Remark The concept of a 1-parameter subgroup is different from that of a 1-parameter group in Section 3.5.

Let ρ be a 1-parameter subgroup of a Lie group G. Associate with ρ the tangent vector $(\mathrm{d}\rho/\mathrm{d}t)(0)$ to ρ at e.

Theorem 3.24 *The correspondence* $\rho \mapsto (\mathrm{d}\rho/\mathrm{d}t)(0)$ *gives a bijective correspondence between* 1-*parameter subgroups of* G *and vectors in* $T_e(G)$.

Proof *Injectivity.* Let ρ be an arbitrary 1-parameter subgroup. Put $v = (\mathrm{d}\rho/\mathrm{d}t)(0) \in T_e(G)$. Then we have

$$\frac{\mathrm{d}\rho(t_0+t)}{\mathrm{d}t} = \frac{\mathrm{d}\rho(t_0)\cdot\rho(t)}{\mathrm{d}t} = \rho(t_0)_* \frac{\mathrm{d}\rho(t)}{\mathrm{d}t}$$

so that

$$\frac{\mathrm{d}\rho}{\mathrm{d}t}(t_0) = \rho(t_0)_* \frac{\mathrm{d}\rho}{\mathrm{d}t}(0) = \rho(t_0)_* v.$$

That is, the tangent vector to ρ at $\rho(t_0)$ is given by $\rho(t_0)_* v$. This means that $\rho(t)$ is an integral curve of the left invariant vector field defined by v satisfying $\rho(0) = e$. It follows from Lemma 3.18 that ρ is uniquely determined by v. Thus the correspondence $\rho \mapsto v$ is injective.

Surjectivity. Given $v \in T_e(G)$, consider the left invariant vector field X_v defined by v. Then it follows from Lemma 3.17 that there exist $\varepsilon > 0$ and an integral curve $l(t, e)$ of X_v through e for $|t| < \varepsilon$. On a manifold, it is impossible to make ε large in general, as observed in Example 2 in Section 3.5. On a Lie group, however, the group structure enables us to reduce the global existence problem to a local one as follows.

For arbitrary $t \in \mathbb{R}$, choose a sufficiently large positive integer a such that $|t|/a < \varepsilon$, and set

$$l(t, e) = l\left(\frac{t}{a}, e\right)^a$$

where the right-hand side denotes the power of $l(t/a, e)$ with exponent a with respect to the multiplication of G.

The proof that this does not depend on the choice of a is similar to that of Theorem 3.19. However, while the product is given by the composition of maps in Theorem 3.19, the product is now given by the multiplication of elements of G.

Next we show that it is a homomorphism.

For arbitrary $t_1, t_2 \in \mathbb{R}$, choose a positive integer a such that

$$|t_1|/a, |t_2|/a, |t_1 + t_2|/a < \varepsilon.$$

Then it follows from Lemma 3.23 that

$$\begin{aligned} l(t_1 + t_2, e) &= l\left(\frac{t_1 + t_2}{a}, e\right)^a = \left\{l\left(\frac{t_1}{a}, e\right) \cdot l\left(\frac{t_2}{a}, e\right)\right\}^a \\ &= l\left(\frac{t_1}{a}, e\right)^a \cdot l\left(\frac{t_2}{a}, e\right)^a = l(t_1, e) \cdot l(t_2, e), \end{aligned}$$

that is, $l(t, e)$ is a homomorphism in t. Here we used the fact that $l(t_1/a, e)$ and $l(t_2/a, e)$ commute by Lemma 3.23. To see that $l(t, e)$ is smooth in t, we need only show it locally, and we see that it is smooth around t by fixing a positive integer a for each t.

Thus there exists a 1-parameter subgroup $l(t, e)$ such that the tangent vector to the curve $l(t, e)$ at e is the given v, that is, the correspondence is surjective. □

Corollary 3.25 *An arbitrary left invariant vector field of a Lie group generates a 1-parameter group of diffeomorphisms.*

Proof Given a left invariant vector field X on a Lie group G, let $\rho: \mathbb{R} \to G$ be the corresponding 1-parameter subgroup of G obtained uniquely by Lemma 3.22 and Theorem 3.24. Then we define a map $l: \mathbb{R} \times G \to G$ by

$$l(t, g) = g\rho(t).$$

Obviously $l(t, g)$ is smooth in both variables t and g, and

$$\begin{aligned} l(0, g) &= g\rho(0) = ge = g \\ l(t_2, l(t_1, g)) &= l(t_2, g\rho(t_1)) = (g\rho(t_1))\rho(t_2) \\ &= g(\rho(t_1)\rho(t_2)) = g\rho(t_1 + t_2) = l(t_1 + t_2, g) \\ &= l(t_2 + t_1, g), \end{aligned}$$

that is, l is a 1-parameter group of diffeomorphisms. Furthermore we have

$$\begin{aligned} \left.\frac{\mathrm{d}l(t, g)}{\mathrm{d}t}\right|_{t=0} &= \left.\frac{\mathrm{d}g\rho(t)}{\mathrm{d}t}\right|_{t=0} = g_* \frac{\mathrm{d}\rho}{\mathrm{d}t}(0) \\ &= g_* X(\rho(0)) = g_* X(e) = X(g) \end{aligned}$$

which means that l is generated by X. □

Remark In this corollary, we do not assume that G is compact.

Definition 3.10 *The exponential map.* Define $\exp\colon T_e(G) \to G$ as follows. Given $v \in T_e(G)$, let ρ_v be the corresponding 1-parameter subgroup of G of Theorem 3.24. Then we set

$$\exp(v) = \rho_v(1).$$

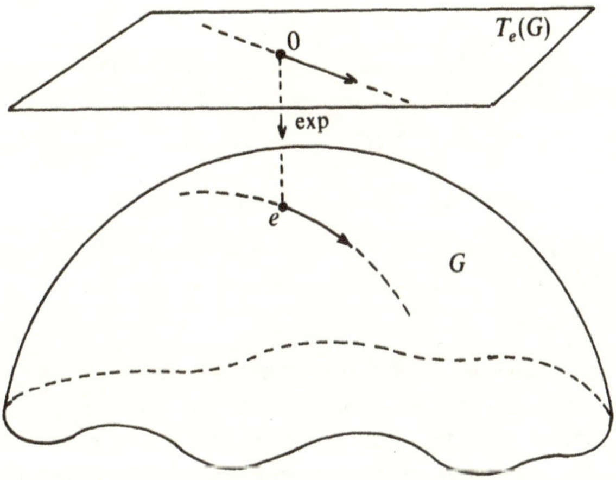

Fig. 3.10 The exponential map.

The exponential map is illustrated in Fig. 3.10.

Remark Making use of the group structure, the map exp can be extended to a map

$$\exp\colon T(G) \to G$$

as follows. For $v \in T_g(G)$, we set

$$\exp(v) = g \exp(g_*^{-1} v).$$

Lemma 3.26 *For $v \in T_e(G)$ and $t \in \mathbb{R}$, we have*

$$\exp(tv) = \rho_v(t).$$

Hence for an integer a, we have

$$\exp(av) = (\exp(v))^a.$$

Proof In view of Lemma 3.13, for each fixed $t_0 \in \mathbb{R}$, we have

$$\left.\frac{\mathrm{d}\rho_v(t_0 t)}{\mathrm{d}t}\right|_{t=0} = t_0 \frac{\mathrm{d}\rho_v}{\mathrm{d}t}(0) = t_0 v,$$

that is, $\rho_v(t_0 t)$ is the 1-parameter subgroup corresponding to the vector $t_0 v \in T_e(G)$. Hence by uniqueness, we have

$$\rho_v(t_0 t) = \rho_{t_0 v}(t).$$

In consequence, we have the required relation

$$\exp(t_0 v) = \rho_{t_0 v}(1) = \rho_v(t_0).$$

Since ρ_v is a homomorphism, we have $\rho_v(-1) = \rho_v(1)^{-1}$ and, for any integer a, we have

$$\exp(av) = \rho_v(a) = \rho_v(1)^a = (\exp(v))^a. \qquad \square$$

Lemma 3.27 *The map* $\exp\colon T_e(G) \to G$ *is smooth and its differential*

$$\exp_*\colon T_0(T_e(G)) = T_e(G) \to T_e(G)$$

is the identity map. Here we identify the tangent space $T_0(T_e(G))$ of $T_e(G)$ at 0 with $T_e(G)$ itself, since $T_e(G)$ is a vector space.

Proof As we observed at the beginning of this section, $X_v(g)$ depends smoothly on both variables $g \in G$ and $v \in T_e(G)$. Accordingly it follows from the general theory of differential equations that, for each fixed $v_0 \in T_e(G)$, there is a number $\varepsilon > 0$ and an open neighbourhood U of v_0 such that, for any $v \in U$, there exists an integral curve

$$\rho_v\colon (-\varepsilon, \varepsilon) \to G, \qquad \rho_v(0) = e$$

of X_v, and $\rho_v(t)$ is a smooth function of $t \in (-\varepsilon, \varepsilon)$ and $v \in U$.

On the other hand, it follows from Theorem 3.24 that, for arbitrary $t \subset \mathbb{R}$ and $v \in T_e(G)$, there exists a unique solution. By combining this with the result above, one verifies easily that the solution depends smoothly on both variables t and v globally. In particular, $\exp(v) = \rho_v(1)$ is smooth in v.

Given $v \in T_e(G)$, we put $l_v(t) = tv$ for $t \in \mathbb{R}$ which is a smooth curve through 0 in $T_e(G)$. As before we denote by $[l_v]$ the tangent vector to the curve $l_v(t)$ at 0. Then we identify $T_e(G)$ with $T_0(T_e(G))$ by this canonical correspondence $v \mapsto [l_v]$.

By making use of Lemma 3.26, we compute

$$\begin{aligned} \exp_*[l_v] &= [\exp tv] = [\rho_{tv}(1)] \\ &= [\rho_v(t)] = v, \end{aligned}$$

that is, $\exp_*$ is the identity under the above identification. $\square$

Remark $\exp\colon T_e(G) \to G$ is neither injective nor surjective in general, even if G is connected.

Proposition 3.28 *Given a homomorphism $f\colon G_1 \to G_2$ of Lie groups, the following diagram is commutative:*

$$\begin{array}{ccc} T_e(G_1) & \xrightarrow{f_*} & T_e(G_2) \\ \Big\downarrow{\scriptstyle \exp} & & \Big\downarrow{\scriptstyle \exp} \\ G_1 & \xrightarrow{f} & G_2 \end{array}$$

Proof For $v \in T_e(G_1)$, we denote by ρ_v the corresponding 1-parameter subgroup of G_1. Then the composite map $f \cdot \rho_v\colon \mathbb{R} \to G_2$ is a 1-parameter subgroup of G_2 and satisfies

$$\frac{df \cdot \rho_v}{dt}(0) = f_* \frac{d\rho_v}{dt}(0) = f_*(v),$$

that is, $f \cdot \rho_v$ is the 1-parameter subgroup of G_2 corresponding to $f_*(v)$. Hence we have

$$\exp f_*(v) = \rho_{f_*(v)}(1) = f \cdot \rho_v(1) = f \cdot \exp(v). \qquad \square$$

Example 1 We consider $G = GL(n, \mathbb{R})$. We begin by recalling Example 3 in Section 3.2. Let $M(n, \mathbb{R})$ be the set of $n \times n$ real matrices. We identify $M(n, \mathbb{R})$ with the n^2-dimensional Euclidean space $\mathbb{R}^{n^2}$ so that $M(n, \mathbb{R})$ is a manifold. Since $GL(n, \mathbb{R})$ is an open submanifold of $M(n, \mathbb{R})$, there is a canonical identification $T_e(GL(n, \mathbb{R}))$ with $T_e(M(n, \mathbb{R}))$. Moreover, since $M(n, \mathbb{R})$ is a real vector space, there is a canonical identification of $T_e(M(n, \mathbb{R}))$ with $M(n, \mathbb{R})$ as shown in the proof of Lemma 3.27. Thus we have a canonical identification

$$T_e(GL(n, \mathbb{R})) = M(n, \mathbb{R}).$$

For our purpose, we shall give the correspondence explicitly in the following.

Since $GL(n, \mathbb{R})$ is open in $M(n, \mathbb{R}) = \mathbb{R}^{n^2}$, the matrix entries x_{ij}, $1 \le i, j \le n$, for any $X = (x_{ij}) \in GL(n, \mathbb{R})$ are coordinates of a single chart covering the group $GL(n, \mathbb{R})$. Therefore, given a smooth curve $l(t) = (x_{ij}(t))$ through e in $GL(n, \mathbb{R})$, the tangent vector $(dl/dt)(0)$ to $l(t)$ at e is obtained by differentiating each entry at 0, that is,

$$\frac{dl}{dt}(0) = \left(\frac{dx_{ij}}{dt}(0)\right) \in M(n, \mathbb{R}).$$

Here note that in general the matrix so obtained no longer lies in $GL(n, \mathbb{R})$ but in $M(n, \mathbb{R})$.

When $G = GL(n, \mathbb{R})$, e is the $n \times n$ identity matrix I.

We now show that for $A \in M(n, \mathbb{R}) = T_e(GL(n, \mathbb{R}))$,

$$\exp(A) = I + A + \frac{A^2}{2} + \cdots + \frac{A^n}{n!} + \cdots.$$

To see this, put

$$\rho_A(t) = I + At + \frac{A^2t^2}{2} + \cdots + \frac{A^n t^n}{n!} + \cdots.$$

Then it is easily shown that ρ_A is a smooth homomorphism from $\mathbb{R}$ to $GL(n, \mathbb{R})$. Since $(\mathrm{d}\rho_A/\mathrm{d}t)(0) = A$, ρ_A is the 1-parameter subgroup of $GL(n, \mathbb{R})$ corresponding to A. Thus we have

$$\exp(A) = \rho_A(1) = I + A + \frac{A^2}{2} + \cdots + \frac{A^n}{n!} + \cdots.$$

Remark The term exp originates from this example.

Theorem 3.29 *The map* exp *is a diffeomorphism of an open neighbourhood of* 0 *in $T_e(G)$ with an open neighbourhood of e in G.*

Proof This follows immediately from Lemma 3.27 and the inverse function theorem. □

Lemma 3.30 *Let $T_e(G) = V \oplus W$ be a direct sum decomposition. Then the map $f: V \oplus W \to G$ defined by*

$$f(v, w) = \exp(v)\cdot\exp(w) \quad \textit{for} \quad v \in V, w \in W$$

is a diffeomorphism of an open neighbourhood of 0 *in $T_e(G)$ with an open neighbourhood of e in G.*

Proof Since the product in G is smooth, it follows from Lemma 3.27 that f is smooth. Since $f(v, 0) = \exp(v)$, we have

$$f_* \mid V = \exp_* \mid V = \mathrm{id}$$

by Lemma 3.27. Similarly we have $f_* \mid W = \mathrm{id}$. Hence f_* is the identity on $T_e(G)$ so that we may proceed as in Theorem 3.29. □

Lemma 3.31 *Let $\varphi: U \to \mathbb{R}^n$ be a chart of G at e such that $\varphi(e) = 0$ and $\varphi(U)$ is bounded. Then there exists a number $\varepsilon > 0$ such that $\varphi^{-1}(tv)\cdot\varphi^{-1}(tw)$ lies in U for $|t| < \varepsilon$ and $v, w \in \varphi(U)$, and we have*

$$\lim_{t\to 0} \frac{\varphi(\varphi^{-1}(tv)\cdot\varphi^{-1}(tw))}{t} = v + w$$

where $\varphi^{-1}(tv)\cdot\varphi^{-1}(tw)$ denotes the product in G.

Proof Since the product in G is continuous, there is an open neighbourhood V of e with $V^2 \subset U$. For a positive number s, we denote by $\mathring{D}(s)$ the open disc of radius s in $\mathbb{R}^n$ having the origin as its centre, that is,

$$\mathring{D}(s) = \{v \in \mathbb{R}^n \mid \|v\| < s\}.$$

By assumption, there exist $R > r > 0$ such that

$$\mathring{D}(r) \subset \varphi(V) \subset \varphi(U) \subset \mathring{D}(R).$$

These subsets are illustrated in Fig. 3.11.

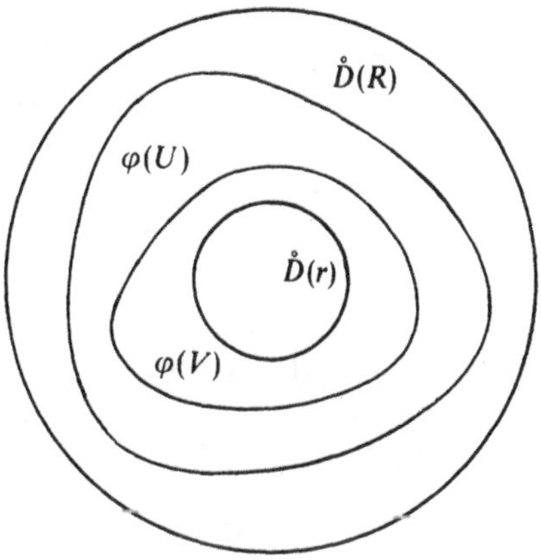

Fig. 3.11 The open discs $\mathring{D}(r)$ and $\mathring{D}(R)$.

Put $\varepsilon = r/R$. Then for any $|t| < \varepsilon$ and $v, w \in \varphi(U)$, we have tv, $tw \in \mathring{D}(r) \subset \varphi(V)$ so that

$$\varphi^{-1}(tv) \cdot \varphi^{-1}(tw) \in V^2 \subset U.$$

For fixed $v, w \in \varphi(U)$, we define a map

$$f\colon (-\varepsilon, \varepsilon) \times (-\varepsilon, \varepsilon) \to \mathbb{R}^n$$

by

$$f(x, y) = \varphi(\varphi^{-1}(xv) \cdot \varphi^{-1}(yw)) - (xv + yw).$$

Then f is smooth and satisfies

$$f(x, 0) = f(0, y) = 0 \qquad \text{for} \quad |x|, |y| < \varepsilon$$

so that we have

$$\frac{\partial f}{\partial x}(0, 0) = \frac{\partial f}{\partial y}(0, 0) = 0.$$

We now put

$$F(t) = f(t, t) \qquad \text{for} \quad |t| < \varepsilon.$$

Then $F(0) = 0$ and

$$\frac{dF}{dt}(0) = \frac{\partial f}{\partial x}(0, 0)\frac{dx}{dt} + \frac{\partial f}{\partial y}(0, 0)\frac{dy}{dt} = 0.$$

This means that

$$\lim_{t \to 0} \frac{f(t, t)}{t} = 0,$$

which is the assertion of the lemma. □

Corollary 3.32 *Given any $v, w \in T_e(G)$, there exists a number $\varepsilon > 0$ such that the function after the limit symbol in the following equation makes sense for $|t| < \varepsilon$ and we have*

$$\lim_{t \to 0} \frac{\exp^{-1}(\exp(tv) \cdot \exp(tw))}{t} = v + w.$$

Proof According to Theorem 3.29, the map exp is a diffeomorphism of an open neighbourhood U of 0 in $T_e(G)$ with $\exp(U)$ which is an open neighbourhood of e in G. Since it is possible to choose a bounded open set as U, we may assume that U is bounded in $T_e(G)$. Then the map

$$\exp^{-1}: \exp(U) \to U$$

is a chart satisfying the conditions of φ in Lemma 3.31. Hence Lemma 3.31 applies to yield Corollary 3.32. □

Theorem 3.33 *If G is an abelian Lie group, then $\exp: T_e(G) \to G$ is a homomorphism.*

Proof According to Corollary 3.32, given $v, w \in T_e(G)$, $\exp^{-1}(\exp(v/a) \cdot \exp(w/a))$ makes sense for a sufficiently large positive integer a and we have

$$\begin{aligned}\lim_{a \to \infty} a \exp^{-1}(\exp(v/a) \cdot \exp(w/a)) &= \lim_{a \to \infty} \frac{\exp^{-1}(\exp(v/a) \cdot \exp(w/a))}{1/a} \\ &= v + w.\end{aligned}$$

Since exp is continuous, this together with Lemma 3.26 yields

$$\begin{aligned}\exp(v + w) &= \exp\left(\lim_{a \to \infty} a \exp^{-1}(\exp(v/a) \cdot \exp(w/a))\right) \\ &= \lim_{a \to \infty} \exp(a \exp^{-1}(\exp(v/a) \cdot \exp(w/a))) \\ &= \lim_{a \to \infty} \{\exp(\exp^{-1}(\exp(v/a) \cdot \exp(w/a)))\}^a \\ &= \lim_{a \to \infty} \{\exp(v/a) \cdot \exp(w/a)\}^a.\end{aligned}$$

Since, however, G is abelian, it follows from Lemma 3.26 again that

$$\{\exp(v/a)\cdot\exp(w/a)\}^a = \{\exp(v/a)\}^a\cdot\{\exp(w/a)\}^a$$
$$= \exp(v)\cdot\exp(w),$$

which is independent of a. Thus we have

$$\exp(v + w) = \exp(v)\cdot\exp(w).$$ □

Corollary 3.34 *If G is a connected abelian Lie group, then* $\exp: T_e(G) \to G$ *is surjective.*

Proof The image of exp is a neighbourhood of e by Theorem 3.29 and is a subgroup of G by Theorem 3.33. Recall now that if G is connected, any neighbourhood of e generates G in general (see Exercise 2 in Chapter 1), and hence the corollary follows. □

Theorem 3.35 *A connected abelian Lie group G has the form $T^a \times \mathbb{R}^b$. In particular, a compact connected abelian Lie group is a torus. Here T^a denotes the torus of dimension a.*

Proof It follows from Theorem 3.33 and Corollary 3.34 that $\exp: T_e(G) \to G$ is a surjective homomorphism. In view of Theorem 3.29 there exists an open neighbourhood U of 0 in $T_e(G)$ such that

$$\operatorname{Ker}\exp \cap U = \{0\},$$

that is, Ker exp is discrete.

We now recall the following well-known results concerning a discrete subgroup Γ of $\mathbb{R}^n$. Under these circumstances, there is a finite subset $\{e_1, \ldots, e_a\}$ of Γ such that $\{e_1, \ldots, e_a\}$ are linearly independent over $\mathbb{R}$ and

$$\Gamma = \left\{\sum_i r_i e_i \,\middle|\, r_i \in \mathbb{Z}\right\}.$$

If we choose vectors $e_{a+1}, \ldots, e_n$ such that $\{e_1, \ldots, e_a, e_{a+1}, \ldots, e_n\}$ is a basis of $\mathbb{R}^n$, then Γ can be written as

$$\Gamma = \{(r_1, \ldots, r_a, 0, \ldots, 0) \mid r_i \in \mathbb{Z}\}$$

with respect to the above basis.

Thus we get an isomorphism

$$G \cong T_e(G)/\operatorname{Ker}\exp \cong T^a \times \mathbb{R}^{n-a}$$

of Lie groups. □

3.7 A closed subgroup of a Lie group is a Lie subgroup

Let G be a Lie group. A subgroup of G which is also a submanifold of G will be called a *Lie subgroup* of G. Evidently, a Lie subgroup of a Lie group is a Lie group.

Example 1 An open subgroup of a Lie group G is a Lie subgroup, because an open set of G is always a submanifold of G. In particular, the identity component G_0 of G is a Lie subgroup, because G_0 is a subgroup of G by Proposition 1.21 and is open in G.

The following theorem, concerning a Lie subgroup, is fundamental.

Theorem 3.36 *A closed subgroup H of a Lie group G is a Lie subgroup.*

Proof The theorem will be proved by a fairly tall stack of four claims, so be prepared! We need only show that H is a submanifold of G. For this, we seek a chart of H at e.

Choose a positive definite inner product in $T_e(G)$, which defines a norm of a vector of $T_e(G)$. Then the concept of convergence of a sequence of vectors is introduced by the norm.

First we define a subset V of $T_e(G)$ as follows:

$$V = \{v \in T_e(G) \mid \exists v_i \in T_e(G),\ \exists a_i \in \mathbb{R} \text{ such that } \exp(v_i) \in H,\ v_i \to 0,\quad a_i v_i \to v\}.$$

Although the definition of V seems a little complicated, V is something like a tangent space of H at e as Fig. 3.12 indicates. Once H is shown to be a Lie

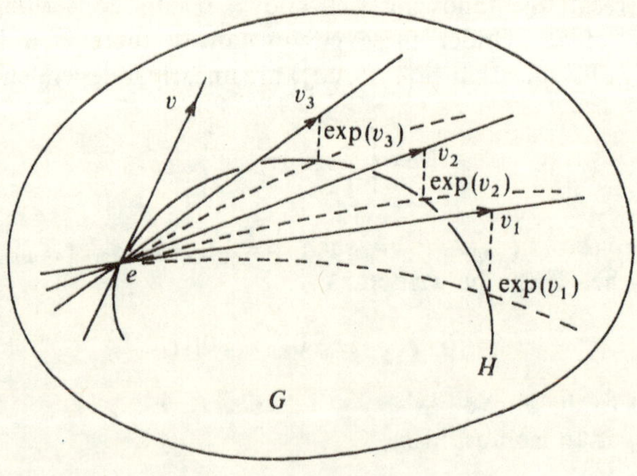

Fig. 3.12 The subset $V \subset T_e(G)$.

subgroup of G in the sequel, V will become the tangent space of H at e publicly.

We begin with the proof of the following claim

Claim 1 $\exp(v) \in H$ for all $v \in V$.

If $v = 0$, then we have trivially $\exp(0) = e \in H$.

Suppose that $v \neq 0$. Let a_i, v_i be as in the definition of V such that $a_i v_i \to v$. Since $v \neq 0$ and $v_i \to 0$, we have $|a_i| \to \infty$. For each a_i, we choose an integer b_i with

$$|a_i - b_i| \leq \tfrac{1}{2}.$$

Then $b_i/a_i \to 1$ and hence we have

$$b_i v_i = \frac{b_i}{a_i} a_i v_i \to 1 \cdot v = v.$$

It turns out that

$$\exp(b_i v_i) \to \exp(v).$$

Since b_i is an integer, it follows from Lemma 3.26 that

$$\exp(b_i v_i) = (\exp(v_i))^{b_i} \in H.$$

Recall now that H is closed by assumption. Hence we can conclude that $\exp(v) \in H$.

Claim 2 V is a vector subspace of $T_e(G)$.

Let $t \in \mathbb{R}$ and $v \in V$. Choose a_i, v_i as in the definition of V such that $a_i v_i \to v$. If we employ ta_i instead of a_i, then $ta_i v_i \to tv$, and hence $tv \in V$.

Let $v, w \in V$. Then there exists a positive integer i_0 such that, for any $i > i_0$, $\exp(v/i) \cdot \exp(w/i)$ lies in the neighbourhood of e in G chosen in Theorem 3.29. Put

$$x_i = \exp^{-1}(\exp(v/i) \cdot \exp(w/i)) \qquad \text{for} \quad i > i_0.$$

Obviously $x_i \to 0$. Combining Claim 1 with the first half of the proof of Claim 2 mentioned above, we have

$$\exp(x_i) = \exp(v/i) \cdot \exp(w/i) \in H.$$

We now apply Corollary 3.32 with $t = 1/i$ for $i > i_0$ and deduce that

$$\lim_{i \to \infty} ix_i = \lim_{i \to \infty} \frac{\exp^{-1}(\exp(v/i) \cdot \exp(w/i))}{1/i} = v + w.$$

Hence by taking i as a_i, we see that $v + w \in V$. This completes the proof of Claim 2.

Let W be a vector subspace of $T_e(G)$ with $T_e(G) = V \oplus W$ and put $f(v, w) = \exp(v) \cdot \exp(w)$ for $v \in V$, $w \in W$. Then it follows from Lemma 3.30 that f is a diffeomorphism of an open neighbourhood of 0 in $T_e(G)$ with an open neighbourhood of e in G. Hence there exist an open neighbourhood V' of 0 in V and an open neighbourhood W' of 0 in W such that $f|(V' + W'): V' + W' \to f(V' + W')$ is a diffeomorphism between an open neighbourhood of 0 in $T_e(G)$ and an open neighbourhood of e in G. Here $V' + W'$ denotes the set consisting of the elements $v' + w'$ for $v' \in V'$, $w' \in W'$.

Claim 3 $\exp(V')$ is a neighbourhood of e in H.

Suppose the claim does not hold. Then there is a sequence $h_i \in H$ such that $h_i \to e$ and $h_i \notin \exp(V')$. Since $f(V' + W')$ is a neighbourhood of e, there is a positive integer i_0 such that $h_i \in f(V' + W')$ for any $i > i_0$. Since the restriction $f|(V' + W')$ is a diffeomorphism, there exist unique $v_i \in V'$ and $w_i \in W'$ for $i > i_0$ such that

$$f(v_i, w_i) = \exp(v_i)\cdot\exp(w_i) = h_i, \quad v_i \to 0, \quad w_i \to 0.$$

Moreover the assumption $h_i \notin \exp(V')$ implies that $w_i \neq 0$. Since both h_i and $\exp(v_i)$ lie in H, $\exp(w_i)$ lies also in H. Then $w_i/\|w_i\|$ is a sequence in the unit sphere $S(W)$. Since $S(W)$ is compact, we can find a convergent subsequence $w_{i_j}/\|w_{i_j}\|$, whose limit we denote by $w \in S(W)$. By taking $1/\|w_{i_j}\|$ as a_j in the definition of V, we see that $w \in V$. This is a contradiction; that is, Claim 3 holds.

Finally, we show that H is a submanifold of G. For simplicity, we denote by f_1 the diffeomorphism

$$f|(V' + W'): V' + W' \to f(V' + W')$$

obtained just before Claim 3. In virtue of Claim 3, there is an open set U' of H such that

$$e \in U' \subset \exp(V') = f_1(V').$$

Then there exists an open set U of $f(V' + W')$ satisfying

$$U' = H \cap U$$

by the definition of the relative topology.

Claim 4 $f_1^{-1}(U)$ is an open neighbourhood of 0 in $T_e(G)$ satisfying

$$f_1^{-1}(U) \cap V = f_1^{-1}(U').$$

For any point x of $f_1^{-1}(U) \cap V$, we have

$$f_1(x) \in U \cap f(V) \subset U \cap H = U'$$

by Claim 1, that is, $x \in f_1^{-1}(U')$. Thus we have

$$f_1^{-1}(U) \cap V \subset f_1^{-1}(U').$$

Conversely, let x be an arbitrary element of $f_1^{-1}(U')$. Then

$$f_1(x) \in U' = U \cap H \subset \exp(V') = f_1(V').$$

Hence $x \in f_1^{-1}(U)$ and there exists $x' \in V'$ such that $f_1(x) = f_1(x')$. But f_1 is bijective, so $x = x' \in V'$. Thus we have shown

$$f_1^{-1}(U) \cap V \supset f_1^{-1}(U) \cap V' \supset f_1^{-1}(U'),$$

which completes the proof of Claim 4.

According to Claim 4, the map

$$\varphi = f_1^{-1} : U \to f_1^{-1}(U)$$

is a chart of G at e and satisfies

$$\varphi(U \cap H) = \varphi(U') = f_1^{-1}(U') = f_1^{-1}(U) \cap V = \varphi(U) \cap V.$$

Given any $h \in H$, the set hU is an open neighbourhood of h in G and the map

$$\varphi \cdot L_{h^{-1}} : hU = L_h(U) \to f_1^{-1}(U)$$

is a chart of G satisfying

$$\begin{aligned}\varphi \cdot L_{h^{-1}}(hU \cap H) &= \varphi \cdot L_{h^{-1}} \cdot L_h(U \cap H) = \varphi(U \cap H) \\ &= \varphi(U) \cap V = \varphi \cdot L_{h^{-1}} \cdot L_h(U) \cap V \\ &= \varphi \cdot L_{h^{-1}}(hU) \cap V.\end{aligned}$$

Thus we have shown that H is a submanifold of G by the charts:

$$\varphi \cdot L_{h^{-1}} : hU \to f_1^{-1}(U) \qquad \text{for} \quad h \in H.$$

As a consequence, we can consider the tangent space $T_e(H)$ of H at e. Last of all, we shall show that $T_e(H)$ coincides with the V mentioned above. Suppose $v \in T_e(H)$. Then we have $\exp(v/i) \in H$ for any positive integer i. Hence, by putting $a_i = i$, $v_i = v/i$, we see that $v \in V$ by definition. That is, the inclusion $T_e(H) \subset V$ holds. On the other hand, it follows from the observation above that

$$\dim T_e(H) = \dim H = \dim V.$$

Thus we have $T_e(H) = V$.

This makes the proof of the theorem complete. □

Remark The reason why we consider U' and U in place of $f_1(V')$ and $f(V' + W')$ in the above proof is that we do not know if $f_1(V')$ is open in H and that the equation

$$f(V' + W') \cap H = f_1(V')$$

does not necessarily hold in general as illustrated in Fig. 3.13.

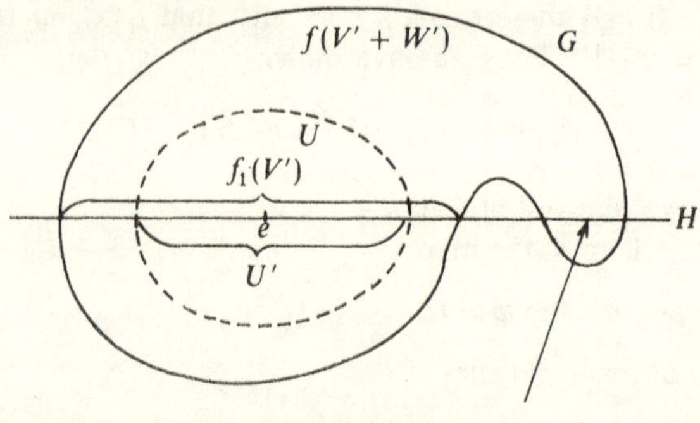

Fig. 3.13 Since we do not know if $f_1(V')$ is open in H, we take U'. Then we take U in order to avoid the part indicated by the arrow.

Example 2 The following subsets of Lie groups $GL(n, \mathbb{R})$ and $GL(n, \mathbb{C})$ are Lie subgroups:

$$O(n) = \{A \in GL(n, \mathbb{R}) \mid A^t A = I\},$$

$$SL(n, \mathbb{R}) = \{A \in GL(n, \mathbb{R}) \mid \det A = 1\},$$

$$SO(n) = O(n) \cap SL(n, \mathbb{R}),$$

$$U(n) = \{A \in GL(n, \mathbb{C}) \mid A^t \bar{A} = I\},$$

$$SL(n, \mathbb{C}) = \{A \in GL(n, \mathbb{C}) \mid \det A = 1\},$$

$$SU(n) = U(n) \cap SL(n, \mathbb{C}).$$

For example, $O(n)$ can be shown to be a Lie subgroup of $GL(n, \mathbb{R})$ as follows. Define a map

$$f\colon GL(n, \mathbb{R}) \to GL(n, \mathbb{R})$$

by $f(A) = A^t A$. Then f is continuous and

$$f^{-1}(I) = O(n).$$

Since the one-point set $\{I\}$ is closed in $GL(n, \mathbb{R})$, $f^{-1}(I)$ is also closed. Moreover it is easy to see that $O(n)$ is a subgroup of $GL(n, \mathbb{R})$. Thus it follows from Theorem 3.36 that $O(n)$ is a Lie subgroup of $GL(n, \mathbb{R})$.

3.8 Differential structures on homogeneous spaces G/H and local cross-sections

Let G be a Lie group and H a closed subgroup of G. Then we shall show in this section that there is a canonical smooth structure on the homogeneous space G/H. Moreover we shall show that the projection $\pi\colon G \to G/H$ has a smooth local cross-section, so that it becomes a principal fibre bundle. These considerations will be indispensable when we investigate the structure of G-manifolds in Chapter 4.

Theorem 3.37 *Let G be a Lie group and H a closed subgroup of G. Then there exists a unique smooth structure on the homogeneous space G/H such that:*

(1) *the projection $\pi\colon G \to G/H$ is smooth;*

(2) *for an arbitrary smooth manifold M, a map $f\colon G/H \to M$ is smooth if and only if $f\cdot\pi\colon G \to M$ is smooth.*

The dimension of G/H is $\dim G - \dim H$.

Proof The proof of this theorem is also fairly hard.

Previously we showed that G/H is Hausdorff. Since G satisfies the second countability axiom, it follows from Corollary 1.56 that its homogeneous space G/H satisfies also the second countability axiom.

Let V and W be the same as in the proof of Theorem 3.36. That is, $V = T_e(H)$ and $T_e(G) = V \oplus W$. However, as opposed to the prior map f, we now define a map

$$f\colon V \oplus W \to G$$

by $f(v, w) = \exp(w)\cdot\exp(v)$. Although the order of the product is reversed, we know, as before, that there is an open neighbourhood V' of 0 in V and an open neighbourhood W' of 0 in W such that $f(V' + W')$ is an open set of G and the restriction

$$f_1 = f\,|\,(V' + W')\colon V' + W' \to f(V' + W')$$

is a diffeomorphism.

Claim 1 There is an open neighbourhood W'' of 0 in W' such that if $\exp(w)H = \exp(w')H$ for $w, w' \in W''$, then $w = w'$.

Suppose the claim does not hold. Then there are sequences $w_i, w_i' \in W'$ and $h_i \in H$ such that $w_i \neq w_i'$, $w_i \to 0$, $w_i' \to 0$ and

$$\exp(w_i) = \exp(w_i')\,h_i.$$

Since $\exp(w_i) \to e$ and $\{\exp(w_i')\}^{-1} \to e$, we have that $h_i \to e$. Hence there exists a positive integer i_0 such that $h_i \in \exp(V')$ for $i > i_0$. Accordingly there exist unique elements v_i of V' with

$$\exp(v_i) = h_i, \qquad \exp(w_i) = \exp(w_i')\cdot\exp(v_i).$$

Since f_1 is injective, we have

$$(0, w_i) = (v_i, w_i'),$$

which is a contradiction. Thus the claim holds.

We now introduce the following maps which we will use frequently in this section and which are illustrated in Fig. 3.14. Let

$$i\colon W'' \to V' + W'' \qquad \text{and} \qquad \pi'\colon V' + W'' \to W''$$

be the inclusion map and the projection map, respectively. Then we define a map f_2 by the composition of the following maps:

$$f_2\colon W'' \xrightarrow{i} V' + W'' \xrightarrow{f_1} G \xrightarrow{\pi} G/H.$$

For each $g \in G$, let

$$\varphi_g\colon W'' \to G/H$$

be the map defined by $\varphi_g = gf_2$. Then we show in the following that G/H is a smooth manifold with charts $(\varphi_g(W''), \varphi_g^{-1})$ for $g \in G$.

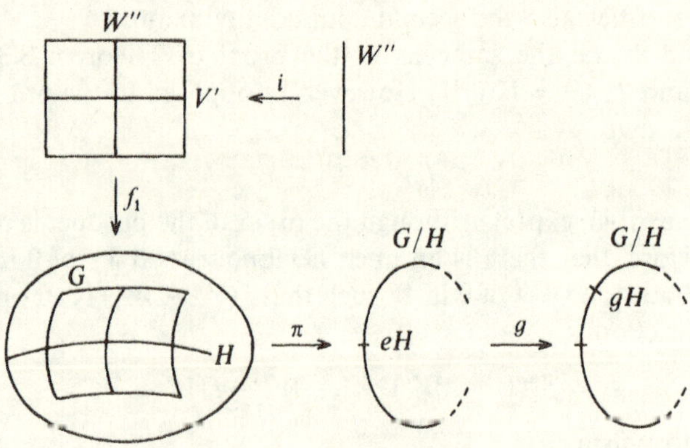

Fig. 3.14

First we show that φ_g is a homeomorphism onto an open set of G/H. To see this, it suffices to show that $\varphi_e = f_2$ is a homeomorphism onto an open set of G/H. The injectivity of f_2 follows from claim 1. Evidently f_2 is continuous. Observe that

$$f_2(W'') = \pi \cdot f_1(W'') = \pi \cdot f_1(V' + W'').$$

Since π is an open map, the last set above is open. We now show that

$$f_2^{-1}\colon f_2(W'') \to W''$$

is continuous, that is, we shall show that f_2 is an open map. Suppose O is an open set of W''. Then $f_2(O)$ is open if and only if $\pi^{-1}(f_2(O))$ is open by the definition of the topology on G/H. Note that we have

$$f_2(O) = \pi \cdot f_1 \cdot i(O) = \pi(\exp(O)) = \exp(O)H.$$

Since $H \supset \exp(V') \neq \varnothing$, we have

$$\pi^{-1}(f_2(O)) = \pi^{-1}(\exp(O)H) = \exp(O)H$$
$$= \exp(O)\cdot\exp(V')H = f_1(V' + O)H$$

where wc usc the symbol $\exp(O)H$ for the subset of G as well as for the subset of G/H. Note that $V' + O$ is open in $V' + W''$. Since $f_1 : V' + W' \to f(V' + W')$ is a homeomorphism onto the open set $f(V' + W')$ of $G, f_1(V' + O)$ is open in G. Hence $f_1(V' + O)H$ is open in G. Thus we have shown that f_2 is a homeomorphism.

Next we shall show that $\{(\varphi_g(W''), \varphi_g^{-1})\}$ is a smooth atlas. Suppose that $g, g' \in G$ satisfy

$$\varphi_g(W'') \cap \varphi_{g'}(W'') \neq \varnothing,$$

that is, there exist w, $w' \in W''$ with

$$gf_2(w) = g' f_2(w').$$

Then we shall show that the map $\varphi_{g'}^{-1}\cdot\varphi_g = f_2^{-1}g'^{-1}gf_2$ is smooth on a neighbourhood of w. We now put

$$h = f_1(w)^{-1}g^{-1}g' f_1(w').$$

It is easy to see that h lies in H, so that we have $\pi\cdot R_h = \pi$ where $R_h\colon G \to G$ denotes the right translation by h in Section 1.2. Note that

$$R_h g'^{-1} g f_1(w) = g'^{-1} g f_1(w) f_1(w)^{-1} g^{-1} g' f_1(w') = f_1(w').$$

It turns out that we can consider the composition $f_1^{-1} R_h g'^{-1} g f_1 \cdot i$ on a neighbourhood of w.

Putting all this together, we compute

$$f_2^{-1} g'^{-1} g f_2 = f_2^{-1} g'^{-1} g\pi\cdot f_1\cdot i = f_2^{-1}\cdot\pi g'^{-1} g f_1\cdot i$$
$$= f_2^{-1}\cdot\pi\cdot R_h g'^{-1} g f_1\cdot i$$
$$= (f_2^{-1}\cdot\pi\cdot f_1)\cdot(f_1^{-1}\cdot R_h g'^{-1} g f_1\cdot i).$$

For the benefit of the reader, we should explain the reason for putting R_h in the above computation: without R_h, we cannot in general consider even locally the composition of $g'^{-1}gf_1\cdot i$ with f_1^{-1}.

The map $f_1^{-1}R_h g'^{-1} g f_1\cdot i$ is obviously smooth on a neighbourhood of w. We now show that

$$f_2^{-1}\cdot\pi\cdot f_1 = \pi'.$$

Given $v \in V'$, $w \in W''$, we have

$$\pi\cdot f_1(v, w) = \exp(w)\cdot\exp(v)H = \exp(w)H.$$

On the other hand, $f_2(w) = \exp(w)H$ by definition. Consequently, since f_2 is bijective, we have

$$f_2^{-1}\cdot\pi\cdot f_1(v, w) = w,$$

that is, we have $f_2^{-1}\cdot\pi\cdot f_1 = \pi'$ which is evidently smooth.

Thus we have shown that $\varphi_{g'}^{-1}\cdot\varphi_g$ is smooth, so that G/H is a smooth manifold with atlas $\{(\varphi_g(W''), \varphi_g^{-1})\}$.

Next we shall show that conditions (1) and (2) hold for this smooth structure.

(1) For any $g\in G$, we shall show that π is smooth on a neighbourhood of g. As charts of G and G/H at g and gH, respectively, we take

$$(gf_1)^{-1}: gf_1(V' + W'') \to V' + W'',$$
$$\varphi_g^{-1}: gf_2(W'') \to W''.$$

By virtue of the above result, we have

$$\varphi_g^{-1}\cdot\pi g f_1 = f_2^{-1}g^{-1}\pi g f_1 = f_2^{-1}\cdot\pi\cdot f_1 = \pi',$$

which means that π is smooth.

(2) If $f\colon G/H \to M$ is smooth, then $f\cdot\pi$ is always smooth in general.

Claim 2 The projection $\pi\colon G \to G/H$ has a smooth local cross-section.

Define a map $s_g\colon gf_2(W'') \to G$ by

$$s_g = gf_1\cdot i\cdot\varphi_g^{-1}.$$

Since $(gf_1)^{-1}\cdot s_g\cdot\varphi_g = i$, s_g is a smooth map. Furthermore we have

$$\begin{aligned}\pi\cdot s_g &= \pi g f_1\cdot i\cdot f_2^{-1}g^{-1} = g\pi\cdot f_1\cdot i\cdot f_2^{-1}g^{-1}\\ &= gf_2\cdot f_2^{-1}g^{-1} = \mathrm{id},\end{aligned}$$

that is, s_g is a smooth local cross-section of π. Thus Claim 2 holds.

On account of Claim 2, the equality

$$f = f\cdot\pi\cdot s_g$$

holds on a neighbourhood of gH. Hence if $f\cdot\pi$ is smooth, then so is f. Thus condition (2) holds.

Finally we shall show uniqueness. Let $(G/H)_0$ denote G/H endowed with another smooth structure. Consider the following commutative diagram:

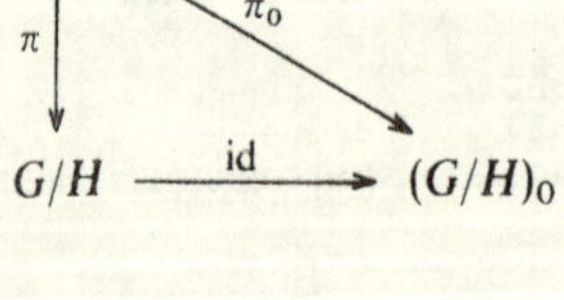

Condition (1) concerning $(G/H)_0$ implies that π_0 is smooth. Hence it follows from condition (2) concerning G/H that id is smooth. Similarly we can deduce that the inverse map is smooth. Hence id is a diffeomorphism.

In view of the above proof, we have

$$\dim G/H = \dim W = \dim G - \dim V = \dim G - \dim H.$$

This completes the proof of the theorem. □

Corollary 3.38 *Even if we replace condition* (2) *of Theorem* 3.37 *with the condition:*

(2)′ *$\pi: G \to G/H$ has a smooth local cross-section,*

the theorem holds.

The proof is contained in the proof of Theorem 3.37.

Corollary 3.39 *$\pi: G \to G/H$ is a smooth principal H-bundle.*

Proof Since $\pi: G \to G/H$ has a local cross-section, it follows from Theorem 2.25 that $\pi: G \to G/H$ is a principal H-bundle. Moreover since these local cross-sections are smooth, it is a smooth fibre bundle. □

Corollary 3.40 *Let G be a Lie group and H a closed subgroup of G. If M is an effective H-manifold, then*

$$G \underset{H}{\times} M \to G/H$$

is a smooth fibre bundle with fibre M associated with the principal H-bundle in the previous corollary. In particular, $G \underset{H}{\times} M$ is a smooth manifold.

Proof In view of Corollary 3.39 $\pi: G \to G/H$ is a smooth principal H-bundle. Hence we have a fibre bundle $G \underset{H}{\times} M \to G/H$ associated with the bundle π by Theorem 2.10. The fact that the associated bundle is smooth is guaranteed by the fact that the transition functions of the associated bundle coincide with those of the original one. □

Remark Even if we remove the effectiveness condition in Corollary 3.40, it follows from the remark immediately after Theorem 2.10, and Corollary 3.41 below that the statement of Corollary 3.40 is also valid where only the structure group is altered.

Corollary 3.41 *Let G be a Lie group and H a closed normal subgroup of G. Then G/H becomes a Lie group so that the projection $\pi: G \to G/H$ is a homomorphism of Lie groups.*

Proof Let $\alpha: G \times G \to G$ denote the map of the product in G. Then α induces a map $\bar{\alpha}: G/H \times G/H \to G/H$ and we have the following commutative diagram:

$$\begin{array}{ccc} G \times G & \xrightarrow{\alpha} & G \\ \downarrow{\scriptstyle \pi \times \pi} & & \downarrow{\scriptstyle \pi} \\ G/H \times G/H & \xrightarrow{\bar{\alpha}} & G/H \end{array}$$

Since there exists a smooth local cross-section of $\pi: G \to G/H$, it follows from the diagram that $\bar{\alpha}$ is also smooth. The map of assigning to each element gH its inverse element $g^{-1}H$ is also shown to be smooth. □

Corollary 3.42 *Let G be a Lie group and H a closed subgroup of G. Then the natural G-action on G/H is smooth.*

Proof Consider the following commutative diagram:

$$\begin{array}{ccc} G \times G & \xrightarrow{\alpha} & G \\ \downarrow{\scriptstyle \mathrm{id} \times \pi} & & \downarrow{\scriptstyle \pi} \\ G \times G/H & \xrightarrow{\varphi} & G/H \end{array}$$

Since $\pi: G \to G/H$ has a smooth local cross-section, it follows from the diagram above that φ is smooth. □

A manifold M is said to be a *homogeneous space* of a Lie group G if there is a transitive smooth G-action on M.

Theorem 3.43 *Let M be a manifold which is a homogeneous space of a compact Lie group G and x_0 be a point of M. Then the map $\bar{f}$ given by*

$$\bar{f}(gG_{x_0}) = gx_0$$

is a G-diffeomorphism from G/G_{x_0} onto M.

Proof By Proposition 1.53, $\bar{f}$ is a homeomorphism from G/G_{x_0} onto M. If we define a map $f: G \to M$ by $f(g) = gx_0$, then f is a smooth G-map and the following diagram commutes:

$$\begin{array}{ccc} G & & \\ \downarrow{\scriptstyle \pi} & \searrow{\scriptstyle f} & \\ G/G_{x_0} & \xrightarrow{\bar{f}} & M \end{array}$$

Since M is Hausdorff, it follows from Corollary 1.38 that G_{x_0} is a closed subgroup of G. Hence G/G_{x_0} is a manifold by Theorem 3.37. Since f is smooth, it follows from (2) of Theorem 3.37 that $\bar{f}$ is smooth.

Claim The differential $\bar{f}_*$ of $\bar{f}$ at eG_{x_0} is injective.

Let u be a vector of $T_{eG_{x_0}}(G/G_{x_0})$ satisfying $\bar{f}_*(u) = 0$. Observe that $\pi_*: T_e(G) \to T_{eG_{x_0}}(G/G_{x_0})$ is surjective, since π has a smooth local cross-section. Accordingly there is a vector v of $T_e(G)$ with $\pi_*(v) = u$. Let $\rho_v: \mathbb{R} \to G$ be the 1-parameter subgroup of G corresponding to v. Then we define a smooth $\mathbb{R}$-action on M:

$$\Phi: \mathbb{R} \times M \to M$$

by $\Phi(t, x) = \rho_v(t)x$. Denote by X the vector field on M defined by the $\mathbb{R}$-action Φ. Since $v = [\rho_v]$ by definition, we have

$$X(x_0) = [f \cdot \rho_v] = f_*(v) = \bar{f}_* \cdot \pi_*(v) = \bar{f}_*(u) = 0.$$

Hence it follows from Corollary 3.21 that x_0 lies in the fixed point set $M^{\mathbb{R}}$ of the action Φ, that is,

$$\rho_v(t)x_0 = \Phi(t, x_0) = x_0, \qquad \text{for any} \quad t \in \mathbb{R}.$$

This means that $\rho_v(t) \in G_{x_0}$ for any $t \in \mathbb{R}$. But the curve $\pi \cdot \rho_v(t)$ represents u, so that we have $u = 0$. Hence $\bar{f}_*$ is injective.

We now put $\dim G/G_{x_0} = n$ and $\dim M = m$. The injectivity of $\bar{f}_*$ implies $n \leq m$.

We next show that $n = m$. To show this, we suppose tentatively that $n < m$. It follows from an application of the inverse function theorem that there exist a chart $\varphi: U \to \mathbb{R}^n$ of G/G_{x_0} at eG_{x_0} with $\varphi(eG_{x_0}) = 0$ and a chart $\psi: V \to \mathbb{R}^m$ of M at x_0 with $\psi(x_0) = 0$ such that $\bar{f}(U) \subset V$ and for each point y of $\varphi(U)$ we have

$$\psi \cdot \bar{f} \cdot \varphi^{-1}(y) = (y, 0) \in \mathbb{R}^n \times \mathbb{R}^{m-n} = \mathbb{R}^m.$$

Since $\bar{f}^{-1}$ is continuous, there exists an open neighbourhood V' of x_0 involved in V such that $\bar{f}^{-1}(V') \subset U$. Then $V' \subset \bar{f}(U)$, so that we have

$$\psi(V') \subset \psi(\bar{f}(U)) \subset \mathbb{R}^n \times \{0\} \subset \mathbb{R}^m.$$

Since $\psi(V')$ is open in $\mathbb{R}^m$, this is a contradiction. Hence the equality $n = m$ holds, that is, $\bar{f}_*$ is a linear isomorphism.

By the inverse function theorem, $\bar{f}^{-1}$ is also smooth on a neighbourhood of x_0. If y_0 is an arbitrary point of M, then there exists $g_0 \in G$ with $g_0 x_0 = y_0$. Since $\bar{f}^{-1}$ is a G-map, we have

$$\bar{f}^{-1}(y) = g_0 \bar{f}^{-1}(g_0^{-1} y)$$

in general. Note that both G-actions on M and on G/G_{x_0} are smooth. As a consequence, $\bar{f}^{-1}$ is smooth on a neighbourhood of y_0. Hence $\bar{f}^{-1}$ is globally smooth.

Thus we have shown that $\bar{f}$ is a G-diffeomorphism. □

Remark In fact, the compactness assumption of G in the theorem is unnecessary. When we showed the continuity of $\bar{f}^{-1}$ in Section 1.6, we used the compactness assumption. Since, however, a Lie group is a locally compact space satisfying the second countability axiom, it is actually possible to show without the compactness assumption that f is open and hence $\bar{f}$ is open.

3.9 Semi-direct products of Lie groups and examples of homogeneous spaces

In Section 1.3, we introduced semi-direct products of topological groups. In the following, we shall discuss semi-direct products of Lie groups.

Given a Lie group N, let $\mathrm{Aut}(N)$ denote the set of all Lie group automorphisms of N. Then $\mathrm{Aut}(N)$ forms a group with respect to the composition of maps. We now define the concept that a homomorphism

$$\theta\colon G \to \mathrm{Aut}(N)$$

of groups is smooth as follows. Given θ as above, we define a map $\theta'\colon G \times N \to N$ by

$$\theta'(g, n) = \theta(g)(n) \qquad \text{for} \quad g \in G, n \in N.$$

If θ' is a smooth map from the product manifold $G \times N$ to N, we say that θ is smooth.

Given a smooth homomorphism $\theta\colon G \to \mathrm{Aut}(N)$, we define a new Lie group

$$N \underset{\theta}{\rtimes} G \qquad \text{(or briefly } N \rtimes G\text{)}$$

as follows.

The manifold structure on $N \underset{\theta}{\rtimes} G$ is given by the product manifold $N \times G$, that is, as a manifold we set

$$N \underset{\theta}{\rtimes} G = N \times G.$$

The group operation is given within it as in the case of topological groups in Section 1.3. Then it is easy to verify by making use of the smoothness assumption of θ that the group operation is smooth. Thus $N \underset{\theta}{\rtimes} G$ becomes a Lie group, which is called the *semi-direct product of N and G with respect to θ.*

Let A and G be Lie groups and $j: A \to G$ be a surjective homomorphism of Lie groups. Then a homomorphism $s: G \to A$ of Lie groups satisfying $j \cdot s = \mathrm{id}$ is called a *splitting* for j as in the case of topological groups.

Proposition 3.44 *Results similar to Proposition* 1.27 *and Corollary* 1.28 *hold in the case of Lie groups. Moreover the isomorphisms of Example* 2 *in Section* 1.3 *hold as Lie groups.*

Remark Since $\operatorname{Ker} j$ of the map $j: A \to G$ in the proposition is a closed subgroup of A, it follows from Theorem 3.36, the last remark of Section 3.8, and Corollary 3.39 that $j: A \to G$ is a principal $\operatorname{Ker} j$ bundle. If there exists a splitting $s: G \to A$ for j, it follows from Theorem 2.12 that the principal $\operatorname{Ker} j$ bundle is trivial, that is, $A \cong \operatorname{Ker} j \times G$ as bundles. However A is not isomorphic to $\operatorname{Ker} j \times G$ in general, but is isomorphic to $\operatorname{Ker} j \rtimes G$ as Lie groups. Example 2 in Section 1.3 provides such an example.

Proposition 3.45 *Let G be a Lie group, H a closed subgroup of G, and K a closed subgroup of H. Then the projection $p: G/K \to G/H$ is a smooth fibre bundle with fibre H/K. If K is a closed normal subgroup of H, then the above is a smooth principal H/K-bundle.*

Proof This is shown by combining Corollary 3.39 and Exercise 4 in Chapter 2. □

The remainder of the section is devoted to giving various examples of homogeneous spaces of Lie groups. We begin by preparing some notation.

In this section we write the elements of $\mathbb{R}^n$ as column vectors and $GL(n, \mathbb{R})$ acts on $\mathbb{R}^n$ by matrix multiplication. We define a positive definite inner product $(\ ,\ )$ in $\mathbb{R}^n$ by

$$(x, y) = \sum_{i=1}^{n} x_i y_i$$

for $x = \begin{bmatrix} x_1 \\ \vdots \\ x_n \end{bmatrix}$, $y = \begin{bmatrix} y_1 \\ \vdots \\ y_n \end{bmatrix} \in \mathbb{R}^n$ as usual. In the case of the complex vector space $\mathbb{C}^n$, we set

$$(x, y) = \sum_{i=1}^{n} x_i \bar{y}_i$$

for $x = \begin{bmatrix} x_1 \\ \vdots \\ x_n \end{bmatrix}$, $y = \begin{bmatrix} y_1 \\ \vdots \\ y_n \end{bmatrix} \in \mathbb{C}^n$. In either case, the length $\|x\|$ of x is defined by

$$\|x\| = \sqrt{(x, x)}.$$

If we put

$$e_i = \begin{bmatrix} 0 \\ \vdots \\ 0 \\ 1 \\ 0 \\ \vdots \\ 0 \end{bmatrix} (i) \qquad 1 \le i \le n,$$

then $\{e_1, \ldots, e_n\}$ forms a basis for $\mathbb{C}^n$ as well as for $\mathbb{R}^n$, which we call the *standard basis of* $\mathbb{R}^n$ (resp. $\mathbb{C}^n$).

Example 1 If we restrict the natural $GL(n, \mathbb{R})$ action on $\mathbb{R}^n$ to the subgroup $O(n)$, then the unit sphere

$$S^{n-1} = \{x \in \mathbb{R}^n \mid \|x\| = 1\}$$

in $\mathbb{R}^n$ is an invariant submanifold of $\mathbb{R}^n$, because the $O(n)$-action preserves the lengths of vectors.

We now show that $O(n)$ acts on S^{n-1} transitively. For any $x_1 \in S^{n-1}$, we can find vectors $x_2, \ldots, x_n$ of $\mathbb{R}^n$ by the Gram–Schmidt process so that $\{x_1, x_2, \ldots, x_n\}$ is an orthonormal basis for $\mathbb{R}^n$, that is,

$$(x_i, x_j) = \delta_{ij} \qquad (= \text{Kronecker delta}).$$

Regarding each vector of $\mathbb{R}^n$ as an $n \times 1$ matrix, we define an $n \times n$ matrix

$$A = [x_1 \quad x_2 \quad \ldots \quad x_n]$$

by lining them up. Then, as is well known, A belongs to $O(n)$ and satisfies

$$Ae_1 = x_1.$$

Hence $O(n)$ acts on S^{n-1} transitively. The isotropy group of this action at e_1 is given by

$$O(n)_{e_1} = \left\{ \begin{bmatrix} 1 & 0 \ldots 0 \\ 0 & \\ \vdots & A' \\ 0 & \end{bmatrix} \middle| \; A' \in O(n-1) \right\}$$

which is identified with $O(n-1)$.

When we regard $O(n)$ as a subset of $\mathbb{R}^{n^2} (= M(n, \mathbb{R}))$, it is closed and is included in the sphere of radius $\sqrt{n}$. Accordingly $O(n)$ is a compact Lie group. It follows from Theorem 3.43 that

$$O(n)/O(n-1) \approx S^{n-1}$$

where $\approx$ stands for $O(n)$-diffeomorphism.

Example 2 Even if we restrict the $O(n)$ action on S^{n-1} in Example 1 to the $SO(n)$ action, it is transitive for $n \geq 2$. This will be seen as follows. For any $x_1 \in S^{n-1}$, we construct a matrix $A = [x_1 \ldots x_n]$ as in Example 1. Then we define a new matrix B by

$$B = [x_1 \quad x_2 \quad \ldots \quad (\det A)x_n].$$

It is trivial to see that $B \in SO(n)$ and satisfies $Be_1 = x_1$. Furthermore it can be shown as above that the isotropy group at e_1 is identified with $SO(n-1)$. Thus we have

$$SO(n)/SO(n-1) \approx S^{n-1} \qquad \text{for} \quad n \geq 2.$$

Proposition 3.46 *Let $O(n+1)$ (resp. $O(n)$) act on S^n (resp. $\mathbb{R}^n$) in the natural manner. Then we have an isomorphism*

$$T(S^n) \cong O(n+1) \underset{O(n)}{\times} \mathbb{R}^n$$

of $O(n+1)$-vector bundles. In particular, this isomorphism holds without taking $O(n+1)$-actions into consideration. If $n \geq 1$, then $O(n)$ and $O(n+1)$ can be replaced by $SO(n)$ and $SO(n+1)$, respectively.

Proof Since S^n and $O(n+1)/O(n)$ are $O(n+1)$-diffeomorphic by Example 1 above, it follows from Theorem 3.15 that there is an isomorphism

$$T(S^n) \cong O(n+1) \underset{O(n)}{\times} T_{eO(n)}(O(n+1)/O(n))$$

of $O(n+1)$-vector bundles. Accordingly we need only investigate the isotropy representation. But we can identify this with the natural $O(n)$-action on $\mathbb{R}^n$ as Fig. 3.15 indicates.

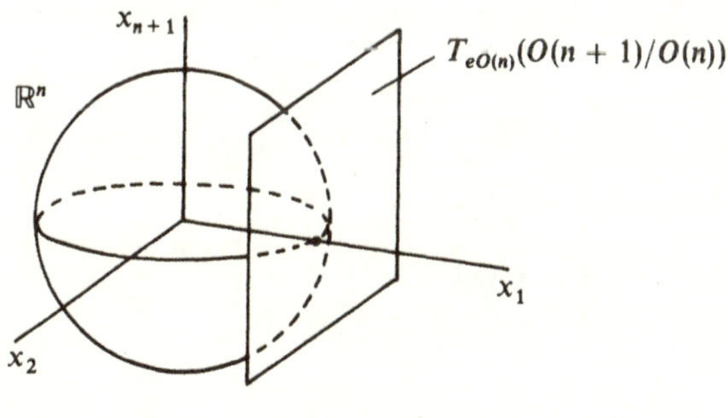

Fig. 3.15

A similar proof is valid for $SO(n)$ and $SO(n+1)$. □

Corollary 3.47 *$T(S^n) \oplus \underline{\mathbb{R}}$ is a trivial vector bundle, where $\underline{\mathbb{R}}$ denotes the trivial line bundle $S^n \times \mathbb{R} \to S^n$.*

Proof On account of Proposition 3.46, we have an isomorphism

$$T(S^n)\oplus \underline{\mathbb{R}} \cong O(n+1) \underset{O(n)}{\times} (\mathbb{R}\oplus\mathbb{R}^n)$$

of $O(n+1)$-vector bundles. Here $O(n)$ acts trivially on $\mathbb{R}$. Hence the $O(n)$-action on $\mathbb{R}\oplus\mathbb{R}^n = \mathbb{R}^{n+1}$ is nothing but the restriction of the natural $O(n+1)$-action on $\mathbb{R}^{n+1}$ to the $O(n)$-action with respect to the inclusion $O(n)\subset O(n+1)$ stated in Example 1 above. Accordingly it follows from Corollary 2.30 that $O(n+1) \underset{O(n)}{\times} \mathbb{R}^{n+1}$ and $(O(n+1)/O(n))\times\mathbb{R}^{n+1} = S^n\times\mathbb{R}^{n+1}$ are isomorphic as $O(n+1)$-vector bundles. □

Example 3 Phenomena similar to Examples 1 and 2 above occur for the unitary group $U(n)$ and the special unitary group $SU(n)$. That is, the natural $U(n)$- (resp. $SU(n)$-) action on the complex n-space $\mathbb{C}^n$ induces a transitive action on the unit sphere

$$S^{2n-1} = \{x\in\mathbb{C}^n \mid \|x\| = 1\}.$$

In the following, we shall show the transitivity. For any $x_1\in S^{2n-1}$, we can find vectors $x_2,\ldots,x_n$ of $\mathbb{C}^n$ by the Gram–Schmidt process in a complex vector space so that $\{x_1, x_2,\ldots,x_n\}$ is an orthonormal basis for $\mathbb{C}^n$, that is,

$$(x_i, x_j) = \delta_{ij}.$$

As in Examples 1 and 2, we set

$$A = [x_1 \quad x_2 \quad \ldots \quad x_n]$$

and

$$B = [x_1 \quad x_2 \quad \ldots \quad (\det A)^{-1}\, x_n].$$

Then one verifies similarly that

$$A\in U(n),\ Ae_1 = x_1 \qquad \text{and} \qquad B\in SU(n),\ Be_1 = x_1$$

and

$$U(n)_{e_1} \cong U(n-1) \qquad \text{and} \qquad SU(n)_{e_1} \cong SU(n-1).$$

Thus we have diffeomorphisms:

$$\begin{aligned} S^{2n-1} &\approx U(n)/U(n-1) \\ &\approx SU(n)/SU(n-1) \qquad \text{for} \qquad n\geq 2. \end{aligned}$$

Example 4 *Stiefel manifolds* $V_n(\mathbb{R}^{n+k})$. An *orthonormal n-frame* in $\mathbb{R}^{n+k}$ is an n-tuple $\mathfrak{f} = (f_1,\ldots,f_n)$ of orthonormal vectors of $\mathbb{R}^{n+k}$. That is, $\mathfrak{f}$ satisfies

$$(f_i, f_j) = \delta_{ij}.$$

Let $V_n(\mathbb{R}^{n+k})$ denote the set of all orthonormal n-frames in $\mathbb{R}^{n+k}$. Given $A\in O(n+k)$ and $\mathfrak{f} = (f_1,\ldots,f_n)\in V_n(\mathbb{R}^{n+k})$, we put,

$$A\circ\mathfrak{f} = (Af_1,\ldots,Af_n)$$

where Af_i denotes the matrix product of A with f_i written as a column vector ($(n+k) \times 1$ matrix). Since the natural $O(n+k)$-action on $\mathbb{R}^{n+k}$ preserves the inner product $(\ ,\ )$, $A \circ \mathfrak{f}$ lies also in $V_n(\mathbb{R}^{n+k})$. Then one verifies easily that the map $(A, \mathfrak{f}) \to A \circ \mathfrak{f}$ defines an $O(n+k)$-action on $V_n(\mathbb{R}^{n+k})$ in the sense of Section 1.1.

Next we shall show that the action is transitive. Given an arbitrary element $\mathfrak{f} = (f_1, \ldots, f_n)$ of $V_n(\mathbb{R}^{n+k})$, we can find vectors $f_{n+1}, \ldots, f_{n+k}$ of $\mathbb{R}^{n+k}$ such that $\{f_1, \ldots, f_n, f_{n+1}, \ldots, f_{n+k}\}$ forms an orthonormal basis of $\mathbb{R}^{n+k}$. Regarding f_i as column vectors, we define a matrix

$$A = [f_1 \quad \cdots \quad f_{n+k}]$$

by lining them up. Then A belongs to $O(n+k)$ and satisfies

$$Ae_i = f_i \qquad \text{for} \quad i = 1, \ldots, n+k$$

where $\{e_1, \ldots, e_{n+k}\}$ is the standard basis of $\mathbb{R}^{n+k}$. Hence for $\mathfrak{e} = (e_1, \ldots, e_n)$, we have

$$A \circ \mathfrak{e} = (f_1, \ldots, f_n) = \mathfrak{f},$$

that is, the action is transitive. The isotropy group at $\mathfrak{e} = (e_1, \ldots, e_n)$ is given by

$$O(n+k)_{\mathfrak{e}} = \left\{ \begin{pmatrix} I_n & 0 \\ 0 & B \end{pmatrix} \middle| B \in O(k) \right\}$$

where I_n denotes the unit matrix of degree n. Evidently $O(n+k)_{\mathfrak{e}}$ is a closed subgroup of $O(n+k)$. Since $O(n+k)_{\mathfrak{e}}$ is canonically isomorphic to $O(k)$, it is often written as $O(k)$. Then by the assignment $AO(k) \mapsto A \circ \mathfrak{e}$, we have a bijective correspondence f from the set $O(n+k)/O(k)$ onto $V_n(\mathbb{R}^{n+k})$.

According to Theorem 3.37, the homogeneous space $O(n+k)/O(k)$ is a smooth manifold. Hence we can put a smooth structure on $V_n(\mathbb{R}^{n+k})$ so that f is a diffeomorphism, and $V_n(\mathbb{R}^{n+k})$ becomes a homogeneous space of $O(n+k)$. The manifold $V_n(\mathbb{R}^{n+k})$ so obtained is called the *Stiefel manifold* of orthonormal n-frames in $\mathbb{R}^{n+k}$.

In the following we shall show that $V_n(\mathbb{R}^{n+k})$ is realized as an orbit of an $O(n+k)$-action on $\mathbb{R}^{n(n+k)}$. Let $\varphi\colon O(n+k) \times \mathbb{R}^{n(n+k)} \to \mathbb{R}^{n(n+k)}$ be the map defined by

$$(A, \mathfrak{a}) \mapsto (Aa_1, \ldots, Aa_n)$$

where $a_i \in \mathbb{R}^{n+k}$ and $\mathfrak{a} = (a_1, \ldots, a_n) \in \mathbb{R}^{n(n+k)}$. The map φ defines a smooth $O(n+k)$-action on $\mathbb{R}^{n(n+k)}$, which is in fact the direct sum of n copies of the natural $O(n+k)$-action on $\mathbb{R}^{n+k}$. Note that each element $\mathfrak{f} = (f_1, \ldots, f_n)$ of $V_n(\mathbb{R}^{n+k})$ can be regarded as a point of $\mathbb{R}^{n(n+k)}$. It follows that $V_n(\mathbb{R}^{n+k})$ can be regarded as a subset of $\mathbb{R}^{n(n+k)}$. Then it is clear that the orbit of $\mathfrak{e} = (e_1, \ldots, e_n)$ under the action φ is nothing but $V_n(\mathbb{R}^{n+k})$.

As will be shown in Section 4.1, each orbit of a smooth G-manifold is in general a G-invariant submanifold. Hence it follows from Corollary 4.4 that $V_n(\mathbb{R}^{n+k})$ is $O(n+k)$-diffeomorphic to the orbit.

For $k \geq 1$, it is similarly shown that $V_n(\mathbb{R}^{n+k})$ is a homogeneous space of $SO(n+k)$ as follows. Since the $O(n+k)$-action on $O(n+k)/O(k)$ is smooth by Corollary 3.42, the action restricted to $SO(n+k)$ is also smooth. Since $k \geq 1$, it is easily verified that the action is also transitive. The isotropy group of this action at $\mathbb{e}$ is given by

$$SO(n+k)_{\mathbb{e}} \cong SO(k).$$

Hence we have an $SO(n+k)$-diffeomorphism

$$SO(n+k)/SO(k) \approx O(n+k)/O(k)$$

by Theorem 3.43.

Example 5 *Grassmann manifolds* $G_n(\mathbb{R}^{n+k})$. An n-dimensional subspace of $\mathbb{R}^{n+k}$ will be called an *n-plane* in $\mathbb{R}^{n+k}$. Let $G_n(\mathbb{R}^{n+k})$ be the set of all n-planes in $\mathbb{R}^{n+k}$. Given $A \in O(n+k)$ and $W \in G_n(\mathbb{R}^{n+k})$, we put

$$AW = \{Ax \mid x \in W\}$$

where Ax denotes the matrix product of A with x written as a column vector. Since AW is an n-plane in $\mathbb{R}^{n+k}$, AW lies also in $G_n(\mathbb{R}^{n+k})$. Then one easily verifies that the correspondence $(A, W) \mapsto AW$ defines an $O(n+k)$-action on $G_n(\mathbb{R}^{n+k})$ in the sense of Section 1.1.

Next we shall show that the action is transitive. Let W_0 be the n-plane spanned by $e_1, \ldots, e_n$ where $\{e_1, \ldots, e_{n+k}\}$ denotes the standard basis of $\mathbb{R}^{n+k}$. Given any $W \in G_n(\mathbb{R}^{n+k})$, choose an orthonormal n-frame $\mathfrak{f} = (f_1, \ldots, f_n)$ in $\mathbb{R}^{n+k}$ which spans W. As shown in Example 4 above, there exists $A \in O(n+k)$ with $A \circ \mathbb{e} = \mathfrak{f}$, so that we have $AW_0 = W$. Thus we see that the action is transitive.

Next we investigate the isotropy group at W_0. It is easy to see that

$$O(n+k)_{W_0} = \left\{ \begin{pmatrix} B & 0 \\ 0 & C \end{pmatrix} \middle| B \in O(n), C \in O(k) \right\}$$

which is evidently a closed subgroup of $O(n+k)$. Since $O(n+k)_{W_0}$ is canonically isomorphic to $O(n) \times O(k)$, it is often written as $O(n) \times O(k)$. Then by the assignment $A(O(n) \times O(k)) \mapsto AW_0$, we have a bijective correspondence f from the set $O(n+k)/(O(n) \times O(k))$ onto $G_n(\mathbb{R}^{n+k})$.

According to Theorem 3.37, the homogeneous space $O(n+k)/(O(n) \times O(k))$ is a smooth manifold. Hence we can put a smooth structure on $G_n(\mathbb{R}^{n+k})$ so that f is a diffeomorphism, and $G_n(\mathbb{R}^{n+k})$ becomes a homogeneous space of $O(n+k)$. The manifold so obtained is called the *Grassmann manifold* of n-planes in $\mathbb{R}^{n+k}$.

We now investigate the relations between $V_n(\mathbb{R}^{n+k})$ and $G_n(\mathbb{R}^{n+k})$. There is a canonical map

$$q\colon V_n(\mathbb{R}^{n+k}) \to G_n(\mathbb{R}^{n+k})$$

which maps each orthonormal n-frame to the n-plane which it spans.

On the other hand, there is a natural projection

$$O(n+k)/O(k) \to O(n+k)/(O(n)\times O(k)).$$

Evidently this map coincides with the map q above under the identifications mentioned before. Thus we have the following commutative diagram

$$\begin{array}{ccc} O(n+k) & & \\ \downarrow{\scriptstyle \pi} & \searrow^{\pi'} & \\ V_n(\mathbb{R}^{n+k}) = O(n+k)/O(k) & \xrightarrow{\quad q \quad} & G_n(\mathbb{R}^{n+k}) = O(n+k)/(O(n)\times O(k)) \end{array}$$

where all the maps are the natural projections. It follows from the property of the smooth structure on homogeneous spaces of Theorem 3.37 that π' is smooth, and hence q is also smooth. Furthermore by virtue of Proposition 3.45, q is the projection map of a smooth principal $O(n)$-bundle.

Finally we have the following diffeomorphism which can be proved in the same manner as Example 4:

$$G_n(\mathbb{R}^{n+k}) \approx SO(n+k)/(SO(n+k)\cap(O(n)\times O(k))).$$

Example 6 In Examples 4 and 5, we dealt with the real Euclidean vector space $\mathbb{R}^{n+k}$. Even if we consider the complex Euclidean space $\mathbb{C}^{n+k}$ instead of $\mathbb{R}^{n+k}$, similar results hold in this case too. That is, the *complex Stiefel manifold* $V_n(\mathbb{C}^{n+k})$ is defined and satisfies:

$$\begin{aligned} V_n(\mathbb{C}^{n+k}) &\approx U(n+k)/U(k) \\ &\approx SU(n+k)/SU(k) \qquad \text{for} \qquad k \geq 1. \end{aligned}$$

Similarly the *complex Grassman manifold* $G_n(\mathbb{C}^{n+k})$ is defined and satisfies:

$$\begin{aligned} G_n(\mathbb{C}^{n+k}) &\approx U(n+k)/(U(n)\times U(k)) \\ &\approx SU(n+k)/(SU(n+k)\cap(U(n)\times U(k))). \end{aligned}$$

As in the real case, we see that the projection map

$$q\colon V_n(\mathbb{C}^{n+k}) \to G_n(\mathbb{C}^{n+k})$$

is smooth and is a smooth principal $U(n)$-bundle by Proposition 3.45.

Remark For the special case $n = 1$, there are canonical diffeomorphisms: $V_1(\mathbb{R}^{k+1}) \approx S^k$, $V_1(\mathbb{C}^{k+1}) \approx S^{2k+1}$, $G_1(\mathbb{R}^{k+1}) \approx \mathbb{R}P^k$, and $G_1(\mathbb{C}^{k+1}) \approx \mathbb{C}P^k$.

Exercises for Chapter 3

1. Show that there are canonical diffeomorphisms $\mathbb{R}P^1 \approx S^1$ and $\mathbb{C}P^1 \approx S^2$.

2. Consider the subset $N = \{(x, 0) \mid x \geq 0\} \cup \{(0, y) \mid y \geq 0\}$ of $\mathbb{R}^2$. Is N a submanifold of $\mathbb{R}^2$ with respect to the natural smooth structure on $\mathbb{R}^2$?

3. Consider the complex projective space $\mathbb{C}P^n = \{[z_0, \ldots, z_n]\}$ and the map $f: \mathbb{C}P^n \to \mathbb{C}P^n$ defined by $[z_0, \ldots, z_n] \mapsto [\bar{z}_0, \ldots, \bar{z}_n]$. Show that f defines a $\mathbb{Z}_2$-action and determine the fixed point set of the action.

4. Given a smooth curve $l: \mathbb{R} \to M$ and a real number s, define a smooth curve $l_s: \mathbb{R} \to M$ by $l_s(t) = l(st)$. Then show that $[l_s] = s[l]$.

5. Find the 1-parameter group generated by the vector field $X(x) = 2x + 1$ on $M = \mathbb{R}$.

6. Find 1-parameter subgroups of T^2.

7. Show that there are the following diffeomorphisms:

$$G_n(\mathbb{R}^{n+k}) \approx G_k(\mathbb{R}^{n+k}), \quad G_n(\mathbb{R}^{n+1}) \approx \mathbb{R}P^n,$$

$$G_n(\mathbb{C}^{n+k}) \approx G_k(\mathbb{C}^{n+k}), \quad G_n(\mathbb{C}^{n+1}) \approx \mathbb{C}P^n.$$

4 Structure of G-manifolds

In this chapter, we elucidate the structure of G-manifolds for G, a compact Lie group, building on the foundation of the previous chapters. This chapter together with Chapter 5 constitutes the principal part of the book.

Initially, we prove the slice theorem which describes the conditions of neighbourhoods of orbits in terms of G-vector bundles. The slice theorem is, so to speak, a local structure theorem, from which we deduce the global structure theorem and the decomposition theorem of G-manifolds.

As another application of the slice theorem, we shall show that there are only a finite number of orbit types in a compact G-manifold. Furthermore, we shall prove the existence of what is called the principal orbit type; that is, a maximum orbit type for a compact Lie group action on a connected manifold. As an application, we shall discuss, for instance, the dimension of compact Lie groups which act effectively on a connected manifold.

4.1 G-embeddings and orbits

If a smooth map $f: M \to N$ between smooth manifolds satisfies a certain condition, M can be regarded as a submanifold of N. Similar comments apply to G-manifolds. In particular, we shall show that the orbit $G(x)$ of each point x of a G-manifold M is a G-invariant submanifold of M. We start by giving definitions.

Let M and N be smooth manifolds and $f: M \to N$ be a smooth map. If for each point x of M, the differential

$$f_*: T_x(M) \to T_{f(x)}(N)$$

is injective, then f is called an *immersion*. If an immersion f is a homeomorphism of M into N, f is called an *embedding*. Here a homeomorphism of M into N means that f is a homeomorphism of M onto its image $f(M)$ with the relative topology of N. If an immersion f is simply injective, f is called an *embedding in the weak sense*.

We will now take G-actions into consideration. Let M and N be G-manifolds and $f: M \to N$ be a smooth G-map. If f is an immersion, an embedding, and an embedding in the weak sense without taking G-actions into consideration, then we say that f is a *G-immersion*, a *G-embedding*, and a *G-embedding in the weak sense*, respectively.

Example 1 Figure 4.1 illustrates an immersion, an embedding, and an embedding in the weak sense.

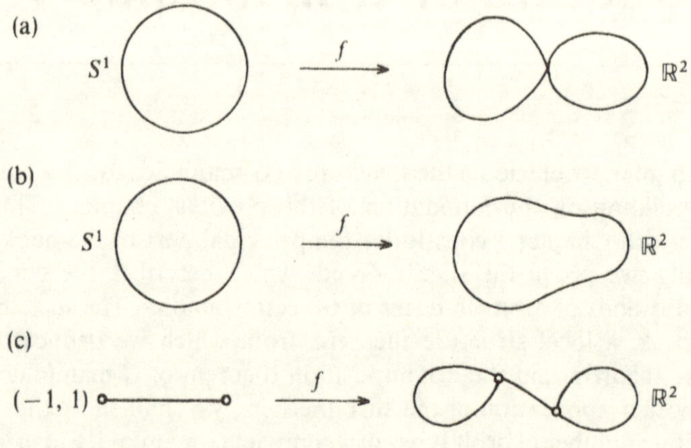

Fig. 4.1 (a) $f: S^1 \to \mathbb{R}^2$ immersion; (b) $f: S^1 \to \mathbb{R}^2$ embedding; (c) $f: (-1, 1) \to \mathbb{R}^2$ embedding in the weak sense.

Theorem 4.1 *If $f: M \to N$ is an embedding, then the image is a submanifold of N. Accordingly $f: M \to f(M)$ is a diffeomorphism.*

Proof By applying the inverse function theorem, we deduce that for each fixed $x_0 \in M$, there exist a chart $\varphi: U \to \mathbb{R}^m$ of M at x_0 with $\varphi(x_0) = 0$ and a chart $\psi: V \to \mathbb{R}^n$ of N at $f(x_0)$ with $\psi(f(x_0)) = 0$ such that:

(1) $f(U) \subset V$;

(2) for an arbitrary point x of $\varphi(U)$,

$$\psi \cdot f \cdot \varphi^{-1}(x) = (x, 0) \in \mathbb{R}^m \times \mathbb{R}^{n-m} = \mathbb{R}^n$$

holds (Fig. 4.2).

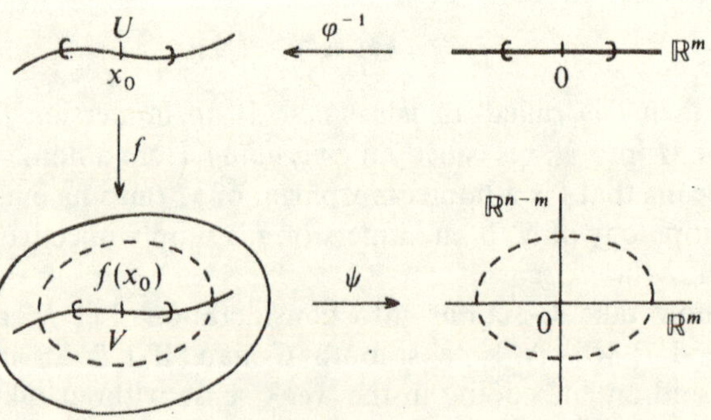

Fig. 4.2

We now denote by $\mathring{D}^a(\varepsilon)$ the open disc of radius ε in $\mathbb{R}^a$ having the origin as its centre. Choose a sufficiently small positive number ε with

$$\varphi(U) \supset \mathring{D}^m(\varepsilon), \qquad \psi(V) \supset \mathring{D}^n(\varepsilon)$$

and put

$$U(\varepsilon) = \varphi^{-1}(\mathring{D}^m(\varepsilon)), \qquad V(\varepsilon) = \psi^{-1}(\mathring{D}^n(\varepsilon)).$$

Then it follows from the above observation that

$$f(U(\varepsilon)) \subset V(\varepsilon) \qquad \text{and} \qquad \psi \cdot f(U(\varepsilon)) = \mathring{D}^n(\varepsilon) \cap \mathbb{R}^m$$

where we identified $\mathbb{R}^m$ with $\mathbb{R}^m \times \{0\}$ ($\subset \mathbb{R}^m \times \mathbb{R}^{n-m} = \mathbb{R}^n$). Although we are apt to hastily conclude that this completes the proof, this is not actually the case. This is because f may map some points outside of $U(\varepsilon)$ into $V(\varepsilon)$ as illustrated in Fig. 4.1(c) (see also Example 2 below). This possibility is, however, eliminated by the assumption that $f: M \to f(M)$ is a homeomorphism as follows.

Since $f^{-1} | f(M)$ is continuous, $f(U(\varepsilon))$ is open in $f(M)$, that is, there exists an open set W of N with

$$f(U(\varepsilon)) = f(M) \cap W.$$

If we put $V' = V(\varepsilon) \cap W$ and $\psi' = \psi \mid V'$, then (V', ψ') is a chart of N at $f(x_0)$. Note that the equality

$$f(U(\varepsilon)) = f(M) \cap V'$$

holds, since $f(U(\varepsilon)) \subset V(\varepsilon)$.

Claim $\psi'(f(M) \cap V') = \psi'(V') \cap \mathbb{R}^m$.

We have $\psi'(f(M) \cap V') = \psi'(f(U(\varepsilon))) = \mathring{D}^n(\varepsilon) \cap \mathbb{R}^m = \psi'(V') \cap \mathbb{R}^m$. Here only the last equality is not obvious. But $V(\varepsilon) \supset V'$ implies $\mathring{D}^n(\varepsilon) \cap \mathbb{R}^m \supset \psi'(V') \cap \mathbb{R}^m$. Since $f(U(\varepsilon)) = f(M) \cap V'$, we have $\psi'(f(U(\varepsilon))) \subset \psi'(V')$, and hence

$$\mathring{D}^n(\varepsilon) \cap \mathbb{R}^m = \psi'(f(U(\varepsilon))) \subset \psi'(V') \cap \mathbb{R}^m.$$

Hence the claim holds.

The claim implies that $f(M)$ is a submanifold of N. □

Example 2 Consider Example 4 of Section 1.4 and define $f: \mathbb{R} \to T^2$ by $f(r) = \varphi(r, (1, 1))$. Then f is an embedding in the weak sense. However f is not an embedding, since the image of f is the orbit of the point (1, 1) which is not homeomorphic to $\mathbb{R}$. Therefore it is not true in general that $f(M)$ is a submanifold of N under the condition that f is an embedding in the weak sense. In this case we cannot choose W as desired in the proof of Theorem 4.1. As illustrated in Fig. 4.3, f maps many points outside of $U(\varepsilon)$ into $V(\varepsilon)$ so that the claim in the proof of Theorem 4.1 does not hold.

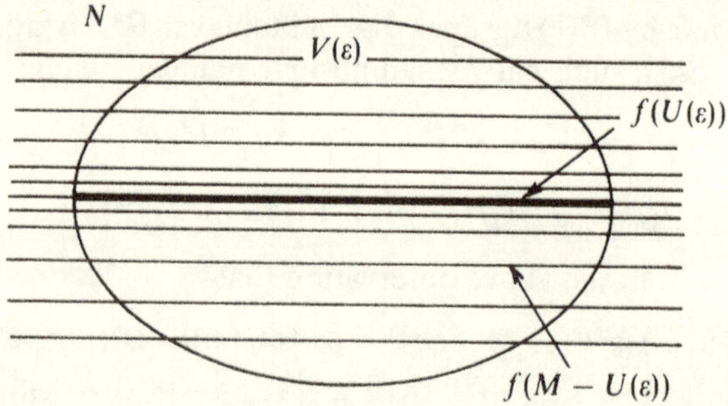

Fig. 4.3 The image of f re-enters $V(\varepsilon)$ infinitely many times.

Lemma 4.2 *If $f: M \sim N$ is an embedding in the weak sense and M is compact, then f is an embedding.*

Proof The continuity of $f^{-1}|f(M)$ follows from Lemma 1.16. □

Theorem 4.3 *Let G be a Lie group and M a G-manifold. For any given $x \in M$, the map $f: G/G_x \to M$ defined by $f(gG_x) = gx$ is an embedding in the weak sense.*

Proof As shown in Section 1.6, f is an injective continuous map. We apply Theorem 3.37 to deduce that G/G_x is endowed with a smooth structure so that f is smooth. Hence it suffices to show that the differential f_* is injective at every point. But the injectivity of f_* at eG_x was shown in the proof of Theorem 3.43 without using the assumptions that G is compact and that the G-action on M is transitive. Since f is a G-map, we have

$$f_{*gG_x} = g_* \cdot f_{*eG_x} \cdot g_*^{-1}$$

for every $g \in G$. Thus f_* is injective at every point. □

Corollary 4.4 *Let G be a compact Lie group and M a G-manifold. Then for every point x of M, the orbit $G(x)$ of x is a G-invariant submanifold of M and the map $f: G/G_x \to G(x)$ is a G-diffeomorphism.*

Proof By combining Lemma 4.2 and Theorem 4.3, we see that f is an embedding. Therefore Theorem 4.1 yields Corollary 4.4. □

Remark Example 2 above shows that if G is not compact, each orbit $G(x)$ is not necessarily a submanifold of M in general.

4.2 Geodesics and the exponential map

The concept of tubular neighbourhoods is basic to the study of differential topology. For this, differential geometric considerations are indispensable; that is, we need notions of geodesics of Riemannian manifolds and the exponential map which is a kind of generalization of the previous exponential map.

In the theory of transformation groups, these ideas are even more important when we prove results such as the slice theorem, the decomposition theorem, and the structure theorems.

Though a detailed discussion is outside the scope of this book, we shall give a general idea of what they are like with as much attention to geometric intuition as possible and shall quote only those theorems which will be needed later. For details, the reader is referred to any of the available books on differential geometry which, conversely, might become easier to understand after reviewing this section.

Given a smooth vector field, an integral curve through each point is obtained locally as has already been shown. Here, however, a different approach will be used to introduce geodesics. This can be done in one of two ways. One way is to define them as a solution of certain differential equations given a priori. The second is to define them from a geometric point of view. Since the study of differential geometry is not the principal objective of this book, we shall introduce and explain geodesics via the second route.

First observe that a manifold M is paracompact, since M is a Hausdorff space satisfying the second countability axiom. Hence for any open covering of M, there exists a smooth partition of unity subordinate to the covering.

Given a smooth vector bundle ξ over a manifold M, there is a metric on ξ as shown in Section 2.8. If we make use of a smooth partition of unity instead of the partition of unity employed there, then we have a smooth metric on ξ. Here we say that a metric $\langle\,,\rangle$ on ξ is smooth if the metric $\langle\,,\rangle_x$ in Section 2.8 depends smoothly on x.

A smooth metric $\langle\,,\rangle$ on the tangent bundle $T(M)$ of a manifold M is called a *Riemannian metric*, and M, together with $\langle\,,\rangle$, is called a *Riemannian manifold*.

Let G be a compact Lie group and M be a G-manifold. A smooth metric on a smooth G-vector bundle ξ over M can be altered into a G-invariant metric by using the invariant integral of Section 2.8. It follows from Corollary 2.35 that the resulting metric is also smooth. The new metric obtained in this way is called a smooth *G-invariant metric* on ξ.

A smooth G-invariant metric $\langle\,,\rangle$ on the tangent G-vector bundle $T(M)$ of a G-manifold M is called a *G-invariant Riemannian metric* on M, and M, together with $\langle\,,\rangle$ is called a *G-Riemannian manifold*.

We now define the notion of the length of a smooth curve l in a Riemannian manifold M. When we denote by $\langle\,,\rangle$ the Riemannian metric, the length $\|v\|$ of a tangent vector v is defined by

$$\|v\| = \sqrt{\langle v, v\rangle}.$$

Given a smooth curve $l\colon (c, d) \to M$, we define, for any a, b with $c < a < b < d$, the length $|l_a^b|$ of this curve from a to b by

$$|l_a^b| = \int_a^b \left\| \frac{\mathrm{d}l}{\mathrm{d}t}(t) \right\| \mathrm{d}t = \int_a^b \|l'(t)\|\, \mathrm{d}t.$$

Using this notion of the length of a curve, we shall give some idea of what geodesics are like. A smooth curve $l\colon (c, d) \to M$ is called a *geodesic* if the following conditions are satisfied.

(1) the length $\|l'(t)\|$ of the velocity vector is constant along $l(t)$ and,
(2) for any $t \in (c, d)$, there exists a positive number ε with $c < t - \varepsilon$, $t + \varepsilon < d$;

such that $|l_{t-\varepsilon}^{t+\varepsilon}|$ is less than or equal to the length of any other smooth curve from $l(t - \varepsilon)$ to $l(t + \varepsilon)$.

Roughly speaking, a geodesic is a curve giving the shortest route locally. The reason why we say 'locally' will be illustrated by the following example (Fig. 4.4). Geodesics on the 2-sphere are precisely the great circles and the larger of the two arcs of a great circle which join two points a and b (which are not at opposite ends of a diameter) is not of minimum length. However in a sufficiently small neighbourhood of each point $l(t)$, the great circle gives the shortest curve.

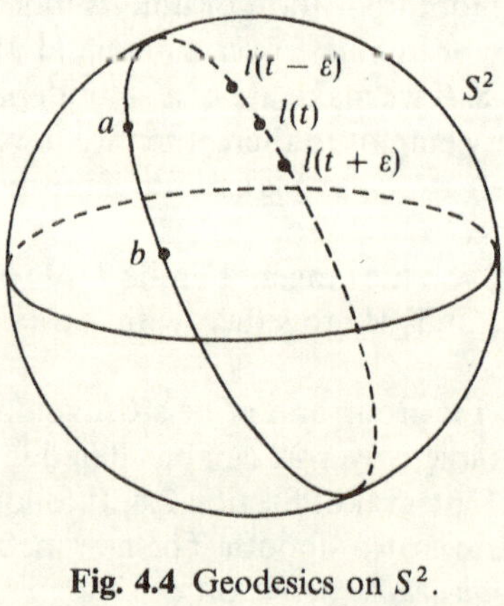

Fig. 4.4 Geodesics on S^2.

The problem of seeking a locally shortest curve is reduced to a variational problem which is expressed as a certain system of second-order differential equations.

The local existence and uniqueness of such a system of differential equations are guaranteed by the theory of differential equations. In fact the problem of solving this system of n second-order equations is reduced to the problem of solving a system of $2n$ first-order equations where $n = \dim M$.

Thus given an arbitrary point x of M and an arbitrary tangent vector v to M at x, there exist a positive number ε and a unique geodesic $l\colon (-\varepsilon, \varepsilon) \to M$ satisfying the conditions $l(0) = x$, $[l] = (\mathrm{d}l/\mathrm{d}t)(0) = v$. The above reduction means that a vector field Z on the tangent bundle $T(M)$ is constructed locally from the initial condition (x, v) so that the projection by $\pi\colon T(M) \to M$ of the integral curve of Z through (x, v) yields the required geodesic.

Let M be a G-Riemannian manifold. Given a smooth curve l in M, we denote by gl the curve defined by $t \mapsto gl(t)$. Since the metric is G-invariant, we have

$$|l_a^b| = |gl_a^b|.$$

In particular, if l is a geodesic, then so is gl.

Next we shall define the exponential map exp which is a generalization of that in Section 3.6.

Given a point x of M and a tangent vector v to M at x, we denote by l_v the geodesic satisfying the initial conditions:

$$l(0) = x, \qquad \frac{\mathrm{d}l}{\mathrm{d}t}(0) = v.$$

Here the domain of the curve l_v is $(-\varepsilon, \varepsilon)$ for some $\varepsilon > 0$. It follows from the uniqueness of the solution of differential equations that

$$l_{sv}(t) = l_v(st) \qquad \text{for } |s| \leq 1.$$

Hence if the length $\|v\|$ is small, then the ε of the domain $(-\varepsilon, \varepsilon)$ of l_v can be large. Accordingly we can take, say, $(-1.1, 1.1)$ as a domain of l_v if v is small enough.

Then the point $l_v(1) \in M$ will be denoted by $\exp(v)$ and is called the *exponential* of the tangent vector v. The geodesic l_v can thus be described by the formula

$$l_v(t) = \exp(tv).$$

It follows from the observation above that $\exp(v)$ is defined only when $\|v\|$ is small enough. In general, $\exp(v)$ is not defined for large vectors v.

We now review the properties of the exponential map.

Theorem 4.5 *There exists an open set W of the tangent bundle $T(M)$ including the zero cross-section (which we identify with M) such that the map* exp *is defined on W with* $\exp|M = \mathrm{id}$ *and is smooth.*

Remark The smoothness of the map exp: $W \to M$ follows from the fact that a solution of a system of differential equations depends smoothly on the initial condition.

For an arbitrary point x of the zero cross-section M, we consider the differential

$$\exp_*: T_x(W) = T_x(T(M)) \to T_x(M)$$

of the exponential map. Obviously we have the direct sum decomposition

$$T_x(T(M)) = T_v \oplus T_h$$

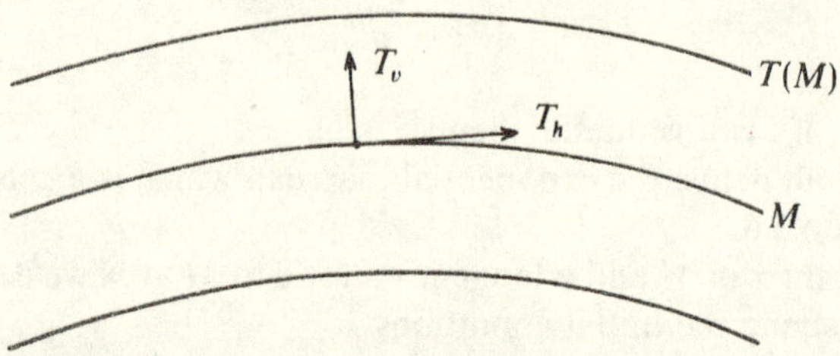

Fig. 4.5

where T_v is the tangent space of the fibre $T_x(M)(\subset T(M))$ at 0 and T_h is the tangent space of the zero section $M(\subset T(M))$ at x. Here v of T_v stands for vertical and h of T_h stands for horizontal as illustrated in Fig. 4.5. Note that there are canonical isomorphisms

$$T_v = T_0(T_x(M)) \cong T_x(M), \qquad T_h \cong T_x(M).$$

We identify both T_v and T_h with $T_x(M)$ via these isomorphisms.

Theorem 4.6 *Under the previous identifications,*

$$\exp_* | T_v = \mathrm{id} \qquad \textit{and} \qquad \exp_* | T_h = \mathrm{id}.$$

Proof Suppose $v_0 \in T_v = T_0(T_x(M))$. Under the previous identification, we regard v_0 not only as an element of T_v but also as an element of $T_x(M)$. Let $l_{v_0}: (-\varepsilon, \varepsilon) \to M$ be the geodesic tangent to v_0 at x, that is,

$$\frac{\mathrm{d}l_{v_0}}{\mathrm{d}t}(0) = v_0, \qquad l_{v_0}(0) = x.$$

Further, we define a smooth curve $v(t)$ in $T_x(M)$ by

$$v(t) = tv_0.$$

The tangent vector to $v(t)$ at 0 is nothing else but v_0 under the previous identification. Then, by making use of the notation in Section 3.3, we have

$$\begin{aligned}\exp_*(v_0) &= \exp_*([v(t)]) = [\exp(v(t))] \\ &= [\exp(tv_0)] = [l_{tv_0}(1)] = [l_{v_0}(t)] = v_0.\end{aligned}$$

Thus the equality $\exp_* | T_v = \mathrm{id}$ holds.

On the other hand, under the identification of the zero section of $T(M)$ with M, we have $T_h = T_x(M)$ and $\exp | M = \mathrm{id}$. It follows that $\exp_* | T_h = \mathrm{id}$. □

As an immediate consequence of the theorem, we have the following corollary.

Corollary 4.7 *The differential*

$$\exp_*\colon T_x(T(M)) = T_v \oplus T_h = T_x(M) \oplus T_x(M) \to T_x(M)$$

is given by $\exp_*(u, v) = u + v$.

4.3 *G*-invariant tubular neighbourhoods

Let M be a manifold and A a submanifold of M. A neighbourhood of A in M is called a tubular neighbourhood if it is diffeomorphic to the normal bundle of A in M. It is a fundamental result in the study of differential topology that there exists a tubular neighbourhood and that uniqueness holds up to isotopy. In this section, we shall generalize this to the case where a compact Lie group acts. Here we shall only show its existence which will be needed later.

Let us formulate it precisely. Given a vector bundle ξ over X equipped with a metric and a continuous map $\varphi\colon X \to \mathbb{R}^+$ (the set of positive numbers), we set

$$D(\xi, \varphi) = \{v \in \xi \mid v \in \xi_x, \|v\| \leq \varphi(x), x \in X\}$$

$$\mathring{D}(\xi, \varphi) = \{v \in \xi \mid v \in \xi_x, \|v\| < \varphi(x), x \in X\}.$$

Regarding a positive number r as the constant function $r\colon X \to \mathbb{R}^+$, we define similarly $D(\xi, r)$ and $\mathring{D}(\xi, r)$. In particular, when $r = 1$, we denote these by $D(\xi)$ and $\mathring{D}(\xi)$, respectively.

Let G be a compact Lie group, M a G-manifold, and A a G-invariant submanifold of M. The restriction $T(M) | A$ of the tangent G-vector bundle $T(M)$ of M has $T(A)$ as a sub-bundle. The quotient bundle

$$\nu = (T(M) | A)/T(A)$$

is called the *normal G-(vector) bundle* of A in M. If M is a G-Riemannian manifold, then $T(A)$ has an orthogonal complement $T(A)^{\perp}$ in $T(M)|A$ and $T(A)^{\perp}$ is canonically isomorphic to ν by Corollary 2.41. Note that ν is canonically endowed with a G-invariant metric by the above isomorphism.

By a *G-invariant tubular neighbourhood* of A in M we mean a G-embedding $f: \nu \to M$ such that the restriction of f to the zero section A of ν is the inclusion of A in M. Then the restriction of f to the unit disc bundle $f|D(\nu): D(\nu) \to M$ is called a *closed G-invariant tubular neighbourhood* of A in M (see Fig. 4.6). In this situation, the image $f(\nu)$ (resp. $f(D(\nu))$) is also called a *G-invariant tubular neighbourhood* (resp. *closed G-invariant tubular neighbourhood*) for brevity. Note that ν is a smooth manifold satisfying $\dim \nu = \dim M$. Hence it follows from the inverse function theorem that $f(\nu)$ is open in M.

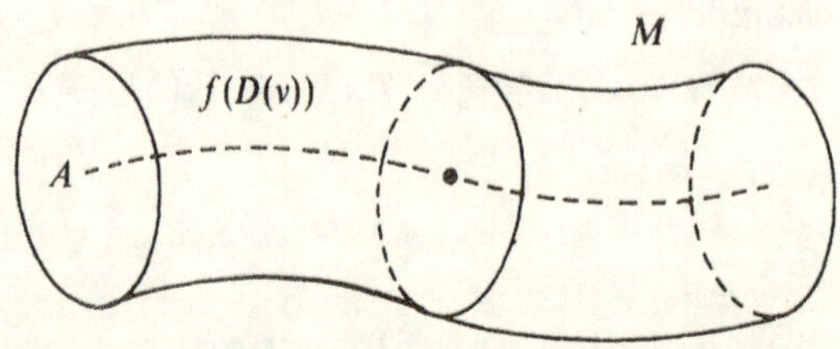

Fig. 4.6 A closed G-invariant tubular neighbourhood.

Theorem 4.8 *Let G be a compact Lie group. For a G-invariant submanifold A of a G manifold M, there exists a G-invariant tubular neighbourhood of A in M.*

Remark Since we make use of this theorem later for the case where A is compact, we recommend that the reader should read the proof of this special case. The essential idea of the proof is included here and the reader can return to it when necessary.

Proof The theorem will be proved by a stack of five claims and a lemma. Take a G-invariant Riemannian metric on M. As shown in the preceding section, there exists an open neighbourhood W of M in $T(M)$ and the map exp is defined on W. Then we first show the following claim.

Claim 1 As W, we can take a G-invariant open neighbourhood and the map $\exp: W \to M$ is G-equivariant.

Since $T(M) - W$ is closed in $T(M)$, it follows from Lemma 1.44 that $G\cdot(T(M) - W)$ is closed in $T(M)$. Hence if we put

$$W' = T(M) - G\cdot(T(M) - W),$$

then W' is a G-invariant open set of $T(M)$ satisfying

$$M \subset W' \subset W.$$

Accordingly it suffices to adopt W' as W. Hereafter we denote W' simply by W. As remarked in the preceding section, each element g of G maps geodesics to geodesics. In consequence, one easily verifies that exp is a G-map.

We now regard ν as the orthogonal complement $T(A)^{\perp}$ of $T(A)$ in $T(M)|A$, so that a G-invariant metric is induced on ν canonically. Then let us consider the following restriction of the above exponential map

$$\exp: \nu \cap W \to M.$$

Note that for any point a of A, there are canonical isomorphisms

$$T_a(\nu \cap W) = T_a(\nu) \cong T_a(A) \oplus \nu_a \cong T_a(M)$$

of vector spaces where ν_a denotes the fibre over a of the bundle ν. If we employ the notation of the preceding section, then $T_a(A)$ and ν_a are subspaces of T_h and T_v, respectively. According to Theorem 4.6, both $\exp_* | T_h$ and $\exp_* | T_v$ are the identity maps, so that the differential

$$\exp_*: T_a(\nu \cap W) = T_a(M) \to T_a(M)$$

is also the identity. Therefore for each point a of A, there is an open neighbourhood U_a of a in $\nu \cap W$ such that the differential

$$\exp_*: T_y(\nu \cap W) \to T_{\exp(y)}(M)$$

is an isomorphism at each point $y \in U_a$. Put

$$U = \bigcup_{a \in A} U_a$$

and change U for a G-invariant open neighbourhood as in the proof of Claim 1. Thus we have shown that there exists a G-invariant open neighbourhood U of A in $\nu \cap W$ on which exp is a G-immersion.

Claim 2 There exists a G-invariant open neighbourhood V of A in U on which exp is a G-embedding.

To prove the claim, we first show the existence of V without considering G-actions and then we change V for a G-invariant open neighbourhood as before. Therefore it suffices to show the following topological lemma without considering G-actions.

Lemma 4.9 *Let X be a locally compact Hausdorff space satisfying the second countability axiom and Y be a Hausdorff space. If a continuous map $f: X \to Y$ is a local homeomorphism and the restriction of f to a locally compact subset A of X is a homeomorphism into Y, then there exists an open neighbourhood V of A on which f is a homeomorphism into Y.*

Proof Since f is a local homeomorphism, we need only find an open neighbourhood V of A on which f is injective. We prove it in three steps.

(1) Suppose A is compact. Since X is Hausdorff, A is closed in X. For an arbitrary point x of $X - A$, there exists an open set U of X such that $\overline{U}$ is compact and

$$x \in U \subset \overline{U} \subset X - A,$$

since X is locally compact.

Let $\{O_i\}$ be a countable open basis of X. Corresponding to the U above, we can choose O_i with $x \in O_i \subset U$. Of course $\overline{O_i} \subset \overline{U}$ and hence $\overline{O_i} \subset X - A$. Since x is an arbitrary point of $X - A$, we have

$$X - A = \bigcup_{\overline{O_i} \subset X - A} \overline{O_i}.$$

Let $\{i_j \mid j = 1, 2, 3, \ldots\}$ be the set of indices i with $\overline{O_i} \subset X - A$ and put $V_j = X - \overline{O_{i_j}}$. Then V_j is an open neighbourhood of A and satisfies

$$\bigcap_j V_j = A.$$

Since A is a compact subset of a locally compact space X, there exists an open neighbourhood W of A in X such that $\overline{W}$ is compact.

Now suppose that for any open neighbourhood V of A in X, the restriction $f \mid V$ is not injective. Then there exist sequences

$$x_n, y_n \in \bigcap_{j \leq n} V_j \cap W$$

such that $x_n \neq y_n$ and $f(x_n) = f(y_n)$. Since both $\{x_n\}$ and $\{y_n\}$ are sequences in the compact set $\overline{W}$, we can choose subsequences x_{nk}, y_{nk} converging to $x, y \in \overline{W}$, respectively. Since f is continuous, we have $f(x) = f(y)$.

We now show that $x, y \in A$. If $x \notin A$, then there exists $O_{i_{j_0}}$ with

$$x \in O_{i_{j_0}} \subset \overline{O_{i_{j_0}}} \subset X - A.$$

Observe that for any $n \geq j_0$,

$$x_n \in \bigcap_{j \leq j_0} V_j \cap W \subset V_{j_0} = X - \overline{O_{i_{j_0}}}.$$

But this contradicts the fact that $x_{nk} \to x$. Hence we have $x \in A$. Similarly we have $y \in A$.

Since f is injective on A, we have $x = y$. But then f is not a local homeomorphism at x. Thus Lemma 4.9 is proved in the case where A is compact.

(2) Let A_0 be a compact subset of A. In the following, we shall show that there exists an open neighbourhood U_0 of A_0 such that $\overline{U_0}$ is compact and f is

a homeomorphism on $\overline{U_0} \cup A$. It follows from (1) above that there exists an open neighbourhood V of A_0 in X such that the restriction $f|V$ is injective. Since A_0 is a compact subset of a locally compact space X, there exists an open set V_0 of X such that

$$A_0 \subset V_0 \subset \overline{V_0} \subset V$$

and $\overline{V_0}$ is compact. Of course the restriction $f|\overline{V_0}$ is injective.

If $f|(U \cup A)$ were not injective for an arbitrary open neighbourhood U of A_0 in V_0, then one verifies in the same manner as (1) above that there would exist a sequence $\{x_n\}$ of $(X - A) \cap V_0$ converging to $a_0 \in A_0$ and a sequence $\{a_n\}$ of A with $f(x_n) = f(a_n)$. Since f is continuous, $f(a_n)$ converges to $f(a_0)$. Since f is a homeomorphism on A, a_n converges also to a_0. But this contradicts the fact that f is a local homeomorphism.

Thus there exists an open neighbourhood U of A_0 in V_0 such that $f|(U \cup A)$ is injective. If we take an open set U_0 of X such that

$$A_0 \subset U_0 \subset \overline{U_0} \subset U \subset V_0$$

and $\overline{U_0}$ is compact as before, then Claim (2) holds.

(3) Since X satisfies the second countability axiom, so does its subspace A. Furthermore A is locally compact, so that there is an ascending sequence of compact sets

$$A_1 \subset A_2 \subset \dots$$

with $A = \bigcup_n A_n$. We prove this first. Let $\{O'_i\}$ be a countable open basis of A. Denote by $\{O'_{i_j}\}$ the subfamily of $\{O'_i\}$ consisting of O'_i whose closure $\overline{O'_i}$ is compact. Put

$$A_n = \bigcup_{j \leq n} \overline{O'_{i_j}}.$$

Then we show that $\{A_n\}$ is the required one. Since A_n is the union of finitely many compact sets, A_n is compact. Obviously $\{A_n\}$ is an ascending sequence. Since A is locally compact, each point a of A has a compact neighbourhood U_a. Then there exists O'_i with $a \in O'_i \subset U_a$. Since $\overline{O'_i} \subset \overline{U_a} = U_a$ and U_a is compact, $\overline{O'_i}$ is also compact. Thus we have $A = \bigcup_n A_n$.

In the following we shall show by induction that there exists an ascending sequence $V_1 \subset V_2 \subset \dots$ of open sets of X such that V_n is an open neighbourhood of A_n, $\overline{V_n}$ is compact, and f is a homeomorphism on $\overline{V_n} \cup A$. In view of (2), there exists an open neighbourhood V_1 of A_1 such that $\overline{V_1}$ is compact and f is a homeomorphism on $\overline{V_1} \cup A$. Suppose now that there exists an ascending sequence $V_1 \subset V_2 \subset \cdots \subset V_n$ of open sets of X such that V_i is an open neighbourhood of A_i, $\overline{V_i}$ is compact, and f is a homeomorphism on $\overline{V_i} \cup A$ for all $i \leq n$. Then $\overline{V_n} \cup A_{n+1}$ is a compact subset of $\overline{V_n} \cup A$. Since f is a homeomorphism on $\overline{V_n} \cup A$, it follows from (2) that there exists an open neighbourhood V_{n+1} of $\overline{V_n} \cup A_{n+1}$ such that $\overline{V_{n+1}}$ is compact and f is a homeomorphism on $\overline{V_{n+1}} \cup \overline{V_n} \cup A = \overline{V_{n+1}} \cup A$. Since V_{n+1} satisfies $V_n \subset V_{n+1}$, the inductive step proceeds. Put

$$V = \bigcup_n V_n.$$

Then V is an open set of X including A. Since $f | V_n$ is injective and $\{V_n\}$ is an ascending sequence, f is injective on V.

This completes the proof of Lemma 4.9 and hence that of Claim 2 as well.

Remark If we take V_n by applying (2) to A_n, then $\{V_n\}$ is not necessarily an ascending sequence. In consequence, it is not guaranteed that f is injective on V.

We now continue the proof of Theorem 4.8. Since A is locally compact, each point a of A has an open neighbourhood U_a in A with $\overline{U_a}$ compact. By Lemma 1.44, $G \cdot U_a$ is open in A and $G \cdot \overline{U_a}$ is compact. Then it is easy to see that there exists $\varepsilon_a > 0$ such that

$$D(v \,|\, G \cdot \overline{U_a}, \varepsilon_a) \subset V.$$

Since A is paracompact, there exists a smooth partition of unity $\{\lambda_\mu\}$ subordinate to the open covering $\{G \cdot U_a \,|\, a \in A\}$. For each μ, we choose a such that

$$\operatorname{supp} \lambda_\mu \subset G \cdot U_a$$

and denote it by $a(\mu)$. Then the map $\varphi \colon A \to \mathbb{R}^+$ defined by $\varphi = \sum_\mu \varepsilon_{a(\mu)} \lambda_\mu$ is smooth.

Claim 3 $D(v, \varphi) \subset V$.

Fix an arbitrary point b of A. Put

$$\varepsilon(b) = \max \{\varepsilon_{a(\mu)} \,|\, \lambda_\mu(b) \neq 0\}.$$

Since there exists μ_0 with $\varepsilon(b) = \varepsilon_{a(\mu_0)}$, we have

$$\varphi(b) \leq \sum_\mu \varepsilon_{a(\mu_0)} \lambda_\mu(b) = \varepsilon_{a(\mu_0)}.$$

Hence if only we note $b \in G \cdot \overline{U_{a(\mu_0)}}$, we see that any element v of v_b with $\|v\| \leq \varphi(b)$ lies in V, which is Claim 3.

By means of the invariant integral, we change φ for a G-invariant map. Here φ is said to be G-invariant if $\varphi(a) = \varphi(ga)$ holds for all $a \in A$, $g \in G$. For this it suffices to put

$$\overline{\varphi}(a) = \int_G \varphi(ga)\,\mathrm{d}g,$$

which is a smooth G-invariant positive function.

Claim 4 $D(\nu, \overline{\varphi}) \subset V$.

Fix an arbitrary point b of A. Since G is compact, there exists

$$\max_{g \in G} \{\varphi(gb)\}$$

which is attained at, say g_0. Then we have

$$\overline{\varphi}(b) \leq \max_{g \in G} \{\varphi(gb)\} = \varphi(g_0 b) \leq \varepsilon_{a(\mu_0)}.$$

Here μ_0 is chosen as above. Since $g_0 b \in G \cdot \overline{U_{a(\mu_0)}}$, we have

$$b \in G \cdot \overline{U_{a(\mu_0)}}.$$

Hence any element v of ν_b with $\|v\| \leq \overline{\varphi}(b)$ lies in V, which is Claim 4.

Since $\overline{\varphi}$ is G-invariant, $D(\nu, \overline{\varphi})$ is a G-invariant subset of ν and $\mathring{D}(\nu, \overline{\varphi})$ is a G-invariant open submanifold of ν. Now we define a map $\psi\colon \nu \to \nu$ by

$$\psi(v) = \frac{\overline{\varphi}(\pi(v))}{\sqrt{1 + \|v\|^2}}\, v$$

where π is the projection $\nu \to A$.

Claim 5 ψ is a G-diffeomorphism onto $\mathring{D}(\nu, \overline{\varphi})$ with $\psi \mid A = \mathrm{id}$.

Clearly $\psi \mid A = \mathrm{id}$. Note that for $t \in \mathbb{R}^+, v \neq 0 \in \nu$,

$$\|\psi(tv)\| = \frac{\overline{\varphi}(\pi(v))}{\sqrt{(1/t\|v\|)^2 + 1}}.$$

Hence for fixed $v \neq 0$, $\|\psi(tv)\|$ is a monotone increasing function of t and $\|\psi(tv)\| \to \overline{\varphi}(\pi(v))$ as $t \to \infty$. Thus we have $\psi(\nu) = \mathring{D}(\nu, \overline{\varphi})$.

It is obvious that ψ is smooth.

We now show that its inverse is also smooth. Observe that ψ changes only the lengths of vectors. Accordingly if we define a map $\psi'\colon \mathring{D}(\nu, \overline{\varphi}) \to \nu$ by

$$\psi'(w) = \frac{w}{\sqrt{\overline{\varphi}(\pi(w))^2 - \|w\|^2}},$$

then it is easy to see that ψ' is smooth and satisfies

$$\psi' \cdot \psi = \mathrm{id} \qquad \text{and} \qquad \psi \cdot \psi' = \mathrm{id}.$$

It follows that ψ is a diffeomorphism onto $\mathring{D}(\nu, \overline{\varphi})$.

For any $g \in G$ and $v \in \nu$, we compute

$$\psi(gv) = \frac{\varphi(\pi(gv))}{\sqrt{1+\|gv\|^2}} gv = \frac{\overline{\varphi}(g\pi(v))}{\sqrt{1+\|v\|^2}} gv = \frac{\overline{\varphi}(\pi(v))}{\sqrt{1+\|v\|^2}} gv$$

$$= g \frac{\overline{\varphi}(\pi(v))}{\sqrt{1+\|v\|^2}} v = g \cdot \psi(v),$$

that is, ψ is a G-map.

Putting all this together, we have that ψ is a G-diffeomorphism onto $\mathring{D}(\nu, \overline{\varphi})$, which completes the proof of Claim 5.

We are now approaching the end of the proof of the theorem. By virtue of Claim 2, $\exp: V \to M$ is a G-diffeomorphism into M. It follows from Claim 4 that $D(\nu, \overline{\varphi}) \subset V$. Furthermore $\psi: \nu \to \mathring{D}(\nu, \overline{\varphi})$ is a G-diffeomorphism by Claim 5. Thus the composite map

$$\exp \cdot \psi: \nu \to M$$

is a G-diffeomorphism into M such that $\exp \cdot \psi \mid A = \mathrm{id}$.

This completes the proof of Theorem 4.8. □

4.4 The slice theorem

Given a G-manifold, the slice theorem describes the conditions of neighbourhoods of each orbit; it is fundamental in the study of the structure of G-manifolds.

Let G be a compact Lie group and M a G-manifold. Since the orbit $G(x)$ of an arbitrary point x of M is a G-invariant submanifold by Corollary 4.4, it follows from Theorem 4.8 that there exists a G-invariant tubular neighbourhood of $G(x)$ in M. That is, there exists an open neighbourhood of $G(x)$ in M which is G-diffeomorphic to the normal G-vector bundle ν of $G(x)$ in M. In view of Corollary 4.4, $G(x)$ is G-diffeomorphic to G/G_x. Accordingly it follows from the observation in Section 2.6 that ν is determined completely by the G_x-action on the fibre ν_x over x, so that ν is isomorphic to

$$G \underset{G_x}{\times} \nu_x \to G/G_x$$

as G-vector bundles. Since G_x is a closed subgroup of G, G_x is a Lie subgroup of G by Theorem 3.36 and the G_x-action on ν_x is smooth. Hence the above bundle isomorphism is a smooth G-vector bundle isomorphism. We have thus established the *slice theorem*.

Theorem 4.10 (The slice theorem) *Let G be a compact Lie group and M a G-manifold. For an arbitrary point x of M, the orbit $G(x)$ of x is a G-invariant submanifold of M. Let ν denote the normal G-vector bundle of $G(x)$ in M. Then* *the fibre v_x over x of v is a representation space of the isotropy group G_x so that v is isomorphic to*

$$G \underset{G_x}{\times} v_x \to G/G_x$$

as smooth G-vector bundles. Moreover there exist a G-invariant open neighbourhood U of $G(x)$ in M and a G-diffeomorphism $f\colon G \underset{G_x}{\times} v_x \to U$ such that the restriction of f to the zero cross-section gives the G-diffeomorphism from G/G_x to $G(x)$ of Corollary 4.4. If we take a G-invariant Riemannian metric on M, then the G_x-action on v_x is given by the G_x-action on the orthogonal complement of $T_x(G(x))$ in $T_x(M)$.

Definition 4.1 The image $f(v_x)$ of v_x under the G-embedding f above is called a *slice* at x and the representation v_x of G_x is called a *slice representation.*

A slice is illustrated in Fig. 4.7.

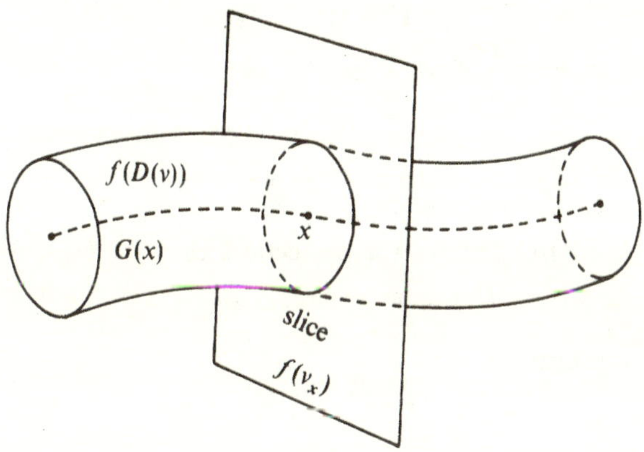

Fig. 4.7 A slice at x.

In fibre bundles and principal bundles in this section and in the next section, we denote the space $U \times F$ giving the local triviality of bundles by the reversed order $F \times U$ and we exchange actions from the left for actions from the right. As we observed in Section 2.2, whichever we employ, the effects are equivalent. However the above change is natural and enables us to understand the situation more easily here. For example, the structure group K acts on the fibre F from the right and K acts freely on the total space of a principal K-bundle from the left (Proposition 2.9).

The remainder of this section is devoted to the study of structures of manifolds with free G-actions carried out by applying the slice theorem. This will be a preparation for the study of structures of general G-manifolds in the next section.

Theorem 4.11 *Let G be a compact Lie group and M a G-manifold. If the action is free, then we can give M/G a smooth structure as a manifold so that the projection $\pi: M \to M/G$ is a smooth principal G-bundle. The smooth structure on M/G has the following properties and any smooth structure with the following properties is unique.*

(1) *$\pi: M \to M/G$ is smooth*

(2) *$h: M/G \to N$ is smooth if and only if $h \cdot \pi$ is smooth where N is a smooth manifold.*

The condition (2) *can be replaced by the condition*:

(2)′ *π has a smooth local cross-section.*

Proof On account of the slice theorem, there exist a vector space V and a G-embedding

$$f: G \times V \to M, \qquad f(e, 0) = x$$

for an arbitrary point x of M. Then the map

$$\bar{f}: (G \times V)/G \to M/G$$

between the orbit spaces induced by f is easily seen to be a homeomorphism into M/G. Moreover if we define a map $j: V \to G \times V$ by $j(v) = (e, v)$ and if $\pi: G \times V \to (G \times V)/G$ denotes the projection, then the composition

$$k = \pi \cdot j: V \to (G \times V)/G$$

is a homeomorphism.

Putting all this together, we have the following commutative diagram:

$$\begin{array}{ccccc} V & \xrightarrow{j} & G \times V & \xrightarrow{f} & M \\ & \searrow^{k} & \downarrow{\scriptstyle \pi} & & \downarrow{\scriptstyle \pi} \\ & & (G \times V)/G & \xrightarrow{\bar{f}} & M/G \end{array}$$

That is, we have the following equalities:

$$\bar{f} \cdot k = \bar{f} \cdot \pi \cdot j = \pi \cdot f \cdot j.$$

We now denote $\bar{f} \cdot k(V)$ by $[V]$ and $\bar{f} \cdot k(v)$ for $v \in V$ by $[v]$. Then $[V]$ is an open set of M/G and $\bar{f} \cdot k: V \to [V]$ is a homeomorphism.

In the following, we shall show that M/G becomes a smooth manifold with charts

$$\psi = (\bar{f} \cdot k)^{-1} : [V] \to V.$$

If we apply the slice theorem to each orbit, then $\{[V]\}$ is an open covering of M/G. Suppose now that two such G-embeddings

$$f_i : G \times V_i \to M, \qquad i = 1, 2$$

obtained by the slice theorem satisfy $[V_1] \cap [V_2] \neq \varnothing$. As we considered $\bar{f}, j, \pi, k, \psi$ corresponding to f, we consider $\bar{f}_i, j_i, \pi_i, k_i, \psi_i$ corresponding to f_i $(i = 1, 2)$. Then it suffices to show that the map

$$\psi_2 \cdot \psi_1^{-1} : \psi_1([V_1] \cap [V_2]) \to \psi_2([V_1] \cap [V_2])$$

is smooth. By definition, we have for $v_i \in V_i$,

$$[v_i] = \bar{f}_i \cdot k_i(v_i) = \bar{f}_i \cdot \pi_i \cdot j_i(v_i) = \pi \cdot f_i \cdot j_i(v_i).$$

Suppose that $[v_1] = [v_2]$ for some $v_i \in V_i$. Then we have

$$\psi_2 \cdot \psi_1^{-1}(v_1) = \psi_2([v_1]) = \psi_2([v_2]) = v_2.$$

Since $f_1 \cdot j_1(v_1)$ and $f_2 \cdot j_2(v_2)$ lie in the same orbit, there exists $g \in G$ with

$$f_2 \cdot j_2(v_2) = g f_1 \cdot j_1(v_1).$$

If $p_2 : G \times V_2 \to V_2$ denotes the projection, then the equality $p_2 g = p_2$ holds trivially.

Claim $\psi_2 \cdot \psi_1^{-1} = p_2 \cdot f_2^{-1} \cdot f_1 \cdot j_1 \mid \psi_1([V_1] \cap [V_2])$.

Since f_i are G-maps, we compute

$$\begin{aligned} p_2 \cdot f_2^{-1} \cdot f_1 \cdot j_1(v_1) &= p_2 \cdot g f_2^{-1} \cdot f_1 \cdot j_1(v_1) = p_2 f_2^{-1} \cdot g f_1 \cdot j_1(v_1) \\ &= p_2 \cdot f_2^{-1} \cdot f_2 \cdot j_2(v_2) = p_2 \cdot j_2(v_2) = v_2, \end{aligned}$$

that is, the claim holds.

Since the maps j_1, f_1, f_2^{-1}, p_2 are all smooth, it follows from the claim above that $\psi_2 \cdot \psi_1^{-1}$ is smooth as well. Therefore $\{([V], \psi)\}$ is a smooth atlas on M/G. Moreover we see that M/G satisfies the second countability axiom by Corollary 1.56 and is Hausdorff by (2) of Proposition 1.58.

Thus we have shown that M/G is a smooth manifold with atlas $\{([V], \psi)\}$.

Next we shall show that $\pi : M \to M/G$ is smooth. For this, we first show that the composition of the following maps

$$\pi^{-1}([V]) = f(G \times V) \xrightarrow{f^{-1}} G \times V \xrightarrow{p} V \xrightarrow{\psi^{-1}} [V]$$

coincides with the map $\pi | \pi^{-1}([V]) : \pi^{-1}([V]) \to [V]$ where p is the projection map. Given $g \in G$, $v \in V$, we have

$$\psi^{-1} \cdot p \cdot f^{-1} \cdot f(g, v) = \psi^{-1} \cdot p(g, v) = \psi^{-1}(v) = \bar{f} \cdot k(v).$$

On the other hand, we have

$$\pi\cdot f(g, v) = \pi\cdot gf(e, v) = \pi\cdot f(e, v) = \bar{f}\cdot k(v).$$

Hence $\psi^{-1}\cdot p\cdot f^{-1} = \pi|\pi^{-1}([V])$ holds. It turns out that

$$p\cdot f^{-1} = \psi\cdot\pi|\pi^{-1}([V]).$$

Since ψ is a chart at $[0]$ and $p\cdot f^{-1}: \pi^{-1}([V]) = f(G\times V)\to V$ is smooth, this means that $\pi|\pi^{-1}([V])$ is smooth. Since this argument is valid for an arbitrary slice, we see that π is smooth, that is, (1) holds.

Next we shall show that $\pi: M\to M/G$ is a smooth principal G-bundle. Let $q_i: G\times V_i\to G$ denote the projections. Define maps

$$\varphi_i: \pi^{-1}([V_i]) = f_i(G\times V_i)\to G\times[V_i]$$

by

$$\varphi_i(x) = (q_i\cdot f_i^{-1}(x), \pi(x))$$

for $x\in\pi^{-1}([V_i]) = f_i(G\times V_i)$, $i = 1, 2$. It follows from (1) above that the φ_i are smooth. Further, the following diagram commutes:

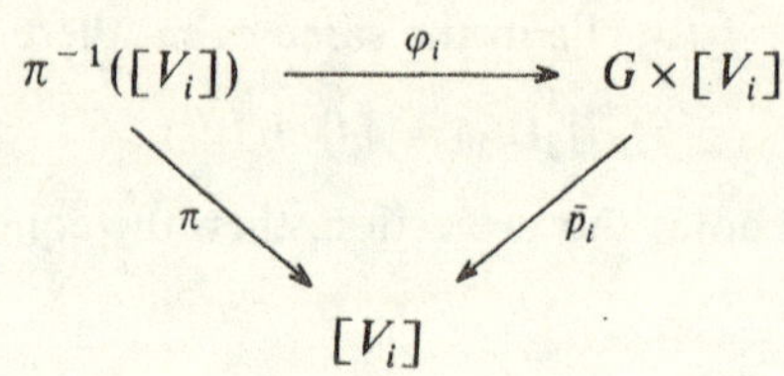

where $\bar{p}_i$ are the projections. Note that

$$\varphi_i\cdot f_i(g, v_i) = (g, [v_i])$$

for $v_i\in V_i$. Since the f_i are bijective, each φ_i is also bijective and the φ_i^{-1} are given by

$$\varphi_i^{-1}(g, [v_i]) = f_i(g, v_i) = f_i(g, \psi_i([v_i])).$$

Consequently we see that the φ_i are diffeomorphisms.

On the other hand, we define a map

$$\theta: [V_1]\cap[V_2]\to G$$

by the composition $q_2\cdot f_2^{-1}\cdot f_1\cdot j_1\cdot\psi_1|[V_1]\cap[V_2]$. Since all the maps in this definition are smooth, θ is smooth as well. For any element v_1 of V_1 with $[v_1]\in[V_1]\cap[V_2]$, we have

$$\theta([v_1]) = q_2\cdot f_2^{-1}\cdot f_1\cdot j_1\cdot\psi_1([v_1]) = q_2\cdot f_2^{-1}\cdot f_1(e, v_1).$$

We now compute the map

$$\varphi_2\cdot\varphi_1^{-1}: G\times([V_1]\cap[V_2])\to G\times([V_1]\cap[V_2]).$$

Making use of the fact that f_i and q_2 are G-maps, we have for $v_1 \in V_1$ with $[v_1] \in [V_1] \cap [V_2]$,

$$\begin{aligned}\varphi_2 \cdot \varphi_1^{-1}(g, [v_1]) &= \varphi_2 \cdot f_1(g, v_1) \\ &= (q_2 \cdot f_2^{-1} \cdot f_1(g, v_1), \pi \cdot f_1(g, v_1)) \\ &= (g q_2 \cdot f_2^{-1} \cdot f_1(e, v_1), \pi \cdot g f_1 \cdot j_1(v_1)) \\ &= (g\theta([v_1]), \pi \cdot f_1 \cdot j_1(v_1)) \\ &= (g\theta([v_1]), [v_1]).\end{aligned}$$

Thus we have shown that $\pi: M \to M/G$ becomes a smooth principal G-bundle with coordinate functions φ_i and transition functions θ.

In particular $\pi: M \to M/G$ has a smooth local cross-section. Accordingly property (2) is verified in the same manner as the proof of Theorem 3.37.

Finally, the uniqueness of the differential structures on M/G with properties (1) and (2) can also be verified in the same manner as the proof of Theorem 3.37. □

Since we need prerequisites from representation theory in order to prove the following theorem, we mention only the result. This will be needed in Chapter 6.

Theorem 4.12 *Let G be a compact Lie group and M a compact G-manifold. Then there exist a representation space V of G and a G-embedding e: $M \to V$.*

Idea of the proof On account of the slice theorem, we know the properties of the neighbourhoods of each orbit. Accordingly these neighbourhoods are G-embedded into some G-representation space by using the results of representation theory. We extend these G-embeddings to G-maps defined on M, which are not necessarily G-embeddings.

Consider this procedure at every orbit. Since M is compact, we can choose finitely many of these G-maps so that M is G-embedded in the direct sum of the corresponding G-representation spaces. □

Example 1 Consider the sphere

$$S^n = \{(x_1, \ldots, x_{n+1}) \in \mathbb{R}^{n+1} \mid \sum x_i^2 = 1\}$$

of Example 2 in Section 3.1. Let g denote the generator of the cyclic group $\mathbb{Z}_2$ of order 2. Then we define a $\mathbb{Z}_2$-action on S^n by

$$g \circ (x_1, \ldots, x_{n+1}) = (-x_1, \ldots, -x_{n+1}).$$

Obviously the action is free and its orbit space $S^n/\mathbb{Z}_2$ becomes a manifold by Theorem 4.11. The projection π: $S^n \to S^n/\mathbb{Z}_2$ is a principal $\mathbb{Z}_2$-bundle and hence π is a regular covering. Moreover $S^n/\mathbb{Z}_2$ is diffeomorphic to $\mathbb{R}P^n$ of

Example 3 in Section 3.1 (Exercise). The bundle $\pi: S^n \to \mathbb{R}P^n$ can be identified with $q: V_1(\mathbb{R}^{n+1}) \to G_1(\mathbb{R}^{n+1})$ of Example 5 in Section 3.9 and is called the *Hopf bundle*. The projection π is called the *Hopf map*.

Let $\mathbb{Z}_2$ act on $\mathbb{R}$ by $g \circ x = -x$. Then the bundle

$$\gamma: S^n \underset{\mathbb{Z}_2}{\times} \mathbb{R} \to \mathbb{R}P^n$$

associated with the Hopf bundle becomes a vector bundle with fibre $\mathbb{R}$, which is called the *Hopf line bundle*.

Example 2 For a positive integer k, we set

$$\zeta = \exp 2\pi\sqrt{-1}/k.$$

Then the cyclic group $\mathbb{Z}_k$ of order k can be written as

$$\{\zeta^i \mid i = 0, 1, \ldots, k-1\}.$$

Given integers $l_1, \ldots, l_n$ relatively prime to k, we define a $\mathbb{Z}_k$-action on

$$S^{2n+1} = \{(z_0, \ldots, z_n) \in \mathbb{C}^{n+1} \mid \sum |z_i|^2 = 1\}$$

by

$$\zeta \circ (z_0, \ldots, z_n) = (z_0\zeta, z_1\zeta^{l_1}, \ldots, z_n\zeta^{l_n}).$$

Obviously the action is free. The orbit space $S^{2n+1}/\mathbb{Z}_k$ is denoted by

$$L(k: l_1, \ldots, l_n)$$

and is called the *lens space*. By virtue of Theorem 4.11, it becomes a manifold and the quotient map $\pi: S^{2n+1} \to L(k: l_1, \ldots, l_n)$ is a principal $\mathbb{Z}_k$-bundle, and hence a regular covering.

Example 3 Set $S^1 = U(1) = \{z \in \mathbb{C} \mid |z| = 1\}$ and define an S^1-action on S^{2n+1} by

$$z \circ (z_0, \ldots, z_n) = (zz_0, \ldots, zz_n), \qquad z \in S^1.$$

Clearly the action is free. By virtue of Theorem 4.11, the orbit space S^{2n+1}/S^1 becomes a manifold and the quotient map $\pi: S^{2n+1} \to S^{2n+1}/S^1$ is a principal S^1-bundle. Furthermore, the orbit space S^{2n+1}/S^1 is diffeomorphic to $\mathbb{C}P^n$ of Example 3 in Section 3.1 (Exercise). The bundle $\pi: S^{2n+1} \to \mathbb{C}P^n$ can be identified with $q: V_1(\mathbb{C}^{n+1}) \to G_1(\mathbb{C}^{n+1})$ of Example 6 in Section 3.9 and is called the *Hopf bundle* as well. The projection π is also called the *Hopf map*. In particular, the special case $\pi: S^3 \to \mathbb{C}P^1 = S^2$ appears frequently.

Let S^1 act on $\mathbb{C}$ by the multiplication of complex numbers. Then the bundle

$$\mathbb{H}: S^{2n+1} \underset{S^1}{\times} \mathbb{C} \to \mathbb{C}P^n$$

associated with the Hopf bundle above becomes a complex vector bundle with fibre $\mathbb{C}$, which is called the *Hopf complex line bundle*.

We denote the conjugate bundle of the Hopf complex line bundle $\mathbb{H}$ by $\mathbb{H}^*$. Here the conjugate bundle $\mathbb{H}^*$ of $\mathbb{H}$ is defined to be the complex line bundle with the same underlying real vector bundle, but with the 'opposite' complex structure.

Proposition 4.13 *There is an isomorphism $T(\mathbb{C}P^n) \oplus \underline{\mathbb{C}} \cong (n+1)\mathbb{H}^*$ of complex vector bundles where $\underline{\mathbb{C}}$ denotes the trivial bundle $\mathbb{C}P^n \times \mathbb{C} \to \mathbb{C}P^n$.*

Proof First we regard $\mathbb{C}P^n$ as the homogeneous space $U(n+1)/(U(n) \times U(1))$. As a chart of $\mathbb{C}P^n$ at the point $[0, \ldots, 0, 1]$, we take the chart $\varphi_n \colon U_n \to \mathbb{C}^n$ of Example 3 in Section 3.1. Then it is easily verified that the isotropy representation $f\colon U(n) \times U(1) \to U(n)$ at this point is given by $(A, z) \mapsto \bar{z}A$. Hence we can apply the complex vector bundle version of Theorem 3.15 and deduce that the complex tangent bundle of $\mathbb{C}P^n$ is given by

$$U(n+1) \underset{U(n) \times U(1)}{\times} \mathbb{C}^n \to U(n+1)/(U(n) \times U(1)).$$

Accordingly the total space of the Whitney sum of this bundle and the trivial complex line bundle $\underline{\mathbb{C}}$ is given by the orbit space of the $U(n) \times U(1)$-action on $U(n+1) \times \mathbb{C}^{n+1}$ defined by

$$(A, z) \circ (x, y) = \left(x \begin{bmatrix} A & 0 \\ 0 & z \end{bmatrix}^{-1}, \begin{bmatrix} \bar{z}A & 0 \\ 0 & 1 \end{bmatrix} y \right)$$

$$\text{for } A \in U(n),\ z \in U(1),\ x \in U(n+1),\ y \in \mathbb{C}^{n+1}.$$

Since elements of $U(n)$ and elements of $U(1)$ commute, the orbit space of the $U(n) \times U(1)$-action coincides with the orbit space of the $U(1)$-action on the orbit space of the $U(n)$-action. Since the $U(n)$-action is given by

$$A \circ (x, y) = \left(x \begin{bmatrix} A & 0 \\ 0 & 1 \end{bmatrix}^{-1}, \begin{bmatrix} A & 0 \\ 0 & 1 \end{bmatrix} y \right),$$

we see, by applying the complex vector bundle version of Corollary 2.30, that its orbit space is identified with $(U(n+1)/U(n)) \times \mathbb{C}^{n+1}$. By paying attention to the correspondence in Proposition 1.98, one verifies easily that the $U(1)$-action on this space is given by

$$\begin{aligned} z \circ ([x], y) &= \left(\left[x \begin{bmatrix} I & 0 \\ 0 & \bar{z} \end{bmatrix} \right], \begin{bmatrix} \bar{z}I & 0 \\ 0 & \bar{z} \end{bmatrix} y \right) \\ &= ([x]\bar{z}, \bar{z}y). \end{aligned}$$

Here $z \in U(1)$, $x \in U(n+1)$, $[x] \in U(n+1)/U(n)$, $y \in \mathbb{C}^{n+1}$, and I denotes the identity matrix of degree n. Thus we see that this bundle is the Whitney sum of $n+1$ copies of $\mathbb{H}^*$. □

Remark In a similar manner to Proposition 4.13, we have the bundle isomorphism

$$T(\mathbb{R}P^n) \oplus \underline{\mathbb{R}} \cong (n+1)\gamma$$

where $\underline{\mathbb{R}}$ denotes the trivial bundle $\mathbb{R}P^n \times \mathbb{R} \to \mathbb{R}P^n$ and γ denotes the Hopf line bundle of the real case.

4.5 Structure theorems for G-manifolds

The purpose of this section is to decompose a G-manifold into disjoint submanifolds by orbit types and to investigate the structure of each submanifold.

In the preceding section, we dealt with the case of free actions by way of preparation. In this section, we first consider the opposite extreme, that is, the fixed point set. By combining the results of both extremes, we investigate G-manifolds with one orbit type. Finally we will examine general G-manifolds.

Theorem 4.14 *Let G be a compact Lie group and M a G-manifold. Then the fixed point set M^G is a closed submanifold of M.*

Proof An arbitrary point x of the fixed point set M^G is a G-invariant submanifold of M. The normal G-vector bundle of $\{x\}$ in M is the tangent space $T_x(M)$ of M at x by definition. Let $f\colon T_x(M) \to M$ be a G-invariant tubular neighbourhood of $\{x\}$ in M. The fixed point set $T_x(M)^G$ is a vector subspace of $T_x(M)$ and the pair $(T_x(M), T_x(M)^G)$ is a pair of Euclidean spaces. Then the map

$$f^{-1}\colon f(T_x(M)) \to T_x(M)$$

is a chart of M at x and satisfies

$$f(T_x(M)) \cap M^G = f(T_x(M)^G),$$

since f is G-equivariant. This means that M^G is a submanifold of M. On the other hand, M^G is closed in M by Theorem 1.46. Thus we have shown that M^G is a closed submanifold of M. □

Remark The hypothesis that G is compact cannot be omitted in Theorem 4.14; an example follows.

Example 1 It is well known that there is a smooth function $\mu\colon \mathbb{R} \to \mathbb{R}$ having the following properties (see Fig. 4.8):

(1) $\mu(x) \geq 0$ for all $x \in \mathbb{R}$;

(2) $\mu^{-1}(\mathbb{R}^+) = (-1, 1)$ where $\mathbb{R}^+ = \{x \in \mathbb{R} \mid x > 0\}$.

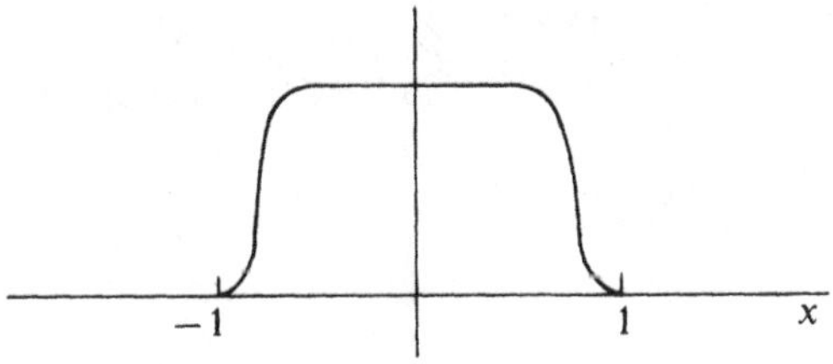

Fig. 4.8

As a manifold M, we take $\mathbb{R}^2$ and define a smooth vector field X on $\mathbb{R}^2$ by

$$X(x, y) = \mu(x^2 + y^2)xy(x, y).$$

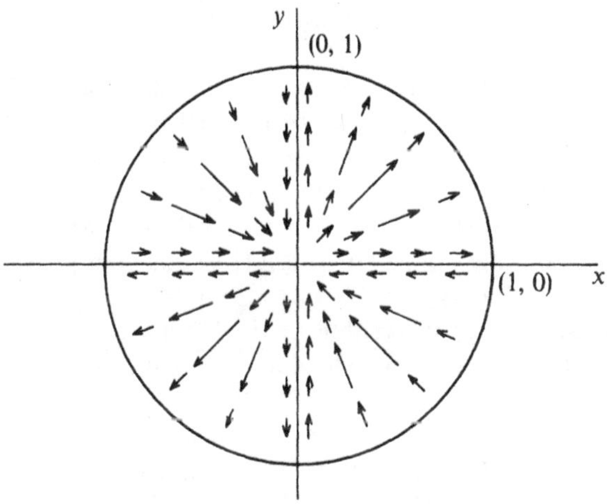

Fig. 4.9

Obviously X is a smooth vector field which vanishes outside of the unit circle. This is illustrated in Fig. 4.9. Since the unit disc is compact, it follows from Theorem 3.19 that X generates a unique 1-parameter group of diffeomorphisms of M. The smooth $\mathbb{R}$-action obtained in this way satisfies

$$M^{\mathbb{R}} = \{(x, y) \in M \mid X(x, y) = 0\}$$

by Corollary 3.21. The right-hand side of this equality consists of the x-axis, the y-axis, the outside of the unit circle, and the unit circle itself (see Fig. 4.10).

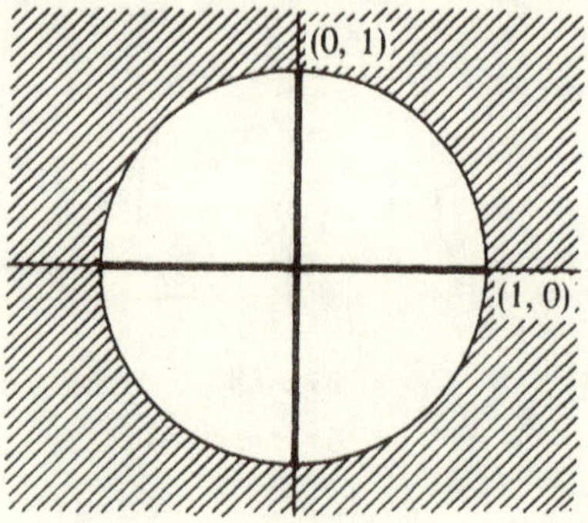

Fig. 4.10

Then $M^{\mathbb{R}}$ cannot be a manifold at the points (0, 0), (1, 0), (− 1, 0), (0, 1), (0, − 1) even if we admit the boundaries and corners in the definition of manifolds.

Thus the fixed point set $M^{\mathbb{R}}$ is not a submanifold of M, which means that the compactness assumption cannot be omitted in Theorem 4.14.

Remark If G is a compact Lie group and M is a G-manifold, then the fixed point set M^G is a G-invariant closed submanifold of M by Theorem 4.14. Then the normal bundle ν of M^G in M is a G-vector bundle over the trivial G-space M^G. This is the type of G-vector bundle which we dealt with in Theorem 2.38 and it can be written as the Whitney sum of G-vector sub bundles indexed by the set of irreducible representations.[68]

Lemma 4.15 *Let G be a Lie group, H a compact subgroup of G, and M a G-manifold. Then $N(H)/H$ becomes a Lie group and the G-action on M induces a smooth $N(H)/H$-action on M^H so that M^H is an $N(H)/H$-manifold.*

Proof It follows from the above theorem that M^H is a submanifold of M. Since H is compact, hence a closed subgroup of G, it follows from Proposition 1.10 that $N(H)$ is also a closed subgroup of G. Accordingly $N(H)$ becomes a Lie group by Theorem 3.36 and hence $N(H)/H$ becomes a Lie group as well by Corollary 3.41. As shown in Section 1.5, M^H is an $N(H)/H$-space. Therefore it suffices to show that the $N(H)/H$-action is smooth.

Evidently the restriction of the G-action on M to $N(H)$ is smooth. In view of Lemma 1.50, M^H is $N(H)$-invariant. Hence M^H is an $N(H)$-manifold. Denote by

$$\varphi: N(H) \times M^H \to M^H$$

the restricted $N(H)$-action. Then φ is smooth and we have the following commutative diagram

$$\begin{array}{ccc} N(H)\times M^H & \xrightarrow{\ \varphi\ } & M^H \\ \downarrow{\scriptstyle \pi\times \mathrm{id}} & & \downarrow{\scriptstyle \mathrm{id}} \\ N(H)/H\times M^H & \xrightarrow{\ \overline{\varphi}\ } & M^H \end{array}$$

where $\overline{\varphi}$ denotes the $N(H)/H$-action induced by φ. According to Corollary 3.38, the projection $\pi\colon N(H)\to N(H)/H$ has a smooth local cross-section. Hence it follows from the above commutative diagram that $\overline{\varphi}$ is smooth. □

Lemma 4.16 *Let G be a group, H a subgroup of G, and X an H-space. Then the isotropy group $G_{[g,x]}$ at $[g,x]$ of the induced G-space $G\underset{H}{\times}X$ defined in Section 1.9 is given by*

$$G_{[g,x]}=gH_xg^{-1}.$$

Proof Since an arbitrary element g' of $G_{[g,x]}$ satisfies

$$[g,x]=g'\circ[g,x]=[g'g,x],$$

there exists $h\in H$ with $g'g=gh^{-1}$, $x=hx$. Hence we have $g'=gh^{-1}g^{-1}\in gH_xg^{-1}$, that is, $G_{[g,x]}\subset gH_xg^{-1}$ holds. Conversely, for any $h\in H_x$ we have

$$(ghg^{-1})\circ[g,x]=[gh,x]=[g,hx]=[g,x],$$

that is, $G_{[g,x]}\supset gH_xg^{-1}$ holds. □

Lemma 4.17 *Let G be a compact group, H a closed subgroup of G and X be a T_1-space as well as an H-space. Then we have*

$$\left(G\underset{H}{\times}X\right)(H)=G\underset{H}{\times}X^H=G/H\times X^H$$

where $(G\underset{H}{\times}X)(H)$ is the union of the orbits of type G/H.

Proof Suppose $[g,x]\in(G\underset{H}{\times}X)(H)$. Then it follows from the preceding lemma that

$$(H)=(G_{[g,x]})=(gH_xg^{-1})=(H_x).$$

That is, there exists $g_0\in G$ with $H=g_0H_xg_0^{-1}$. Hence we have $H_x\subset H=g_0H_xg_0^{-1}$. Since X is a T_1-space, it follows from Theorem 1.37 that H_x is a closed subgroup of H. Moreover H is a closed subgroup of G by assumption, so that H_x is a closed subgroup of G. Furthermore, since G is compact, it follows from Corollary 1.70 that

$$H_x = g_0 H_x g_0^{-1} = H,$$

that is, $x \in X^H$. Thus we have

$$\left(G \underset{H}{\times} X\right)(H) \subset G \underset{H}{\times} X^H = G/H \times X^H.$$

Conversely, for an arbitrary $[g, x] \in G \underset{H}{\times} X^H$, we have

$$G_{[g,x]} = gH_x g^{-1} = gHg^{-1},$$

that is, $[g, x] \in (G \underset{H}{\times} X)(H)$. Thus the reverse inclusion holds. □

We are now ready to state the structure theorem in the special case where there is precisely one orbit type.

Theorem 4.18 (The structure theorem 1) *Let G be a compact Lie group and M a G-manifold. Suppose that every orbit in M has type G/H. Then the orbit space M/G is a smooth manifold and the projection π: $M \to M/G$ is a smooth fibre bundle with fibre G/H and structure group $N(H)/H$. A result similar to Theorem 4.11 holds for the smooth structure on M/G. Moreover, π_1: $M^H \to M^H/(N(H)/H)$ is a smooth principal $N(H)/H$-bundle. Regarding G/H as a right $N(H)/H$-space and M^H as a left $N(H)/H$-space, we consider the following bundle:*

$$G/H \underset{N(H)/H}{\times} M^H \to M^H/(N(H)/H)$$

associated with π_1. Then this bundle is isomorphic to π: $M \to M/G$ as smooth fibre bundles with G-actions. In particular, M is G-diffeomorphic to $G/H \underset{N(H)/H}{\times} M^H$, which is G-diffeomorphic to $G \underset{N(H)}{\times} M^H$ as well.

Proof By assumption, each orbit contains a point x with $G_x = H$. It follows from the slice theorem that there exists an H-representation space V such that $G \underset{H}{\times} V$ is G-diffeomorphic to a G-invariant tubular neighbourhood of the orbit. Since the isotropy type of the isotropy group at every point of $G \underset{H}{\times} V$ is (H) by assumption, we have

$$G \underset{H}{\times} V = \left(G \underset{H}{\times} V\right)(H).$$

Accordingly if we take V as X in Lemma 4.17, then we have

$$G \underset{H}{\times} V = G \underset{H}{\times} V^H.$$

From this, we shall show that $V = V^H$. Note that V^H is a vector subspace of V and both $G \underset{H}{\times} V^H$ and $G \underset{H}{\times} V$ are G-vector bundles over G/H. Hence the inclusion

$$i\colon G \underset{H}{\times} V^H \to G \underset{H}{\times} V$$

is a bijective bundle homomorphism, that is, i is a bundle isomorphism. Since the fibres of them are V^H and V, respectively, the equality $V^H = V$ holds. Thus we have shown the following claim.

Claim 1 Every slice representation V of a G-manifold with one orbit type, say G/H, is trivial and a G-invariant tubular neighbourhood of each orbit is G-diffeomorphic to $G/H \times V$.

Next we show that M/G becomes a manifold. Let

$$f\colon G/H \times V \to M$$

be a G-invariant tubular neighbourhood obtained above. The map

$$\bar{f}\colon (G/H \times V)/G \to M/G$$

between the orbit spaces induced by f is a homeomorphism into M/G. Define a map $j\colon V \to G/H \times V$ by $j(v) = (eH, v)$ and denote by $\pi\colon G/H \times V \to (G/H \times V)/G$ the projection map. Evidently the composition

$$k = \pi \cdot j\colon V \to (G/H \times V)/G$$

is a homeomorphism.

Putting all this together, we have the following commutative diagram:

$$\begin{array}{ccccc}
V & \xrightarrow{\ j\ } & G/H \times V & \xrightarrow{\quad f\quad } & M \\
 & \searrow^{k} & \downarrow{\scriptstyle\pi} & & \downarrow{\scriptstyle\pi} \\
 & & (G/H \times V)/G & \xrightarrow{\quad \bar{f}\quad } & M/G
\end{array}$$

That is, we have the following equalities:

$$\bar{f} \cdot k = \bar{f} \cdot \pi \cdot j = \pi \cdot f \cdot j.$$

We now denote the image $\bar{f} \cdot k(V)$ by $[V]$ and $\bar{f} \cdot k(v)$ for $v \in V$ by $[v]$. Since $[V]$ coincides with $\pi \cdot f(G/H \times V)$ and π is an open map, it follows that $[V]$ is an open set of M/G and $\bar{f} \cdot k\colon V \to [V]$ is a homeomorphism.

In the following, we shall show that M/G becomes a smooth manifold with charts

$$\psi = (\bar{f} \cdot k)^{-1}\colon [V] \to V.$$

If we apply the slice theorem to each orbit, then $\{[V]\}$ is an open covering of M/G. Suppose that two such G-embeddings

$$f_i\colon G/H \times V_i \to M, \qquad i = 1, 2$$

obtained by the slice theorem satisfy $[V_1] \cap [V_2] \neq \emptyset$. As we considered $\bar{f}, j, \pi, k, \psi$ corresponding to f, we consider $\bar{f}_i, j_i, \pi_i, k_i, \psi_i$ corresponding to f_i $(i = 1, 2)$. Then it can be shown by the same argument as in the case of free actions in Theorem 4.11 that the map

$$\psi_2 \cdot \psi_1^{-1}\colon \psi_1([V_1] \cap [V_2]) \to \psi_2([V_1] \cap [V_2])$$

is smooth. Hence $\{([V], \psi)\}$ is a smooth atlas on M/G. Moreover we see that M/G satisfies the second countability axiom by Corollary 1.56 and is Hausdorff by (2) of Proposition 1.58.

Thus we have shown that M/G is a smooth manifold with atlas $\{([V], \psi)\}$. Further, it can be shown in a similar fashion to the proof of Theorem 4.11 that the quotient map $\pi\colon M \to M/G$ is smooth.

Next we shall show that $\pi\colon M \to M/G$ is a smooth fibre bundle with fibre G/H and structure group $N(H)/H$. Let $q_i\colon G/H \times V_i \to G/H$ denote the projections. Define maps

$$\varphi_i\colon \pi^{-1}([V_i]) = f_i(G/H \times V_i) \to G/H \times [V_i]$$

by

$$\varphi_i(x) = (q_i \cdot f_i^{-1}(x), \pi(x))$$

for $x \in \pi^{-1}([V_i]) = f_i(G/H \times V_i)$, $i = 1, 2$. Since π is smooth as shown above, the φ_i are smooth. Further, the following diagram commutes:

$$\begin{array}{ccc} \pi^{-1}([V_i]) & \xrightarrow{\varphi_i} & G/H \times [V_i] \\ & {\scriptstyle \pi}\searrow \quad \swarrow{\scriptstyle \bar{p}_i} & \\ & [V_i] & \end{array}$$

where the $\bar{p}_i$ are the projections. Note that

$$\varphi_i \cdot f_i(gH, v_i) = (gH, [v_i])$$

for $v_i \in V_i$, $i = 1, 2$. Since the f_i are bijective, each φ_i is also bijective and the φ_i^{-1} are given by

$$\varphi_i^{-1}(gH, [v_i]) = f_i(gH, v_i) = f_i(gH, \psi_i([v_i])).$$

Consequently we see that the φ_i are diffeomorphisms.

On the other hand, we define a map

$$\theta\colon [V_1] \cap [V_2] \to G/H$$

by the composition $q_2 \cdot f_2^{-1} \cdot f_1 \cdot j_1 \cdot \psi_1 | ([V_1] \cap [V_2])$. Since all the maps in the above definition are smooth, θ is smooth as well.

In the following we show that the image of θ is included in $N(H)/H$. If elements v_i of V_i satisfy $[v_1] = [v_2]$, then we have

$$\theta([v_1]) = q_2 \cdot f_2^{-1} \cdot f_1 \cdot j_1 \cdot \psi_1([v_1]) = q_2 \cdot f_2^{-1} \cdot f_1(eH, v_1).$$

Accordingly we have

$$f_2^{-1} \cdot f_1(eH, v_1) = (\theta([v_1]), v_2).$$

Put $\theta([v_1]) = gH$. Since the f_i are G-diffeomorphisms into M, we have

$$H = G_{(eH, v_1)} = G_{(gH, v_2)} = G_{[g, v_2]} = gH_{v_2}g^{-1} = gHg^{-1}$$

where we identified $G/H \times V_2$ with $G \times_H V_2$ in the third equality. Thus we have shown that $g \in N(H)$, that is, the image of θ is included in $N(H)/H$. We now regard θ as the map:

$$\theta\colon [V_1] \cap [V_2] \to N(H)/H$$

obtained by restricting the range to $N(H)/H$.

Next we compute the map

$$\varphi_2 \cdot \varphi_1^{-1}\colon G/H \times ([V_1] \cap [V_2]) \to G/H \times ([V_1] \cap [V_2]).$$

By making use of the fact that both f_i and q_2 are G-maps, we have for $v_1 \in V_1$ with $[v_1] \in [V_1] \cap [V_2]$,

$$\begin{aligned}
\varphi_2 \cdot \varphi_1^{-1}(gH, [v_1]) &= \varphi_2 \cdot f_1(gH, v_1) \\
&= (q_2 \cdot f_2^{-1} \cdot f_1(gH, v_1), \pi \cdot f_1(gH, v_1)) \\
&= (gq_2 \cdot f_2^{-1} \cdot f_1(eH, v_1), \pi \cdot gf_1(eH, v_1)) \\
&= (g\theta([v_1]), \pi \cdot f_1 \cdot j_1(v_1)) \\
&= (gH\theta([v_1]), [v_1]).
\end{aligned}$$

Thus we have shown that $\pi\colon M \to M/G$ becomes a smooth fibre bundle with fibre G/H, coordinate functions φ_i, structure group $N(H)/H$, and transition functions θ.

Since $\pi\colon M \to M/G$ is a smooth fibre bundle as shown above, π has a smooth local cross-section. Accordingly, a result similar to the statement concerning smooth structure in Theorem 4.11 is verified.

Next we show that $\pi_1\colon M^H \to M^H/(N(H)/H)$ is a smooth principal $N(H)/H$-bundle. In view of Lemma 4.15, M^H is an $N(H)/H$-manifold.

Claim 2 For any point x of M^H, the equality $G_x = H$ holds, and hence the $N(H)/H$-action on M^H is free.

Since every orbit in M has type G/H by assumption, there exists $g \in G$ with $G_{gx} = H$. Since $G_{gx} = gG_xg^{-1}$, we have $G_x = g^{-1}Hg$. On the other hand,

$x \in M^H$ means that $G_x \supset H$. Accordingly we have $g^{-1}Hg = G_x \supset H$. Since G is compact and H is a closed subgroup of G, it follows from Corollary 1.70 that $G_x = g^{-1}Hg = H$, which is just the claim.

Thus $\pi_1\colon M^H \to M^H/(N(H)/H)$ becomes a smooth principal $N(H)/H$-bundle by virtue of Theorem 4.11.

We now show that two orbit spaces $M^H/(N(H)/H)$ and M/G are canonically diffeomorphic. To see this, consider the following commutative diagram:

$$\begin{array}{ccc} M^H & \xrightarrow{\quad i \quad} & M \\ \Big\downarrow{\scriptstyle \pi_1} & & \Big\downarrow{\scriptstyle \pi} \\ M^H/(N(H)/H) & \xrightarrow{\quad \bar{i} \quad} & M/G \end{array}$$

where i is the inclusion and $\bar{i}$ is the map between the orbit spaces induced by i. It follows from Theorem 4.14 that the inclusion i gives a diffeomorphism onto its image. Since each G-orbit in M contains an H-fixed point, $\bar{i}$ is surjective. If we suppose that two points x and y of M^H satisfy $\pi \cdot i(x) = \pi \cdot i(y)$, then there exists $g \in G$ with $y = gx$. Then it follows from Claim 2 above that

$$H = G_y = G_{gx} = gG_xg^{-1} = gHg^{-1}.$$

Hence we have $g \in N(H)$, so that $\pi_1(x) = \pi_1(y)$, which means that $\bar{i}$ is injective.

Since π_1 has a smooth local cross-section, $\bar{i}$ is smooth. To see that $\bar{i}^{-1}$ is smooth, it suffices to show that π has a smooth local cross-section whose image is included in M^H. For this, by making use of the previous notation, we define a map $s\colon [V] \to \pi^{-1}([V])$ by

$$s([v]) = f \cdot j \cdot \psi([v]) = f(eH, v) \qquad \text{for } [v] \in [V].$$

Then it is smooth and satisfies

$$\pi \cdot s([v]) = \pi \cdot f \cdot j \cdot \psi([v]) = \psi^{-1} \cdot \psi([v]) = [v],$$

that is, s is a smooth local cross-section of π. Since f is a G-diffeomorphism into M and satisfies $G_{(eH, v)} = H$, the image of s is included in M^H. Thus we have shown that $\bar{i}$ is a diffeomorphism.

Next we shall show that the fibre bundles $\pi'\colon G/H \times_{N(H)/H} M^H \to M^H/(N(H)/H)$ and $\pi\colon M \to M/G$ are isomorphic as smooth fibre bundles with G-actions. Here $\pi'\colon G/H \times_{N(H)/H} M^H \to M^H/(N(H)/H)$ is the bundle with fibre G/H associated with $\pi_1\colon M^H \to M^H/(N(H)/H)$. In the following, we shall make use of the notation of Section 4.4 freely.

Consider the following commutative diagram:

$$\begin{array}{ccc} G/H \underset{N(H)/H}{\times} M^H & \xrightarrow{\tilde{\imath}} & M \\ \Big\downarrow{\scriptstyle \pi'} & & \Big\downarrow{\scriptstyle \pi} \\ M^H/(N(H)/H) & \xrightarrow{\bar{\imath}} & M/G \end{array}$$

where $\tilde{\imath}$ is the G-map defined by $\tilde{\imath}([gH, x]) = gx$. Let

$$f': N(H)/H \times V' \to M^H$$

be an $N(H)/H$-invariant tubular neighbourhood of an $N(H)/H$-orbit in M^H and let

$$\varphi': \pi_1^{-1}([V']) \to N(H)/H \times [V']$$

denote the coordinate function of the principal bundle π_1: $M^H \to M^H/(N(H)/H)$ obtained from f' as above. Then the corresponding coordinate function of π' associated with π_1 is given by the composition of the following maps as defined in Section 2.3:

$$\Phi: \pi'^{-1}([V']) \xrightarrow{\varphi'_1} G/H \underset{N(H)/H}{\times} \pi_1^{-1}([V']) \xrightarrow{\varphi'_2} G/H \underset{N(H)/H}{\times} (N(H)/H \times [V'])$$
$$\xrightarrow{\varphi'_3} \left(G/H \underset{N(H)/H}{\times} N(H)/H \right) \times [V'] \xrightarrow{\varphi'_4} G/H \times [V'].$$

Let f: $G/H \times V \to M$ be a G-invariant tubular neighbourhood of a G-orbit in M and let

$$\varphi: \pi^{-1}([V]) \to G/H \times [V]$$

denote the coordinate function of the bundle π: $M \to M/G$ obtained from f. As we considered j, ψ corresponding to f, we consider j', ψ' corresponding to f'. Then we compute the composition

$$\varphi \cdot \tilde{\imath} \cdot \Phi^{-1}: G/H \times ([V'] \cap \bar{\imath}^{-1}([V])) \to G/H \times (\bar{\imath}([V']) \cap [V]).$$

For any $g \in G$ and $v' \in V'$ with $[v'] \in [V'] \cap \bar{\imath}^{-1}([V])$, we have

$$\begin{aligned} \varphi \cdot \tilde{\imath} \cdot \Phi^{-1}(gH, [v']) &= \varphi \cdot \tilde{\imath}([gH, f'(eH, v')]) \\ &= \varphi(gf'(eH, v')) = (q \cdot f^{-1} \cdot gf'(eH, v'), \bar{\imath}([v'])) \\ &= (gq \cdot f^{-1} \cdot f'(eH, v'), \bar{\imath}([v'])) = (gH\theta([v']), \bar{\imath}([v'])). \end{aligned}$$

Here q: $G/H \times V \to G/H$ denotes the projection and

$$\theta: [V'] \cap \bar{\imath}^{-1}([V]) \to G/H$$

is the smooth map defined by

$$\theta([v']) = q \cdot f^{-1} \cdot f'(eH, v') = q \cdot f^{-1} \cdot f' \cdot j' \cdot \psi'([v']).$$

As in the previous case, we can show that the image of θ is included in $N(H)/H$. Hence $\tilde{i}$ is a smooth bundle map preserving the G-actions. Since the map $\bar{i}$ between the base spaces is a diffeomorphism, it is easily seen that $\tilde{i}^{-1}$ is also a smooth bundle map.

Thus we have shown that $\tilde{i}$ is a smooth bundle isomorphism preserving the G-actions. In particular, $\tilde{i}$ gives a G-diffeomorphism between $G/H \underset{N(H)/H}{\times} M^H$ and M.

According to Corollary 1.92, there is a canonical G-homeomorphism

$$F\colon G \underset{N(H)}{\times} M^H \to G/H \underset{N(H)/H}{\times} M^H.$$

Then we consider the following commutative diagram:

$$\begin{array}{ccc} G \times M^H & \longrightarrow & G \underset{N(H)}{\times} M^H \\ \downarrow & & \downarrow F \\ G \underset{H}{\times} M^H = G/H \times M^H & \longrightarrow & G/H \underset{N(H)/H}{\times} M^H \end{array}$$

Since all the maps except for F in the diagram above are projections to orbit spaces of free actions, it follows from Theorem 4.11 that these maps are smooth and have smooth local cross-sections. Accordingly, both F and F^{-1} are smooth, and hence F is a G-diffeomorphism.

This makes the proof of the theorem complete. □

Theorem 4.19 (The structure theorem 2) *Let G be a compact Lie group, H a closed subgroup of G, and M a G-manifold. The union $M(H)$ of the orbits of type G/H is a G-invariant submanifold of M. The orbit space $M(H)/G$ is a smooth manifold and the projection π: $M(H) \to M(H)/G$ is a smooth fibre bundle with fibre G/H and structure group $N(H)/H$. Furthermore, $\pi_1\colon M(H)^H \to M(H)^H/(N(H)/H)$ is a smooth principal $N(H)/H$-bundle. Regarding G/H as a right $N(H)/H$-space and $M(H)^H$ as a left $N(H)/H$-space, we consider the following bundle:*

$$G/H \underset{N(H)/H}{\times} M(H)^H \to M(H)^H/(N(H)/H)$$

associated with π_1. Then the bundle is isomorphic to π: $M(H) \to M(H)/G$ as smooth fibre bundles with G-actions. In particular, $M(H)$ is G-diffeomorphic to $G/H \underset{N(H)/H}{\times} M(H)^H$, which is G-diffeomorphic to $G \underset{N(H)}{\times} M(H)^H$ as well.

Proof Each orbit of type G/H contains a point x with $G_x = H$. On account of the slice theorem, there exists a representation space V of H such that $G \times_H V$ is G-diffeomorphic to a G-invariant tubular neighbourhood of $G(x)$ in M, and we have

$$\left(G \times_H V\right)(H) = G \times_H V^H = G/H \times V^H$$

by Lemma 4.17. Since V^H is a vector subspace of V, $G \times_H V^H$ is a vector subbundle of $G \times_H V$. We now apply Theorem 2.21 to deduce that $G \times_H V^H$ is a G-invariant submanifold of $G \times_H V$. It turns out that $M(H)$ is a G-invariant submanifold of M. Since we now know that $M(H)$ is a G-manifold, the rest of the statement of the theorem is immediate from the preceding theorem. □

Remark $M(H)$ is not necessarily connected and the dimensions of its components are not necessarily constant. Even if M is compact, $M(H)$ is not compact in general.

4.6 Decomposition and blowing up of G-manifolds

In the preceding section, we investigated the structure of the union $M(H)$ of the orbits of type G/H for a G-manifold M. Although $M(H)$ is a submanifold of M as observed there, $M(H)$ is not necessarily compact in general. The purpose of the present section is to investigate how these $M(H)$ fit together.

We now describe this more precisely. Given a G-manifold M, we carefully order the set of isotropy types so that we decompose M into the union of the disc bundles over compact G-invariant submanifolds of M with corners with one orbit type.

This decomposition theorem will be quite useful when we investigate formulae for Euler characteristics (see Section 5.6). It is also useful in the investigation of equivariant h-cobordisms, equivariant s-cobordisms, and equivariant homotopy types.[3,67,71] Moreover, owing to this decomposition theorem, we have stratifications of G-manifolds with respect to orbit types and it can be shown that G-manifolds have the structures of G-CW complexes.[67]

The union $M(\geq H)$ of the orbits whose isotropy types are greater than or equal to (H) is closed and G-invariant in M, but it is not a submanifold in general. Then by replacing each orbit of $M(\geq H)$ by a slightly larger orbit if necessary, we obtain a manifold, which is something like the resolution of singularities in algebraic geometry. That is, $M(\geq H)$ is approximated by a G-manifold.

In order to state the decomposition theorem, we must define several terms. We extend the definition of manifolds to 'manifolds with boundaries' and further, 'manifolds with corners'. In comparison with these manifolds, the manifolds we have dealt with so far are frequently emphasized as 'manifolds without boundaries'.

First, for $i \geq 1$, we put

$$\mathbb{R}^n_i = \{(x_1, \ldots, x_n) \in \mathbb{R}^n \mid x_{n-i+1} \geq 0, \ldots, x_n \geq 0\}.$$

Then there are natural inclusions

$$\mathbb{R}^n \supset \mathbb{R}^n_1 \supset \mathbb{R}^n_2 \supset \cdots.$$

The manifolds we dealt with so far are covered with copies of open sets of $\mathbb{R}^n$. On the contrary, the manifolds covered with copies of open sets of $\mathbb{R}^n_1$ (resp. $\mathbb{R}^n_i$, $i \geq 2$) are called *manifolds with boundaries* (resp. *manifolds with corners*). These are illustrated in Fig. 4.11.

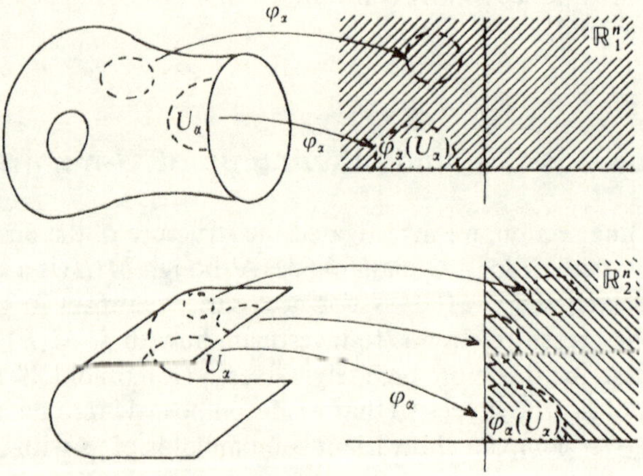

Fig. 4.11 Manifolds with boundaries and manifolds with corners.

The following definitions are important. We define an atlas $\{(U_\alpha, \varphi_\alpha)\}$ to be smooth in this case as follows. When we consider a map

$$\varphi_\beta \cdot \varphi_\alpha^{-1} : \varphi_\alpha(U_\alpha \cap U_\beta) \to \varphi_\beta(U_\alpha \cap U_\beta),$$

$\varphi_\alpha(U_\alpha \cap U_\beta)$ and $\varphi_\beta(U_\alpha \cap U_\beta)$ are open sets of $\mathbb{R}^n_i$, but are not necessarily open in $\mathbb{R}^n$. When the map $\varphi_\beta \cdot \varphi_\alpha^{-1}$ can be extended to a smooth map $f: V \to \mathbb{R}^n$ where V is an open set of $\mathbb{R}^n$ including $\varphi_\alpha(U_\alpha \cap U_\beta)$, we say that the map $\varphi_\beta \cdot \varphi_\alpha^{-1}$ is smooth.

The notions of a smooth map between manifolds with corners and of a smooth G-action on a manifold with corners are defined in the same way.

We now identify $\mathbb{R}^{n-1}$ with $\{(x_1, \ldots, x_n) \in \mathbb{R}^n \mid x_n = 0\}$. The set of points which correspond to points of $\mathbb{R}^{n-1}$ via charts is called the *boundary of M*, denoted by ∂M. It is well known that the definition does not depend on the choice of charts.

Furthermore we identify $\mathbb{R}^{n-i}$ with $\{(x_1, \ldots, x_n) \in \mathbb{R}^n \mid x_{n-i+1} = 0, \ldots, x_n = 0\}$. The set of points which correspond to points of $\mathbb{R}^{n-i}$ via charts is called the *corner of M*.

We denote by $\partial\mathbb{R}^n_i$ the set of boundary points of $\mathbb{R}^n_i$ in $\mathbb{R}^n$ in the sense of a topological space and its subspace. That is, we have

$$\partial\mathbb{R}^n_i = \{x_1, \ldots, x_n\} \in \mathbb{R}^n_i \mid n - i + 1 \leq \exists j \leq n \text{ with } x_j = 0\}$$

and $\mathbb{R}^{n-i} \subset \partial\mathbb{R}^n_i \subset \mathbb{R}^n_i \subset \mathbb{R}^n$.

The sets of points which correspond to points of $\partial\mathbb{R}^n_i$ via charts is called the *boundary of M*, denoted by ∂M (see Fig. 4.12).

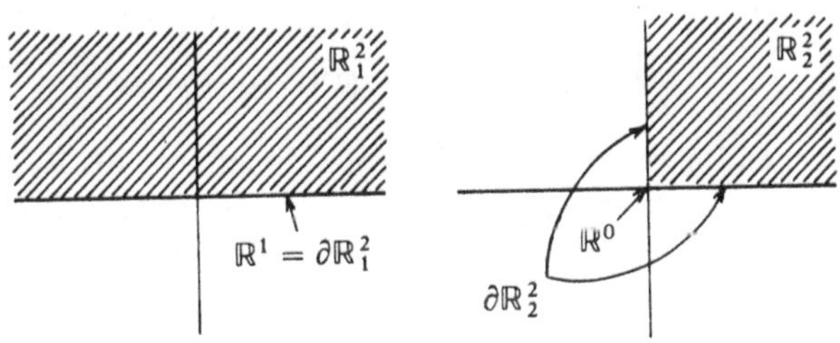

Fig. 4.12 The boundary of M.

We are now ready to describe the decomposition of G-manifolds announced at the beginning of this section.

Let G be a compact Lie group and M a compact G-manifold. As will be shown in the next section, the number of isotropy types occurring in M is finite. Denote by $\{(H_i)\} \mid i = 1, \ldots, n\}$ the set of isotropy types occurring in M. We arrange them such that

$$(H_i) \geq (H_j) \Rightarrow i \leq j$$

by Lemma 1.80.

First we consider $M(H_1)$. Since $M(H_1)$ is closed in M by Corollary 1.78 and is a G-invariant submanifold, it is a compact G-manifold without boundary. Denote by ν_1 the normal G-vector bundle of $M(H_1)$ in M and by $f_1 : \nu_1 \to M$ a G-invariant tubular neighbourhood of $M(H_1)$ in M. Take a G-invariant metric on ν_1 and denote by $\mathring{D}(\nu_1)$ the open unit disc bundle of ν_1 with respect to the metric. Then we put

$$M_2 = M - f_1(\mathring{D}(\nu_1))$$

which is a compact G-invariant submanifold of M with boundary. There no longer exists an orbit in M_2 having the isotropy type (H_1) and the set of isotropy types occurring in M_2 is exactly $\{(H_2), (H_3), \ldots, (H_n)\}$.

Next we consider the union $M_2(H_2)$ of the orbits in M_2 having the isotropy type (H_2). Then $M_2(H_2)$ is generally different from $M(H_2)$ and is a compact G-invariant submanifold of M_2 with boundary. Take a G-invariant Riemannian metric on M_2 such that the metric has the form $\partial M_2 \times I$ in the neighbourhood of ∂M_2 in M_2. The existence of such a metric is seen as follows. In the same way that the existence of a G-invariant tubular neighbourhood was proved, it is shown that there exists a G-invariant neighbourhood of ∂M_2 in M_2 which has the form $\partial M_2 \times I$. This is called a *G-invariant collar neighbourhood* of ∂M_2 in M_2. By making use of this neighbourhood, we derive the desired metric. Denote by $f_2 : v_2 \to M_2$ a G-invariant tubular neighbourhood of $M_2(H_2)$ in M_2. Notice that the notion of G-invariant tubular neighbourhood in the case of 'with boundaries' is similar to that in the case of 'without boundaries' provided we pay attention to the metric near the boundary. Taking a G-invariant metric on v_2, we put

$$M_3 = M_2 - f_2(\mathring{D}(v_2)),$$

which is in general a compact G-invariant submanifold of M_2 with corner. The set of isotropy types occurring in M_3 is exactly $\{(H_3), (H_4), \ldots, (H_n)\}$.

Paying attention to corners, we continue this procedure and have a decomposition

$$M = f_1(D(v_1)) \cup f_2(D(v_2)) \cup \cdots \cup f_n(D(v_n))$$

of M. The last term $f_n(D(v_n))$ is in fact a compact manifold M_n with corner with one isotropy type (H_n), that is, M_n is regarded as a 0-dimensional disc bundle over M_n.

Thus we have the following theorem.

Theorem 4.20 (Decomposition theorem)[65] *Let G be a compact Lie group and M a compact G-manifold. We arrange the set $\{(H_1), \ldots, (H_n)\}$ of isotropy types occurring in M such that*

$$(H_i) \geq (H_j) \Rightarrow i \leq j.$$

Then there exist a compact G-manifold N_i with corner having only the isotropy type (H_i) and a G-vector bundle v_i over N_i for each $i = 1, 2, \ldots, n$ such that we have a decomposition

$$M = D(v_1) \cup D(v_2) \cup \cdots \cup D(v_n)$$

and such that the set of isotropy types occurring in the G-invariant submanifold $D(v_i) \cup D(v_{i+1}) \cup \cdots D(v_n)$ is exactly $\{(H_i), (H_{i+1}), \ldots, (H_n)\}$ or equivalently the isotropy types $\{(H_1), \ldots, (H_i)\}$ occur only in the G-invariant *submanifold*

$$D(v_1) \cup \cdots \cup D(v_i) \qquad \text{for each } i = 1, \ldots, n.$$

Moreover $N_i - \partial N_i$ *is G-diffeomorphic to* $M(H_i)$.

Proof In the explanation preceding the theorem, we need only put $N_1 = M(H_1)$ and $N_i = M_i(H_i)$ for $i \geq 2$. Moreover, since $f_i: v_i \to M_i$ is a G-embedding, $f_i(D(v_i))$ is G-diffeomorphic to $D(v_i)$. By making use of the equivariant collar theorem cited above, one verifies easily that there is a G-diffeomorphism

$$N_i - \partial N_i \cong M(H_i).$$

The rest of the proof is indicated in the above explanation and is left to the reader as an exercise.

Example 1 Let $G = \mathbb{Z}_2 \oplus \mathbb{Z}_2$, and g_1 and g_2 be the generators of $\mathbb{Z}_2 \oplus \{0\}$ and $\{0\} \oplus \mathbb{Z}_2$, respectively. We take

$$M = S^2 = \{(x, y, z) \in \mathbb{R}^3 \mid x^2 + y^2 + z^2 = 1\}$$

and define a G-action on M by

$$g_1 \circ (x, y, z) = (-x, y, z), \quad g_2 \circ (x, y, z) = (x, -y, z).$$

Then the isotropy types occurring in M are

$$H_1 = G, \qquad H_2 = \mathbb{Z}_2 \oplus \{0\}, \qquad H_3 = \{0\} \oplus \mathbb{Z}_2, \qquad H_4 = \{0\} \oplus \{0\}$$

where we have abbreviated parentheses, for G is commutative. It is easy to see that

$$M(H_1) = \{(0, 0, 1), (0, 0, -1)\} = (\text{two points}),$$

$$M(H_2) = \{(0, y, z) \mid y^2 + z^2 = 1, z \neq \pm 1\} \cong S^1 - (\text{two points}),$$

$$M(H_3) = \{(x, 0, z) \mid x^2 + z^2 = 1, z \neq \pm 1\} \cong S^1 - (\text{two points}),$$

$$M(H_4) = M - M(H_1) - M(H_2) - M(H_3).$$

By the way, we can write down the fixed point sets of all the subgroups of G as:

$$M^{H_1} = M(H_1) = \{(0, 0, 1), (0, 0, -1)\},$$

$$M^{H_2} = M(H_1) \cup M(H_2) \cong S^1,$$

$$M^{H_3} = M(H_1) \cup M(H_3) \cong S^1,$$

$$M^{H_4} = M.$$

All these are collected and illustrated in Fig. 4.13. Finally, in Fig. 4.14 we add to Fig. 4.13 the decomposition of M stated in the theorem.

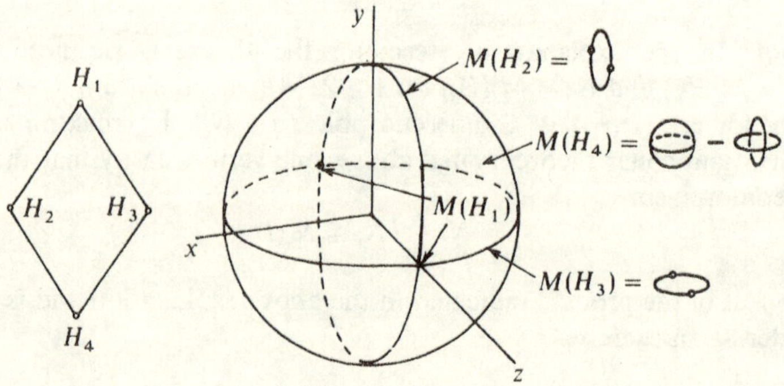

Fig. 4.13

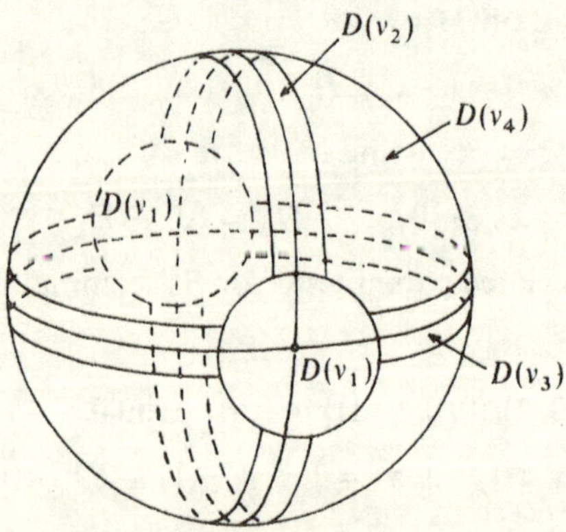

Fig. 4.14

As a clue to investigating a G-manifold M, there are subspaces

$$M^H, \qquad M(H), \qquad M(\geq H), \qquad D(v_i).$$

However they have both strong and weak points. Each of them is indispensable and it is necessary to make use of them by combining them well. We now collect the strong and weak points of them in Table 4.1.

Let G be a compact Lie group, H a closed subgroup of G, and M a compact G-manifold.

Table 4.1

	Pros	Cons
M^H	compact submanifold	not G-invariant
$M(H)$	G-invariant submanifold	not necessarily compact, maps fitting them together are not clear
$M(\geq H)$	compact, G-invariant	not necessarily submanifold
$D(\nu_i)$	compact, G-invariant submanifold, maps fitting them together are easy to see, base space of ν_i has one orbit type	with corners, involves many orbit types

We shall now discuss blowing up.

Construction Let G be a compact Lie group and H a closed subgroup of G. Given a G-manifold M, the H-fixed point set M^H is an $N(H)$-manifold. Since the inclusion map $j: M^H \to M$ is an $N(H)$-map, it follows from Theorem 1.83 that we have the following commutative diagram:

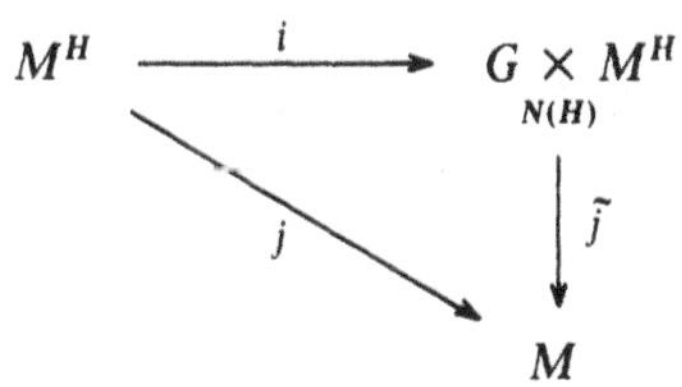

Then we prove the following theorem.

Theorem 4.21 (Blowing up)

(1) $G \underset{N(H)}{\times} M^H \to G/N(H)$ *is a smooth fibre bundle with fibre* M^H. *In particular,* $G \underset{N(H)}{\times} M^H$ *becomes a smooth* G*-manifold and* $\tilde{j}$ *is a smooth* G*-map.*

(2) $\operatorname{Im}(\tilde{j}) = M(\geq H)$.

(3) *For each orbit $G(x)$ in $M(\geq H)$, we have*

$$\tilde{j}^{-1}(G(x)) = G \underset{N(H)}{\times} G(x)^H.$$

(4) *For $x \in M(\geq H)$, there are canonical homeomorphisms:*

$$\left(G \underset{N(H)}{\times} G(x)^H\right)/G \approx G(x)^H/N(H) \approx N(H)\backslash N(H, G_x)^{-1}/G_x$$

where $N(H, G_x)^{-1}$ denotes the subspace $\{a \in G \mid a^{-1}Ha \subset G_x\}$ of G on which $N(H)$ acts from the left and G_x acts from the right by multiplication in G and $N(H)\backslash N(H, G_x)^{-1}/G_x$ denotes the orbit space of this $N(H) - G_x$ action on $N(H, G_x)^{-1}$. The orbit type in $G \underset{N(H)}{\times} G(x)^H$ corresponding to $[a] \in N(H)\backslash N(H, G_x)^{-1}/G_x$ is given by $G/(N(H) \cap aG_xa^{-1})$.

Proof (1) Since H is a closed subgroup of G, it follows from Proposition 1.10 that $N(H)$ is also a closed subgroup of G. Accordingly $G \to G/N(H)$ becomes a smooth principal $N(H)$-bundle by Corollary 3.39. Hence it follows from Corollary 3.40 that we have a smooth fibre bundle

$$G \underset{N(H)}{\times} M^H \to G/N(H)$$

associated with the principal $N(H)$-bundle above. In particular, we see that $G \underset{N(H)}{\times} M^H$ becomes a manifold. This follows also from Theorem 4.11, since the $N(H)$-action on $G \times M^H$ is smooth and free. Moreover it follows from the method of endowing $G \underset{N(H)}{\times} M^H$ with the smooth structure that it is a G-manifold.

Now we consider the following commutative diagram:

$$\begin{array}{ccc}
G \times M^H & \xrightarrow{\mathrm{id} \times j} & G \times M \\
\downarrow \pi & & \downarrow \varphi \\
G \underset{N(H)}{\times} M^H & \xrightarrow{\tilde{j}} & M
\end{array}$$

where φ denotes the G-action on M. Then the composition $\varphi \cdot (\mathrm{id} \times j)$ is a smooth map. Since $N(H)$ is also a compact Lie group, we may apply Theorem 4.11 to the projection π and deduce that $\tilde{j}$ is a smooth G-map as well.

(2) For arbitrary $g \in G$ and $x \in M^H$, we have $\tilde{j}([g, x]) = gj(x) = gx$. Since $G_x \supset H$ by assumption, we have

$$G_{gx} = gG_xg^{-1} \supset gHg^{-1},$$

that is, $gx \in M(\geq H)$. Hence we have

$$\mathrm{Im}(\tilde{j}) \subset M(\geq H).$$

Conversely, for any $x \in M(\geq H)$, there exists $g \in G$ with $gG_xg^{-1} \supset H$ by definition. In consequence, $G_{gx} = gG_xg^{-1} \supset H$, hence $gx \in M^H$. Observe that $\tilde{j}([g^{-1}, gx]) = g^{-1}(gx) = x$. Thus we have

$$\mathrm{Im}(\tilde{j}) \supset M(\geq H).$$

(3) Let $[g, y]$ be an arbitrary element of $\tilde{j}^{-1}(G(x))$ where $g \in G$, $y \in M^H$. Since

$$\tilde{j}([g, y]) = gy \in G(x),$$

there exists $g' \in G$ with $gy = g'x$. If we put $a = g^{-1}g'$, then $y = ax$, and hence

$$y \in G(x) \cap M^H = G(x)^H.$$

Thus we have

$$\tilde{j}^{-1}(G(x)) \subset G \underset{N(H)}{\times} G(x)^H.$$

Conversely, for any given $g \in G$ and $y \in G(x)^H$, we have

$$\tilde{j}([g, y]) = gy \in gG(x) = G(x),$$

that is,

$$\tilde{j}^{-1}(G(x)) \supset G \underset{N(H)}{\times} G(x)^H$$

holds.

(4) According to Proposition 1.90, the space $G(x)^H/N(H)$ of $N(H)$-orbits in $G(x)^H$ and the space $(G \underset{N(H)}{\times} G(x)^H)/G$ of G-orbits in $G \underset{N(H)}{\times} G(x)^H$ are canonically homeomorphic. We now define a map

$$f\colon N(H)\backslash N(H, G_x)\ {}^{1}/G_x \to G(x)^H/N(II)$$

by $f([a]) = N(H)(ax)$ where $[a]$ denotes the class in $N(H)\backslash N(H, G_x)^{-1}/G_x$ represented by $a \in N(H, G_x)^{-1}$. Then it is easy to see that f is well defined. Since $N(H, G_x)^{-1}$ is compact by Lemma 1.63, one verifies easily that f is a homeomorphism. Thus we have the required homeomorphisms.

In consequence, $G \underset{N(H)}{\times} G(x)^H$ can be regarded as the union of G-orbits with indexing set $N(H)\backslash N(H, G_x)^{-1}/G_x$. In view of the above correspondence, the G-orbit type in $G \underset{N(H)}{\times} G(x)^H$ corresponding to $[a] \in N(H)\backslash N(H, G_x)^{-1}/G_x$ is given by $G/G_{[e, ax]}$. For $g \in G$, we have

$$\begin{aligned} g \in G_{[e, ax]} &\Leftrightarrow g[e, ax] = [e, ax] \\ &\Leftrightarrow [g, ax] = [e, ax] \Leftrightarrow g \in N(H) \text{ and } g \in G_{ax}, \end{aligned}$$

that is, $G_{[e, ax]} = N(H) \cap G_{ax}$ holds.

Thus we have

$$G/G_{[e, ax]} = G/(N(H) \cap aG_x a^{-1}). \qquad \square$$

Remark If a G-manifold M has only the orbit type G/H, then we see by Theorem 4.18 that the map $\tilde{j}$ in Theorem 4.21 is a G-diffeomorphism.

4.7 Finiteness of number of orbit types

The purpose of the present section is to show that there are a finite number of orbit types in a compact G-manifold M where G is a compact Lie group. To see this, we first prepare the following lemma.

Lemma 4.22 *Let $\pi: E \to X$ be a fibre bundle with fibre F. If both F and X are compact, then so is the total space E.*

Proof Let $\{W_\lambda\}$ be an arbitrary open covering of E. For each $x \in X$, the fibre $\pi^{-1}(x)$ over x is compact, so that there is a finite subset $\{\lambda_1, \ldots, \lambda_n\}$ of the set $\{\lambda\}$ of indices with

$$\pi^{-1}(x) \subset \bigcup_{i=1}^{n} W_{\lambda_i}.$$

Claim There exists an open neighbourhood U_x of x such that

$$\pi^{-1}(U_x) \subset \bigcup_{i=1}^{n} W_{\lambda_i}.$$

If we make use of the local trivialities, the proof is almost apparent. However we will give a proof of the claim for certainty's sake. Take a coordinate function

$$\varphi: \pi^{-1}(U) \to U \times F$$

with $x \in U$. Then $\pi^{-1}(U) \cap W_{\lambda_i}$ is an open set of $\pi^{-1}(U)$ and

$$\{\varphi(\pi^{-1}(U) \cap W_{\lambda_i}) \mid i = 1, \ldots, n\}$$

is an open covering of $\{x\} \times F$ in $U \times F$. For each $y \in F$, there exist an open neighbourhood U_y of x in U and an open neighbourhood V_y of y in F such that

$$U_y \times V_y \subset \bigcup_{i=1}^{n} \varphi(\pi^{-1}(U) \cap W_{\lambda_i}),$$

since the right-hand side above is an open neighbourhood of (x, y) in $U \times F$. Here notice that the index y of U_y is not a misprint, because x is fixed and such a neighbourhood U_y depends on $y \in F$. Needless to say, $\{V_y \mid y \in F\}$ is an open covering of F. Since F is compact, there are finite points $y_1, \ldots, y_l$ of F

such that

$$F = \bigcup_{j=1}^{l} V_{y_j}.$$

We now put

$$U_x = \bigcap_{j=1}^{l} U_{y_j}.$$

Then U_x is an open set of U and satisfies

$$\begin{aligned} U_x \times F = U_x \times \left(\bigcup_{j=1}^{l} V_{y_j} \right) &= \bigcup_{j=1}^{l} (U_x \times V_{y_j}) \\ &\subset \bigcup_{j=1}^{l} (U_{y_j} \times V_{y_j}) \subset \bigcup_{i=1}^{n} \varphi(\pi^{-1}(U) \cap W_{\lambda_i}). \end{aligned}$$

It turns out that

$$\pi^{-1}(U_x) = \varphi^{-1}(U_x \times F) \subset \bigcup_{i=1}^{n} (\pi^{-1}(U) \cap W_{\lambda_i}) \subset \bigcup_{i=1}^{n} W_{\lambda_i},$$

that is, the claim holds.

The family $\{U_x\}_{x \in X}$ obtained by the claim for each $x \in X$ is an open covering of X. Since X is compact, there are a finite number of points $x_1, \ldots, x_m$ of X such that

$$X = \bigcup_{k=1}^{m} U_{x_k}.$$

As a consequence, we have

$$E = \pi^{-1}(X) = \bigcup_{k=1}^{m} \pi^{-1}(U_{x_k}) \subset \bigcup_{k=1}^{m} \bigcup_{i} W_{k\lambda_i},$$

that is, we see that E is compact. Here $\bigcup_i W_{k\lambda_i}$ means the union of finitely many W_λ which are chosen as at the beginning of the proof for each x_k. □

We are now in a position to state the main theorem of this section.

Theorem 4.23 *Let G be a compact Lie group and M a compact G-manifold. Then there are only finitely many orbit types in M.*

Proof We shall give a proof by induction on the dimension of M. First, since a zero-dimensional compact manifold consists of a finite number of points, there are only finitely many orbit types.

Suppose now that the theorem holds for compact G-manifolds of dimension less than n. Let M be an arbitrary compact G-manifold of dimension n. According to the slice theorem, for each $x \in M$, there exists a G-invariant open neighbourhood v of $G(x)$ in M which is G-diffeomorphic to $G \times_{G_x} V_x$ for some representation space V_x of G_x. By considering this for each point x in M, we get an open covering $\{v\}$ of M. Since M is compact, $\{v\}$ has a finite subcovering $\{v_i \mid i = 1, \ldots, k\}$. Therefore to prove the theorem, it is sufficient to show that there are only finitely many orbit types in each v. That is, we need only investigate G-manifolds of the form $v = G \times_{G_x} V_x$.

It follows from Lemma 4.16 that the isotropy group $G_{[g,v]}$ at a point $[g, v]$ of $G \times_{G_x} V_x$ is given by

$$G_{[g,v]} = g(G_x)_v g^{-1}.$$

Since G_x acts linearly on V_x, we have

$$(G_x)_{tv} = (G_x)_v \qquad \text{for } t \neq 0 \in \mathbb{R}.$$

As a consequence, we have

$$G_{[g,tv]} = G_{[g,v]} \qquad \text{for } t \neq 0 \in \mathbb{R}.$$

By taking a G-invariant metric on v, we denote by $S(v)$ the total space of the unit sphere bundle associated with v. Since the zero cross-section of v is canonically identified with G/G_x, it follows from the above observation that

The set of orbit types in v = The set of orbit types in $S(v) \cup \{G/G_x\}$.

Observe that $S(v)$ is the total space of a fibre bundle with base space G/G_x and fibre the sphere of dimension $(n-1) - \dim G/G_x$. Since both base space and fibre are compact, it follows from Lemma 4.22 that the total space $S(v)$ is compact. Hence $S(v)$ is a compact G-manifold of dimension $n-1$. By the inductive assumption, the number of orbit types occurring in $S(v)$ is finite. Thus we have shown that the number of orbit types occurring in v is finite. □

This theorem can be extended slightly to the case of non-compact G-manifolds.

Corollary 4.24 *Let G be a compact Lie group and M a compact G-manifold. Then there are only finitely many orbit types in the total space E of a smooth G-vector bundle over M.*

Proof By choosing a G-invariant metric on E, we consider the associated sphere bundle $\pi: S(E) \to M$. Since its fibre is compact, it follows from Lemma 4.22 that $S(E)$ is a compact G-manifold. As explained in the proof of the theorem above, we have

$$G_{tv} = G_v \qquad \text{for any} \quad v \in E \quad \text{and} \quad t \neq 0 \in \mathbb{R}.$$

Therefore we have similarly that

The set of orbit types in E = The set of orbit types in $S(E)$
$\cup$ The set of orbit types occurring in the zero cross-section ($=M$).

Since both $S(E)$ and M are compact G-manifolds, it follows from Theorem 4.23 that there are only finitely many orbit types in E. □

Since a G-representation space can be regarded as a G-vector bundle over one point, we have the following corollary.

Corollary 4.25 *There are only finitely many orbit types in a representation space of a compact Lie group.*

4.8 Principal orbits

Among the orbit types, the least one is G/G and the largest one is $G/\{e\}$. Given a G-space X, these orbit types do not always occur in X.

The purpose of this section is to show that there exists a maximum orbit type for orbits in a G-manifold M with M/G connected. Since the set of orbit types occurring in a G-manifold is not linearly ordered in general, it is not obvious that there exists a maximum element. A minimum element does not always exist in general. Furthermore, if M/G is not connected, then even a maximum element does not exist in general.

We begin by preparing the following lemma, whose proof is now easy.

Lemma 4.26 *Let G be a compact Lie group, H a closed subgroup of G, and V a representation space of H. Then any orbit in $G \times_H V$ has type greater than or equal to G/H. Accordingly, in a G-manifold, any orbit has a minimum type locally.*

Proof According to Lemma 4.16,

$$G_{[g,v]} = gH_v g^{-1}$$

for $[g, v] \in G \times_H V$. Hence we have

$$(G_{[g,v]}) = (H_v) \leq (H),$$

that is, $G/G_{[g,v]} \geq G/H$ holds.

For a general G-manifold, the locally minimum property of orbit types is immediate from the slice theorem. □

We are now in a position to state the theorem.

Theorem 4.27 *Let G be a compact Lie group and M a G-manifold. If the orbit space M/G is connected, then there exists a maximum orbit type G/H in M. The union $M(H)$ of the orbits of type G/H is open and dense in M and its orbit space $M(H)/G$ is connected.*

Definition 4.2 The maximum orbit type for orbits in M guaranteed by Theorem 4.27 is called the *principal orbit type* and orbits of this type are called *principal orbits.* The corresponding isotropy groups are called *principal isotropy groups.*

Proof The proof will be given by induction on the dimension of M. If M is of dimension 0, then the orbit space M/G is just a point, since M/G is connected by assumption. That is, M consists of one orbit. Hence the theorem holds evidently in this case.

Suppose that the theorem holds for manifolds of dimension less than n. Let M be a G-manifold of dimension n satisfying the assumption in the theorem. If there exists an orbit $G(x)$ with $\dim G(x) = n$, then $G(x)$ is open in M, because $G(x)$ is a submanifold of M with the same dimension as M. Therefore the point $G(x)/G$ is open in M/G. On the other hand, M/G is Hausdorff by Proposition 1.58, so that each point of M/G is closed in M/G. Since M/G is connected by assumption, it turns out that $M = G(x)$. Conversely, if $M = G(x)$, then we have trivially that $\dim G(x) = n$. Thus the following equivalence is shown.

There exists an orbit $G(x)$ of dimension n if and only if M is a homogeneous space.

The theorem obviously holds in the case where this equivalent condition is satisfied.

Next we shall deal with the case where $\dim G(x) < n$ for every orbit $G(x)$. It follows from the slice theorem that, for any $x \in M$, there exist a G_x-representation space V_x of dimension greater than or equal to 1 and a G-embedding

$$f\colon \nu = G \underset{G_x}{\times} V_x \to M, \qquad f([e, 0]) = x.$$

Then our provisional target is to prove the theorem for the G-manifold ν. By taking a G-invariant metric on ν, we consider the unit sphere bundle $S(\nu)$ associated with ν. Then note that $S(\nu) = G \underset{G_x}{\times} S(V_x)$ and $\dim S(\nu) = n - 1$. To make use of the inductive assumption, we hope $S(\nu)/G$ will be connected. Unfortunately this is not always the case. Such an undesirable case occurs when $\dim V_x = 1$ and G_x acts trivially on V_x, because $S(\nu)/G$ and $S(V_x)/G_x$ are homeomorphic by Proposition 1.90. In this case, we have $\nu = G/G_x \times \mathbb{R}$. Hence ν has only the orbit type G/G_x, so that G/G_x is the principal orbit. Since $\nu(G_x) = \nu$, $\nu(G_x)$ is open and dense in ν. Moreover $\nu(G_x)/G$ is connected, because $\nu(G_x)/G = \mathbb{R}$. Thus the theorem holds for this kind of ν.

Next we deal with the case where $S(\nu)/G$ is connected. Then there exists a principal orbit in $S(\nu)$ by the inductive assumption. Let G/H denote the principal orbit type in $S(\nu)$.

In the following, we shall show that $\nu(H)$ is open and dense in ν and that $\nu(H)/G$ is connected. For this, we divide the following proof into two cases.

(1) In the case where $\nu = \nu(H)$, $\nu(H)$ is open and dense in ν. Since ν/G $(=\nu(H)/G)$ is homeomorphic to V_x/G_x by Proposition 1.90, $\nu(H)/G$ is connected.

(2) For the case where $\nu \neq \nu(H)$, we first show that $\nu(H)$ is included in ν – (zero section). As shown in the proof of Theorem 4.23, the set of orbit types occurring in ν is the union of those in $S(\nu)$ and $\{G/G_x\}$. Let G/K be an arbitrary orbit type occurring in ν. Then it follows from Lemma 4.26 that G/K is greater than or equal to G/G_x. Thus we have

$$G/G_x \leq G/K \leq G/H.$$

Hence if $G/G_x = G/H$, then we have $\nu = \nu(H)$, which is a contradiction. Accordingly, the zero section $(=G/G_x)$ of ν is not included in $\nu(H)$, that is, $\nu(H)$ is included in ν – (zero section).

Let $\mathbb{R}^+$ denote the set of positive numbers on which G acts trivially. Consider the following commutative diagram:

$$\begin{array}{ccc} \mathbb{R}^+ \times S(\nu) & \xrightarrow{f_1} & \nu - (\text{zero section}) \\ \uparrow{\scriptstyle \mathrm{id} \times i_1} & & \uparrow{\scriptstyle i_2} \\ \mathbb{R}^+ \times S(\nu)(H) & \xrightarrow{f_2} & \nu(H) \end{array}$$

where i_1 and i_2 are the inclusion maps and f_1 is the map defined by $f_1(r, y) = ry$ for $r \in \mathbb{R}^+$, $y \in S(\nu)$ and f_2 is the map defined by the restriction $f_1 \mid \mathbb{R}^+ \times S(\nu)(H)$ of f_1. On the other hand, we define a map

$$f'_1 : \nu - (\text{zero section}) \to \mathbb{R}^+ \times S(\nu)$$

by $f'_1(v) = (\|v\|, v/\|v\|)$ and put

$$f'_2 = f'_1 \mid \nu(H) : \nu(H) \to \mathbb{R}^+ \times S(\nu)(H).$$

Evidently these are all continuous G-maps and satisfy

$$f'_1 \cdot f_1 = f_1 \cdot f'_1 = \mathrm{id} \qquad \text{and} \qquad f'_2 \cdot f_2 = f_2 \cdot f'_2 = \mathrm{id}.$$

Accordingly, f_i and f'_i are all G-homeomorphisms. Since $S(\nu)(H)$ is open and dense in $S(\nu)$ by the inductive assumption, $\mathbb{R}^+ \times S(\nu)(H)$ is open and dense in $\mathbb{R}^+ \times S(\nu)$. Hence it follows from the above diagram that $\nu(H)$ is open and dense in ν – (zero section). Evidently ν – (zero section) is open and dense in ν.

Combining the above results, we can conclude that $v(H)$ is open and dense in v. Furthermore, the G-homeomorphism f_2 induces a homeomorphism

$$\bar{f}_2\colon (\mathbb{R}^+ \times S(v)(H))/G \to v(H)/G,$$

between orbit spaces. Since $(\mathbb{R}^+ \times S(v)(H))/G = \mathbb{R}^+ \times (S(v)(H)/G)$ and since $S(v)(H)/G$ is connected by the inductive assumption, we can conclude that $v(H)/G$ is connected.

Thus we have shown that the theorem holds for v.

Next we shall show that the theorem holds for a general G-manifold M of dimension n. On account of the slice theorem, M is covered with open sets which are the images of G-embeddings of the form

$$f\colon v = G \underset{G_x}{\times} V_x \to M.$$

Then we shall show that the principal orbit types obtained above for v all coincide. Suppose now that the principal orbit type depends on v. Let G/H be the principal orbit type of one of the v. We divide the set of v into two classes: the set $\{v_\lambda\}$ of v whose principal orbit type is G/H; and the set $\{v_\mu\}$ of v whose principal orbit type is different from G/H. Then we have an open covering:

$$M = \bigcup_\lambda f_\lambda(v_\lambda) \cup \bigcup_\mu f_\mu(v_\mu)$$

of M. Since both $f_\lambda(v_\lambda)$ and $f_\mu(v_\mu)$ are open in M, $f_\lambda(v_\lambda)/G$ and $f_\mu(v_\mu)/G$ are open in M/G. In consequence, we have an open covering:

$$M/G = \bigcup_\lambda f_\lambda(v_\lambda)/G \cup \bigcup_\mu f_\mu(v_\mu)/G$$

of M/G. Since M/G is connected by assumption, we have

$$\left(\bigcup_\lambda f_\lambda(v_\lambda)/G\right) \cap \left(\bigcup_\mu f_\mu(v_\mu)/G\right) \neq \varnothing.$$

That is, there exist λ_0 and μ_0 such that

$$f_{\lambda_0}(v_{\lambda_0})/G \cap f_{\mu_0}(v_{\mu_0})/G \neq \varnothing.$$

Since both $f_{\lambda_0}(v_{\lambda_0})$ and $f_{\mu_0}(v_{\mu_0})$ are G-invariant, we have

$$f_{\lambda_0}(v_{\lambda_0}) \cap f_{\mu_0}(v_{\mu_0}) \neq \varnothing.$$

If we put $U = f_{\lambda_0}(v_{\lambda_0}) \cap f_{\mu_0}(v_{\mu_0})$, then U is open in M, and hence is open in both $f_{\lambda_0}(v_{\lambda_0})$ and $f_{\mu_0}(v_{\mu_0})$ as well. As shown previously, the union of principal orbits of $f_{\lambda_0}(v_{\lambda_0})$ is dense in $f_{\lambda_0}(v_{\lambda_0})$, so that there exists a principal orbit of $f_{\lambda_0}(v_{\lambda_0})$ in U. Moreover, since the union of principal orbits of $f_{\lambda_0}(v_{\lambda_0})$ is open in $f_{\lambda_0}(v_{\lambda_0})$, there exists a G-invariant open set W in U which has only the orbit type G/H.

On the other hand, the union of principal orbits of $f_{\mu_0}(v_{\mu_0})$ is dense in $f_{\mu_0}(v_{\mu_0})$. Since W is open in U, and hence open in $f_{\mu_0}(v_{\mu_0})$, there exists a principal orbit of $f_{\mu_0}(v_{\mu_0})$ in W. This is a contradiction.

Thus we have shown that the principal orbit types for all the v coincide.

Next we shall show that $M(H)$ is open in M. The open covering

$$M = \bigcup_{\lambda} f_\lambda(v_\lambda)$$

of M induces a covering

$$M(H) = \bigcup_{\lambda} f_\lambda(v_\lambda(H))$$

of $M(H)$. As shown above, $v_\lambda(H)$ is open in v_λ so that $f_\lambda(v_\lambda(H))$ is open in $f_\lambda(v_\lambda)$. Since $f_\lambda(v_\lambda)$ is open in $M, f_\lambda(v_\lambda(H))$ is also open in M. Thus we see that $M(H)$ is open in M.

Let x be an arbitrary point of M and W be an arbitrary open neighbourhood of x in M. Choose λ_0 with $x \in f_{\lambda_0}(v_{\lambda_0})$. Then the intersection $W \cap f_{\lambda_0}(v_{\lambda_0})$ is a non-empty open set of $f_{\lambda_0}(v_{\lambda_0})$. Note that $f_{\lambda_0}(v_{\lambda_0}(H)) = f_{\lambda_0}(v_{\lambda_0})(H)$. As observed previously, $f_{\lambda_0}(v_{\lambda_0}(H))$ is dense in $f_{\lambda_0}(v_{\lambda_0})$, so that we have

$$W \cap M(H) \supset W \cap f_{\lambda_0}(v_{\lambda_0})(H) = (W \cap f_{\lambda_0}(v_{\lambda_0})) \cap f_{\lambda_0}(v_{\lambda_0})(H) \neq \varnothing,$$

that is, $M(H)$ is dense in M.

Finally we shall show that $M(H)/G$ is connected. Each G-embedding $f_\lambda : v_\lambda \to M$ induces a map $\bar{f}_\lambda : v_\lambda/G \to M/G$ between orbit spaces. Then $\bar{f}_\lambda$ is a homeomorphism onto an open set of M/G and the open covering

$$M(H) = \bigcup_{\lambda} f_\lambda(v_\lambda(H))$$

of $M(H)$ mentioned above induces an open covering

$$M(H)/G = \bigcup_{\lambda} f_\lambda(v_\lambda(H))/G = \bigcup_{\lambda} \bar{f}_\lambda(v_\lambda(H)/G)$$

of $M(H)/G$. Previously, we showed that $v_\lambda(H)/G$ is connected.

Suppose now that $M(H)/G$ is not connected. Then there exist non-empty open subsets U_1 and U_2 of $M(H)/G$ such that

$$M(H)/G = U_1 \cup U_2, \qquad U_1 \cap U_2 = \varnothing.$$

Since each $\bar{f}_\lambda(v_\lambda(H)/G)$ is connected, it is included entirely in either U_1 or U_2. Then we divide the index set $\{\lambda\}$ into two parts by setting

$$\Lambda_i = \{\lambda \mid \bar{f}_\lambda(v_\lambda(H)/G) \subset U_i\} \qquad i = 1, 2.$$

According to this division, we write the open covering $M/G = \bigcup_\lambda f_\lambda(v_\lambda)/G$ of M/G as

$$M/G = \bigcup_{\lambda \in \Lambda_1} f_\lambda(v_\lambda)/G \cup \bigcup_{\lambda \in \Lambda_2} f_\lambda(v_\lambda)/G.$$

Since M/G is connected, there exist $\lambda_1 \in \Lambda_1$ and $\lambda_2 \in \Lambda_2$ such that

$$f_{\lambda_1}(v_{\lambda_1})/G \cap f_{\lambda_2}(v_{\lambda_2})/G \neq \varnothing.$$

Hence we have

$$f_{\lambda_1}(v_{\lambda_1}) \cap f_{\lambda_2}(v_{\lambda_2}) \neq \varnothing.$$

Since $f_{\lambda_i}(v_{\lambda_i}(H))$ are open and dense in $f_{\lambda_i}(v_{\lambda_i})$ for $i = 1, 2$, we can prove that

$$f_{\lambda_1}(v_{\lambda_1}(H)) \cap f_{\lambda_2}(v_{\lambda_2}(H)) \neq \varnothing,$$

in the same manner as in the proof of the fact that the principal orbit types for all the v coincide. Therefore we have

$$f_{\lambda_1}(v_{\lambda_1}(H))/G \cap f_{\lambda_2}(v_{\lambda_2}(H))/G \neq \varnothing.$$

It turns out that

$$\varnothing \neq \bar{f}_{\lambda_1}(v_{\lambda_1}(H)/G) \cap \bar{f}_{\lambda_2}(v_{\lambda_2}(H)/G) \subset U_1 \cap U_2 = \varnothing,$$

which is a contradiction. Since this is a consequence of the assumption that $M(H)/G$ is not connected, we can conclude that $M(H)/G$ is connected.

This completes the proof of the theorem. □

The following theorem is a characterization of the principal orbit type.

Theorem 4.28 *Let G be a compact Lie group, H a closed subgroup of G and M a G-manifold with M/G connected. If the union $M(H)$ of the orbits of type G/H is dense in M or a non-empty open set of M, then G/H is the principal orbit type.*

Proof Let G/H_0 be the principal orbit type of M. It follows from Theorem 4.27 that $M(H_0)$ is a non-empty open set of M. If $M(H)$ is dense in M, then

$$M(H_0) \cap M(H) \neq \varnothing,$$

that is, the equality $(H_0) = (H)$ holds.

If $M(H)$ is a non-empty open set, then by exchanging the roles of H_0 and H we reach the same conclusion. □

Example 1 The hypothesis that M/G is connected in the theorem cannot be omitted. For example, let $S^1 = \{z \in \mathbb{C} \mid |z| = 1\}$ and M be the disjoint union of two copies S^1_1 and S^1_2 of S^1. Put $G = S^1 \times S^1$. Then we define a G-action on M by

$$(t_1, t_2) \circ x_1 = t_1 x_1 \qquad \text{for} \quad (t_1, t_2) \in S^1 \times S^1, x_1 \in S^1_1,$$

$$(t_1, t_2) \circ x_2 = t_2 x_2 \qquad \text{for} \quad (t_1, t_2) \in S^1 \times S^1, x_2 \in S^1_2.$$

The isotropy group G_{x_1} at $x_1 \in S^1_1$ is given by $G_{x_1} = \{1\} \times S^1$ and the isotropy group G_{x_2} at $x_2 \in S^1_2$ is given by $G_{x_2} = S^1 \times \{1\}$. Since there is no relation of order between (G/G_{x_1}) and (G/G_{x_2}), there is no maximum orbit type in M.

Corollary 4.29 *Let G be a compact Lie group and M a G-manifold with M/G connected. Then the orbit $G(x)$ of a point x in M is principal if and only if the slice representation of G_x is trivial.*

Proof Let $f: G \times_{G_x} V_x \to M$ be a G-embedding arising from the slice theorem.

First suppose that G_x acts trivially on V_x. Then we have $G \times_{G_x} V_x = (G/G_x) \times V_x$ so that all the orbits in this space have the type G/G_x. Since the union of principal orbits is dense in M in general, there exists a principal orbit in the open set $f(G \times_{G_x} V_x)$. Hence $G(x)$ $(=f(G/G_x))$ is nothing but the principal orbit.

The converse is shown by *reductio ad absurdum*. Suppose that G_x acts non-trivially on V_x. Then there exists a point v in V_x such that $(G_x)_v \subsetneqq G_x$. Hence it follows from Lemma 4.16 that

$$G_{[g,v]} = g(G_x)_v g^{-1} \subsetneqq gG_x g^{-1}.$$

If $G(x)$ were a principal orbit, then the equality $(G_{[g,v]}) = (G_x)$ would hold, that is, there would exist $g_0 \in G$ such that

$$G_{[g,v]} = g_0 G_x g_0^{-1}.$$

Therefore we would have

$$G_x \subsetneqq g_0^{-1} g G_x (g_0^{-1} g)^{-1}. \tag{4.1}$$

Since G_x is a closed subgroup of a compact group G satisfying $G_x \subset g_0^{-1} g G_x (g_0^{-1} g)^{-1}$, it follows from Corollary 1.70 that

$$G_x = g_0^{-1} g G_x (g_0^{-1} g)^{-1}.$$

But this would contradict (4.1) above. Accordingly $G(x)$ is not a principal orbit. □

4.9 Dimension of compact Lie groups which act effectively on manifolds

The purpose of this section is to show that dimensions of compact Lie groups which act effectively on a given manifold are bounded. This will be a result in which the concept of the principal orbit is particularly appropriate for the proof. We begin with an example.

Example 1 As shown previously, $SO(n+1)$ acts naturally on the n-dimensional sphere

$$S^n = \{(x_1, \ldots, x_{n+1}) \in \mathbb{R}^{n+1} \mid \sum x_i^2 = 1\}.$$

Evidently this action is effective. It is easy to see that the tangent space $T_I(SO(n+1))$ to $SO(n+1)$ at the unit matrix I is given by $\{A \in M(n+1, \mathbb{R}) \mid A + {}^tA = 0 \text{ (zero matrix)}\}$. It follows that $\dim T_I(SO(n+1)) = n(n+1)/2$. As a consequence, we see that the dimension of $SO(n+1)$ is also $n(n+1)/2$.

The following theorem asserts that the example above attains the maximum dimension.

Theorem 4.30 *Let G be a compact Lie group and M a connected G-manifold of dimension n. If its action is effective, then we have*

$$\dim G \leq \frac{n(n+1)}{2}.$$

Proof We first remark that we need only prove the theorem in the case where G is connected, because, if a G-action is effective, then the restriction of G to the identity component is also effective. The proof will proceed by induction on the dimension n of M.

Since a zero-dimensional connected manifold M is nothing but a point, the unit group only acts effectively on M. Hence the theorem holds trivially in this case.

Suppose next that the theorem holds for manifolds of dimension less than n. Let M be an arbitrary connected G-manifold of dimension n. Since M is connected, and hence M/G is connected, there exists a principal orbit in M by Theorem 4.27. Let G/H be its orbit type. Since H is a closed subgroup of a compact Lie group G, it follows from Theorem 3.36 that H is a compact Lie subgroup of G and satisfies

$$\dim G - \dim H = \dim G/H \leq \dim M = n.$$

Hence we have $\dim G \leq n + \dim H$.

Let H_0 denote the identity component of H. Notice that $\dim H_0 = \dim H$. Hence if we show that H_0 acts effectively on a connected manifold of dimension less than n, then we have

$$\dim G \leq n + \frac{(n-1)n}{2} = \frac{n(n+1)}{2}$$

by the inductive assumption, which is the assertion of the theorem.

Suppose that an element g of G satisfies $(G/H)^g = G/H$. Since the union of principal orbits is dense in M, we have $M^g = M$. Hence g is the identity by assumption. Thus G acts effectively on the connected manifold G/H. In particular, H_0 acts effectively on G/H. If H_0/H_1 is the principal orbit type of this H_0-action on G/H, then similarly H_0 acts effectively on H_0/H_1.

We now consider the following two cases.

(1) In the case where $\dim H_0/H_1 < \dim G/H$, since $\dim G/H \leq n$, we have $\dim H_0/H_1 \leq n-1$. Hence H_0 acts effectively on a connected manifold of dimension less than n, so that the inductive step proceeds.

(2) In the case where $\dim H_0/H_1 = \dim G/H$, since H_0/H_1 is a submanifold of G/H with the same dimension, H_0/H_1 is open in G/H. Since H_0/H_1 is compact, it is closed in G/H as well. Hence we have $H_0/H_1 = G/H$, because G/H is connected. This means that the H_0-action on G/H is transitive. But the equality $H_0eH = eH$ holds, so G/H has to be one point. As observed above, the G-action on G/H is effective, and this happens only when G is the unit group. Thus case (2) occurs only when G is the unit group. In this case, H_0 is the unit group as well, so that H_0 acts effectively on any manifold.

This makes the inductive proof of the theorem complete. □

The hypothesis that M is connected in the theorem cannot be omitted. An example follows.

Example 2 Let G be the product $S^1 \times \cdots \times S^1$ of n copies of S^1 and M the disjoint union $S_1^1 \cup \cdots \cup S_n^1$ of n copies of S^1. Then we define a G-action on M by

$$(t_1, \ldots, t_n) \circ x_i = t_i x_i$$

for $(t_1, \ldots, t_n) \in S^1 \times \cdots \times S^1$ and $x_i \in S_i^1$. Evidently this action is effective. But $\dim G = n$ and $\dim M = 1$, so that the inequality

$$n \leq \frac{1(1+1)}{2} = 1$$

does not hold for $n \geq 2$.

In the following we shall give an example showing that the compactness assumption of G is also indispensable in the theorem.

Example 3 Take a smooth function $f: \mathbb{R} \to \mathbb{R}$ such that $f(x) \geq 0$ for all x, $f(\frac{1}{2}) > 0$ and $f(x) = 0$ for all $x \notin (\frac{1}{4}, \frac{3}{4})$ (see Fig. 4.15).

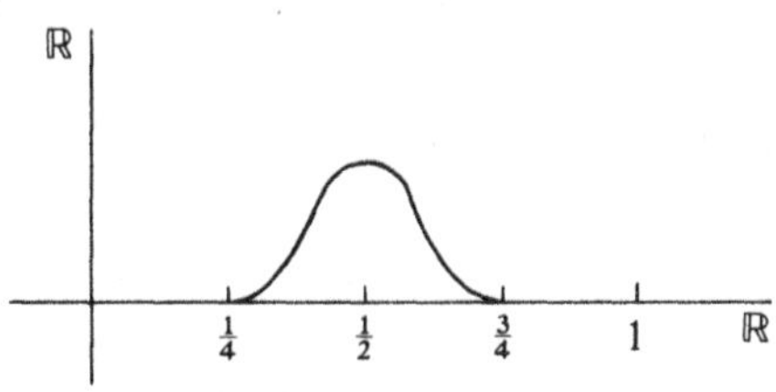

Fig. 4.15

Define a vector field X on $\mathbb{R}$ by

$$X(x) = f(x).$$

Since the vector field X vanishes outside of the compact set $[\frac{1}{4}, \frac{3}{4}]$, it follows from Theorem 3.19 that X generates a 1-parameter group, which we denote by

$$\varphi\colon \mathbb{R} \times \mathbb{R} \to \mathbb{R}.$$

If $t > 0$, then $\varphi(t, \frac{1}{2}) > \frac{1}{2}$. If $t < 0$, then $\varphi(t, \frac{1}{2}) < \frac{1}{2}$. Hence the action is effective. We now define an $\mathbb{R}^n$-action on $\mathbb{R}$

$$\Phi\colon \mathbb{R}^n \times \mathbb{R} \to \mathbb{R}$$

by

$$\Phi(t_1, \ldots, t_n, x) = \begin{cases} x & \text{if } x \notin [1, n+1] \\ \varphi(t_i, x - i) + i & \text{if } x \in [i, i+1] \\ & (i = 1, \ldots, n). \end{cases}$$

One verifies easily that Φ defines a smooth $\mathbb{R}^n$-action. If $(t_1, \ldots, t_n) \neq (0, \ldots, 0)$, then there exists i with $t_i \neq 0$. Then

$$\Phi(t_1, \ldots, t_n, i + 1/2) = \varphi(t_i, 1/2) + i.$$

Since $\varphi(t_i, 1/2) \neq 1/2$ for $t_i \neq 0$, we see that the action Φ is effective. Since n can be an arbitrary positive integer, a non-compact Lie group of any dimension can act effectively on the one-dimensional connected manifold $\mathbb{R}$.

Let M be a connected manifold. Then we set

$$N(M) = \max\{\dim G \mid G \text{ is a compact Lie group which acts effectively on } M\},$$

which is called the *degree of symmetry* of M.

The main theorem of this section means that $N(M) \leq n(n+1)/2$ where $n = \dim M$. It is known that there are higher dimensional manifolds M with $N(M) = 0$.

Exercises for Chapter 4

1. Given $n+1$ integers $a_0, \ldots, a_n$, let S^1 act on $\mathbb{C}P^n$ by $\varphi\colon S^1 \times \mathbb{C}P^n \to \mathbb{C}P^n$, defined by $\varphi(z, [z_0, \ldots, z_n]) = [z^{a_0} z_0, \ldots, z^{a_n} z_n]$. Determine the fixed point set, the principal orbits, and the isotropy groups of this action, and apply the decomposition theorem to this action.

2. Let M be a compact S^1-manifold. Show that there exists a positive integer k such that the equality $M^{S^1} = M^{\mathbb{Z}_n}$ holds for any integer $n > 1$ with $(n, k) = 1$. Here $\mathbb{Z}_n$ is regarded as a subgroup of S^1 in a natural way. The condition 'any integer $n > 1$ with $(n, k) = 1$' may be replaced by 'any integer n with $n > k$'.

3. Let M be a compact S^1-manifold without fixed point. Show that there exists a positive integer k such that the restricted $\mathbb{Z}_n$-action is free for any integer $n > 1$ with $(n, k) = 1$.

4. Let G be a compact commutative Lie group and let M be a G-manifold such that the action is effective and M/G is connected. Show that the principal isotropy group is the trivial group $\{e\}$.

5. Let G be a finite group and let M be a connected G-manifold such that the action is effective. Show that the principal isotropy group is the trivial group $\{e\}$.

6. Let G be a finite group and let M be a connected G-manifold with one orbit type such that the action is effective. Show that the action is free.

7. Consider the diagonal actions of the orthogonal group $O(n)$ on $\mathbb{R}^n \oplus \mathbb{R}^n$, $\mathbb{R}^n \oplus \mathbb{R}^n \oplus \mathbb{R}^n, \ldots$. Determine the isotropy types of these actions.

8. Repeat Exercise 7 above after replacing $\mathbb{R}^n$ by $\mathbb{C}^n$ and $O(n)$ by $U(n)$.

9. Consider the Brieskorn variety

$$W_k^{2n-1} = \{(z_0, \ldots, z_n) \in \mathbb{C}^{n+1} \mid z_0^k + z_1^2 + \cdots + z_n^2 = 0, \\ |z_0|^2 + \cdots + |z_n|^2 = 1\}.$$

Let $O(n)$ act on the coordinates $z_1, \ldots, z_n$ as a subgroup of $U(n)$. Show that W_k^{2n-1} is invariant under this action and determine the isotropy types occurring in this action.

10. Define a map $f\colon M = \mathbb{C}P^2 \to \mathbb{C}P^2$ by $f([z_0, z_1, z_2]) = [\bar{z}_0, -\bar{z}_2, \bar{z}_1]$. Show that f defines a $\mathbb{Z}_4$-action. Determine the subspaces $M(\{e\})$, $M(\mathbb{Z}_2)$, $M(\mathbb{Z}_4)$ in the sense of Section 4.6 and apply the decomposition theorem to this action.

5 Algebraization

We have studied transformation groups in a geometric way. In this chapter we change entirely the method employed so far and consider, so to speak, algebraization of geometric structures. Specifically, we shall study the algebraic aspects of transformation groups. The essence of transformation groups will be abstracted, recognized, and understood via algebraization. As a means of algebraization, we introduce the concept of G-complexes for G, a finite group, and study first the relationships among the Euler characteristics of G-complexes, orbit spaces, and various kinds of fixed point sets. The Möbius inversion formula plays an important role in this study. When G is a compact Lie group, similar formulae for Euler characteristics of G-manifolds are also obtained by making use of the decomposition theorem in Section 4.6 (see Section 5.6).

Next we shall investigate the homological relationships between a G-complex and its orbit space by means of a transfer map. Then we shall introduce the celebrated Smith theory which gives us the flavour of this direction. Also, we shall present the Lefschetz fixed point theorem and apply it to the study of G-complexes and G-manifolds.

Finally, we shall discuss a necessary condition for a finite group to act freely on a sphere along the lines of the Smith theory.

For an action of the compact group G, we derive considerable information by restricting G to various kinds of its subgroups. Among the several types of homology groups, we adopt those of G-complexes which seem to be easy to understand for beginners.

I would like to remark in advance that every G-manifold can be given the structure of a G-complex for G a finite group.

5.1 *G*-complexes

As a first step towards abstracting algebra from geometric objects, we introduce the concept of complexes and triangulations. First of all, we will briefly review simplicial complexes. Since it is convenient to treat them in an abstract way, we adopt the following method.

Let X be a finite set. The set consisting of all the subsets of X is called the *power set* of X and is denoted by $\mathscr{P}(X)$. Notice that the empty set $\varnothing$ and X itself belong to $\mathscr{P}(X)$.

A *simplicial complex* (or briefly *complex*) K *on a vertex set* X is a subset of $\mathscr{P}(X) - \{\varnothing\}$ satisfying:

(1) if $x \in X$, then $\{x\} \in K$;

(2) if $s \in K$ and $s' \subset s$, then $s' \in K$.

Then we say that K is a *simplicial complex made from* X. Elements of X are called *vertices* of K and elements of K are called *simplexes.* We frequently identify the vertex x with the simplex $\{x\}$. When a simplex s consists of a subset $\{x_0, x_1, \ldots, x_k\}$ of X, we write

$$s = (x_0, x_1, \ldots, x_k).$$

A simplex s containing exactly $k + 1$ vertices is called a *k-simplex.* We also say that the *dimension* of s is k and write $\dim s = k$. Then it follows from the above identification that the set of 0-simplices of K is regarded as the set X of vertices of K.

If K is a simplicial complex, we put

$$\dim K = \max\{\dim s \mid s \in K\}.$$

A non-empty subset s' of a simplex s is called a *face* of s (a *proper face* if $s' \neq s$) and is denoted by $s' \subset s$ (or $s \supset s'$).

Let Y be another finite set and L be a simplicial complex made from Y. Generally, a map $f\colon X \to Y$ induces a map $\mathscr{P}(X) \to \mathscr{P}(Y)$, which we denote also by f. If f satisfies $f(K) \subset L$, then f is called a *simplicial map* from K to L.

A *subcomplex* L of a simplicial complex K, denoted by $L \subset K$, is a subset of K that is a simplicial complex.

Given a finite set $X = \{x_1, \ldots, x_n\}$ of n elements, we consider $\mathbb{R}^n$ and identify x_i with the ith unit coordinate vector $e_i = (0, \ldots, 0, 1, 0, \ldots, 0)$ in $\mathbb{R}^n$. Let K be a simplicial complex made from X. For $s = (x_{i_0}, \ldots, x_{i_k}) \in K$, we denote by $|s|$ the convex hull of s, that is, we put

$$|s| = \left\{t_0 x_{i_0} + \cdots + t_k x_{i_k} \mid t_i \geq 0, \sum_{i=0}^{k} t_i = 1\right\}.$$

Then $(t_0, \ldots, t_k)$ is called a *barycentric coordinate* of $t_0 x_{i_0} + \cdots + t_k x_{i_k}$ with respect to vertices $\{x_{i_0}, \ldots, x_{i_k}\}$. We now put

$$|K| = \bigcup_{s \in K} |s|.$$

The topological space $|K|$ endowed with the relative topology of $\mathbb{R}^n$ is called a *topological* (or *geometric*) *realization* of K or *polyhedron* of K.

Let $f\colon K \to L$ be a simplicial map. For an arbitrary point y of $|K|$, there exists $s = (x_{i_0}, \ldots, x_{i_k}) \in K$ with $y \in |s|$ by definition, that is, y can be written as $y = \sum_{j=0}^{k} t_j x_{i_j}$ with $t_j \geq 0$ for all j and $\sum_{j=0}^{k} t_j = 1$. Then we define a map $|f|\colon |K| \to |L|$ by

$$|f|(y) = \sum_{j=0}^{k} t_j f(x_{i_j}).$$

Note that some of the points $\{f(x_{i_j})\}$ may coincide; nevertheless $|f|(y)$ belongs to the set $|f(s)|$. It is easy to see that $|f|$ is well defined and continuous.

A *triangulation* (K, f) of a topological space M consists of a simplicial complex K and a homeomorphism $f\colon |K| \to M$.

The *barycentric subdivision* of a simplicial complex K is the simplicial complex K' defined by

$$K' = \{(s_0, \ldots, s_k) | s_i \in K, s_0 \subset \cdots \subset s_k\},$$

that is, K' is the simplicial complex whose vertices are the simplices of K and whose simplices are the sets $(s_0, \ldots, s_k)$, where the s_i are simplices of K such that $s_0 \subset \cdots \subset s_k$. This is in fact defined by regarding the barycentre of each simplex of K as a vertex of K'. It is evident that K' is a simplicial complex.

Example 1 We consider the simplicial complex K consisting of a 2-simplex and its proper faces. Then K' denotes the barycentric subdivision of K. In Fig. 5.1 we indicate the barycentric subdivision K' of K by accumulating their simplices and recommend the reader to draw them accurately.

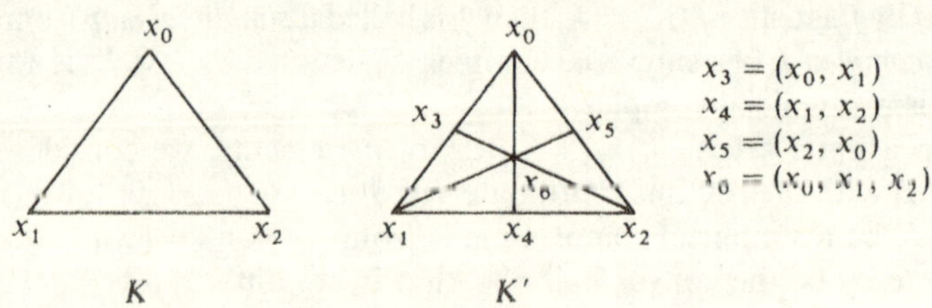

Fig. 5.1 The barycentric subdivision K' of K.

One verifies easily that there is a canonical homeomorphism

$$|K'| \approx |K|.$$

Next we take G-actions into consideration. Let G be a finite group with the discrete topology and K a (simplicial) complex. We say that a G-action on K is *simplicial* if the map

$$g\colon K \to K$$

induced by the action is simplicial for every $g \in G$. Precisely speaking, let X be the set of vertices of K. A G-action on X induces a G-action on the power set $\mathscr{P}(X)$ of X in general. If K is G-invariant, then the induced G-action on K is called a *simplicial G-action.* Observe that each $g \in G$ preserves the dimension of simplices of K. The simplicial complex K, together with such an action is called a (*simplicial*) *G-complex*.

Lemma 5.1 *Given a G-complex K, a continuous G-action is induced on its polyhedron* $|K|$.

Proof Each transformation $g: K \to K$ is simplicial by definition. Hence g induces a continuous map $|g|: |K| \to |K|$ as observed above. Evidently this defines a continuous G-action on $|K|$. □

A *G-triangulation* (K, f) of a G-space M consists of a G-complex K and a G-homeomorphism $f: |K| \to M$ where $|K|$ is given the G-action induced by the G-action on K according to Lemma 5.1.

We now prepare the following technical definition and lemmas which will be needed to derive formulae for Euler characteristics and transfer theorems, and to develop the Smith theory.

Following G. Bredon,[15] we consider the following two properties concerning a G-complex K:

(P_1) For any $g \in G$ and simplex s of K, g leaves $s \cap gs$ pointwise fixed.

(P_2) If $g_0, g_1, \ldots, g_k$ are elements of G and $(x_0, \ldots, x_k)$ and $(g_0x_0, \ldots, g_kx_k)$ are both simplices of K, then there exists an element g of G such that $gx_i = g_ix_i$ for all i. Here overlaps of some of x_i are allowed.

Lemma 5.2 *Let K be a G-complex having the property* (P_1) *and* $s = (x_0, \ldots, x_k)$ *be a simplex of K. If there exists* $g \in G$ *with* $gx_i = x_j$, *then we have* $x_i = x_j$.

Proof Since $x_j \in s \cap gs$, it follows from property (P_1) that $gx_j = x_j$. Hence we have $x_j = g^{-1}x_j = x_i$. □

Lemma 5.3 *Property* (P_2) *implies property* (P_1).

Proof Let K be a G-complex having property (P_2). If $x \in s \cap gs$, then $(x, gx) \subset gs$. In consequence, (x, gx) is a simplex of K. Since (x, x) is also a simplex of K, it follows from the assumption that there exists $g' \in G$ such that $g'x = x$, $g'x = gx$. Hence $gx = g'x = x$, which is property (P_1). □

Definition 5.1 For a G-complex K, we denote by ${}^GK$ the set of simplices of K whose vertices are all fixed by the G-action. On the other hand, we denote by K^G the fixed point set of the G-action on K.

Lemma 5.4 *For a G-complex K,* ${}^GK$ *is a subcomplex of K in general. If K possesses property* (P_1), *then* K^G *coincides with* ${}^GK$. *In consequence* K^G *becomes a subcomplex of K in this case.*

Proof Evidently ${}^GK$ is a subcomplex of K. Note that $K^G \supset {}^GK$ holds in general. Suppose that K possesses property (P_1). If $s \in K^G$, then $gs = s$ for any $g \in G$ by definition. It follows from property (P_1) that g leaves $s \cap gs = s$ pointwise fixed, that is, $s \in {}^GK$. □

Remark K^G does not coincide with ${}^GK$ in general and is not necessarily a subcomplex of K. See Example 2 which will be given later.

Corollary 5.5 *If a G-complex K possesses property* (P_1), *then* $|{}^GK| = |K|^G$.

Proof Obviously we have $|{}^GK| \subset |K|^G$. For any given $y \in |K|^G$, there exist unique $s = (x_0, \ldots, x_k) \in K$ and $t_i > 0$ for $i = 0, \ldots, k$ such that $\sum_{i=0}^k t_i = 1$ and $y = \sum_{i=0}^k t_i x_i$. Then for $g \in G$, gy is given by $\sum_{i=0}^k t_i g x_i$. Since $gy = y$ for any $g \in G$ by assumption, it follows from the uniqueness of the expression in terms of barycentric coordinate that the set of vertices of gs coincides with that of s. That is, $gs = s$, which means $s \in K^G$ by definition. On account of Lemma 5.4, we have $s \in {}^GK$. Thus $y \in |s| \subset |{}^GK|$, that is, the inclusion $|{}^GK| \supset |K|^G$ holds. □

Remark If K does not possess property (P_1), then $|K^G| \neq |K|^G \neq |{}^GK|$ in general. See Example 2 which will be given later.

A G-complex K does not have properties (P_1) and (P_2) in general. However, given a G-complex, we shall construct in a natural way a new G-complex which has both properties (P_1) and (P_2). It is easy to see that the barycentric subdivision K' of K inherits naturally the G-action on K and becomes a G-complex. The barycentric subdivision of K' is called the *second barycentric subdivision* of K and is denoted by K''.

Lemma 5.6 *If K is a G-complex, then the induced G-action on the barycentric subdivision K' possesses property* (P_1).

Proof For any given $g \in G$ and $s' \in K'$, we take an arbitrary vertex s of $s' \cap gs'$, thus a simplex of K. Then $(s, gs) \subset gs'$ clearly holds. Since (s, gs) is a set of vertices included in a simplex gs' of K', it follows from the definition of simplices of K' that either $s \subset gs$ or $s \supset gs$ holds. But G preserves the dimension of simplices of K, so that $s = gs$. □

Lemma 5.7 *If a G-complex K has property* (P_1), *then its barycentric subdivision K' has property* (P_2).

Proof Suppose that K has property (P_1). We shall prove property (P_2) for K' by induction on k which appears in the definition of (P_2). For $k = 0$, we have nothing to prove. Suppose that K' has property (P_2) for simplices of dimension less than k. Let $(s_0, s_1, \ldots, s_k)$ be a k-simplex of K'. By reordering, we may assume that $s_0 \subset \cdots \subset s_k$ as simplices of K. If $(g_0 s_0, \ldots, g_k s_k)$ is a simplex of K', then it follows from the inductive assumption that there exists $g \in G$ with $gs_i = g_i s_i$ for $0 \leq i < k$. Acting by g^{-1} on $(g_0 s_0, \ldots, g_k s_k)$, we have that

$$(s_0, s_1, \ldots, s_{k-1}, g^{-1} g_k s_k)$$

is a simplex of K'. Since these are ordered by ascending (or strictly speaking by non-descending) dimension, we have that

$$s_0 \subset s_1 \subset \cdots \subset s_{k-1} \subset g^{-1} g_k s_k.$$

Hence we have

$$s_{k-1} \subset s_k \cap g^{-1} g_k s_k.$$

It follows from (P_1) that $g^{-1} g_k$ leaves all the vertices of s_{k-1} fixed. Since s_i is a subset of s_{k-1} for $i \leq k-1$, we have

$$g^{-1} g_k s_i = s_i \qquad \text{for} \qquad i \leq k-1.$$

As a consequence, we have

$$g_k s_i = g s_i = g_i s_i \qquad \text{for} \qquad i \leq k-1.$$

Thus $g_k s_i = g_i s_i$ for all i. □

Following Bredon[15] we give the following definition. A simplicial action of G on K is called *regular* if K possesses property (P_2) for the action of each subgroup of G. Then K together with the regular action is called a *regular G-complex*.

By combining Lemmas 5.6 and 5.7, we have the following theorem.

Theorem 5.8 *Let K be an arbitrary G-complex. Then the second barycentric subdivision K″ of K is a regular G-complex.*

The remainder of the section will be devoted to the study of the orbit space of a G-complex. We begin with some preparation.

Let X and Y be finite sets. As explained earlier, a map $f\colon X \to Y$ induces a map $\mathscr{P}(X) \to \mathscr{P}(Y)$, which we denote also by f. Let K be a simplicial complex such that X is the set of vertices of K. Then it is easy to see that the image $f(K)$ becomes a simplicial complex such that $f(X)$ is the set of vertices of $f(K)$. We apply this observation to our situation.

Let K be a G-complex and X be the set of vertices of K. Concerning the induced G-action on X, we consider its orbit space X/G and the orbit map $p\colon X \to X/G$. As observed above, p induces a map $\mathscr{P}(X) \to \mathscr{P}(X/G)$, which we denote by p as well and $p(K)$ becomes a simplicial complex such that X/G is the set of vertices of $p(K)$. However, if K is not regular, then some peculiar phenomena occur as will be seen later. As observed previously, we have the inclusion $X \subset K$. For $s \in K$, we denote $p(s)$ by $\bar{s}$.

Lemma 5.9 *Let K be a G-complex possessing property* (P_1) *and X be the set of vertices of K. If* $s = (x_0, \ldots, x_k)$ *is a k-simplex of K, then* $\{\bar{x}_0, \ldots, \bar{x}_k\}$ *are distinct* $k + 1$ *points of* X/G.

Proof If $\bar{x}_i = \bar{x}_j$, then there exists $g \in G$ with $x_i = gx_j$. Hence we have $x_i = x_j$ by Lemma 5.2. □

Next we consider another kind of orbit space. Let K be a G-complex. Denote by K/G the orbit space of the G-action on K and by $\pi: K \to K/G$ the orbit map. For $s \in K$, we denote $\pi(s)$ by $[s]$. Note that K/G is not a simplicial complex in general and K/G does not coincide with $p(K)$ in general. For more details of this, see Example 2 below.

Under the regularity condition, we investigate K/G in the following.

Lemma 5.10 *Let K be a regular G-complex and X be the set of vertices of K. Then the orbit space K/G becomes a simplicial complex such that the set of vertices of K/G is X/G and K/G is naturally isomorphic to p(K). Moreover the orbit map* $\pi: K \to K/G$ *is a simplicial map preserving dimension of simplexes.*

Proof Suppose that two simplices $s = (x_0, \ldots, x_k)$ and $s' = (x'_0, \ldots, x'_k)$ of K satisfy $[s] = [s']$. That is, there exists $g \in G$ with $s' = gs$, which implies $\bar{s} = \bar{s}'$. Conversely, suppose that $\bar{s} = \bar{s}'$. Then it follows from the regularity condition that there exists $g \in G$ with

$$s' = (x'_0, \ldots, x'_k) = (gx_0, \ldots, gx_k) = gs.$$

Hence we have $[s] = [s']$. Thus K/G is naturally in bijective correspondence with the complex $p(K)$. In consequence, for a simplex $s = (x_0, \ldots, x_k)$ of K, the G-orbit $G(s)$ of s is determined by $([x_0], \ldots, [x_k])$. Note that this does not hold in general without the regularity condition. Accordingly the orbit space K/G is given the structure of a simplicial complex such that X/G is the set of vertices of K/G and K/G is naturally isomorphic to $p(K)$ as complexes. It turns out that the projection $\pi: K \to K/G$ is identified with the projection $p: K \to p(K)$. Consequently π is also simplicial. Further, it follows from Lemma 5.9 that π preserves the dimension of simplices. □

Exercise When K is a regular G-complex, show that $|K|/G$ and $|K/G|$ are naturally homeomorphic.

Example 2 We consider the complex K consisting of a 2-simplex and its proper faces as in Example 1. Denote by K' the barycentric subdivision of K. Figure 5.2 illustrates the topological realization of K and K'.

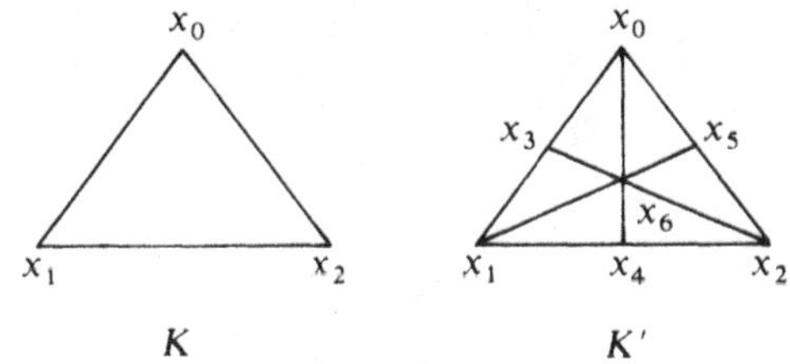

Fig. 5.2 Topological realization of K and K'

Consider the $\mathbb{Z}_3$-actions on K and K' defined by the obvious rotations. Note that K does not have property (P_1), while K' has property (P_1), but does not have property (P_2). Then the following phenomena occur in the G-complexes K and K'.

1. Since the map p takes all 0-simplices, 1-simplices, and the 2-simplex to a 0-simplex, $p(K)$ consists only of one point. On the other hand, the orbit space K/G consists of $[x_0, x_1, x_2]$, $[x_0, x_1]$, and $[x_0]$, so that K/G is not a simplicial complex. The fixed point set K^G of the G-action on K consists only of the 2-simplex, and hence is not a subcomplex of K. On the contrary, we have ${}^GK = \varnothing$. Note that $|K|^G$ coincides with the barycentre of the 2-simplex of the topological realization of K.

2. Using the notation of Figure 5.2, we put $s = (x_0, x_3, x_6)$ and $s' = (x_1, x_3, x_6)$. Any element of G does not take s to s'. That is, $[s] \neq [s']$. However as elements of $p(K')$ we have $\bar{s} = \bar{s}'$. Therefore $K'/G \neq p(K')$. Figure 5.3 illustrates K'/G.

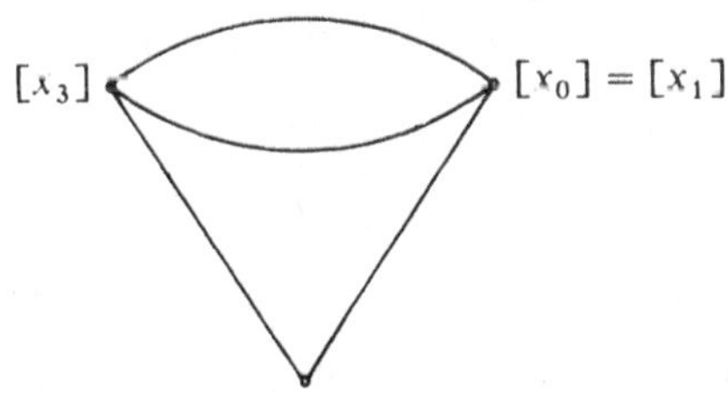

Fig. 5.3 K'/G

Although $[x_0] = [x_1]$, $[x_0, x_3]$ does not coincide with $[x_1, x_3]$. Specifically, the pair $([x_0], [x_3])$ does not define a simplex, hence K'/G is not a simplicial complex. Figure 5.4 shows $p(K')$ and illustrates that $p(K')$ is a simplicial complex. Since K' has property (P_1), we have $K'^G = {}^GK'$ and $|{}^GK'| = |K'|^G$.

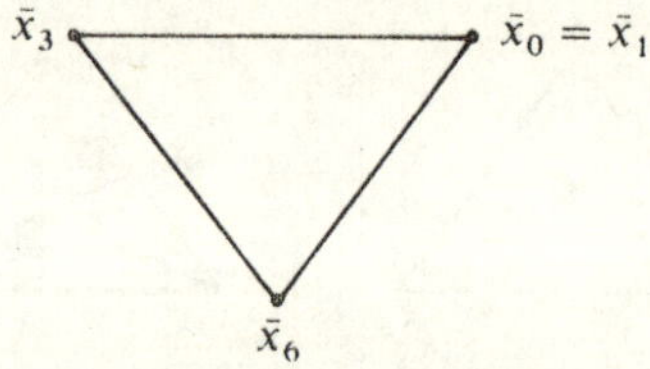

Fig. 5.4 $p(K')$

5.2 Formulae for Euler characteristics and Möbius inversion

In this section, we shall investigate the relationships among the Euler characteristics of a regular G-complex K, of its orbit space K/G, and of its various kinds of fixed point sets GK^H.

We begin with a brief review of the Euler characteristic following Spanier.[119] Let K be a simplicial complex and L be a subcomplex of K. An oriented k-simplex of K is a k-simplex $s \in K$ together with an equivalence class of total orderings of the vertices of s, two orderings being equivalent if they differ by an even permutation of the vertices.

Let $C_k(K, L)$ be the free abelian group generated by the oriented k-simplexes of $K - L$, modulo the relations $\sigma_1 + \sigma_2 = 0$ whenever σ_1 and σ_2 are different oriented k-simplexes corresponding to the same k-simplex of $K - L$.

We define homomorphisms

$$\partial_k: C_k(K, L) \to C_{k-1}(K, L)$$

for $k \geq 1$ as usual and get a chain complex

$$C(K, L) = \{C_k(K, L), \partial_k\}$$

which is called the *oriented chain complex* of (K, L). Then we put

$$Z_k(K, L) = \text{Ker}\ \partial_k, \qquad B_k(K, L) = \text{Im}\ \partial_{k+1}.$$

Then the quotient group

$$Z_k(K, L)/B_k(K, L)$$

is called the (*relative*) kth *homology group* of (K, L) and is denoted by $H_k(K, L)$. If L is empty, $H_k(K, \phi) = H_k(K)$ is called the (*absolute*) kth *homology group* of K. Put

$$b_k = \text{rank}\ H_k(K, L),$$

which is called the kth *Betti number* of (K, L). Then the *Euler characteristic* $\chi(K, L)$ of (K, L) is defined as the alternating sum

$$\chi(K, L) = \sum_{k=0}^{n} (-1)^k b_k \qquad (n = \dim K).$$

To study the Euler characteristic, we first put

$$a_k(K, L) = \text{the number of } k\text{-simplexes of } K - L$$

and recall the following familiar theorem.

Theorem 5.11 (Euler–Poincaré formula)

$$\chi(K, L) = \sum_{k=0}^{n} (-1)^k a_k(K, L).$$

Proof One verifies easily that

$$\operatorname{rank} C_k(K, L) = \operatorname{rank} Z_k(K, L) + \operatorname{rank} B_{k-1}(K, L)$$
$$\operatorname{rank} Z_k(K, L) = \operatorname{rank} B_k(K, L) + \operatorname{rank} H_k(K, L).$$

On the other hand, it follows from the definition that $\operatorname{rank} C_k(K, L) = a_k(K, L)$. By combining these relations, we obtain the formula in the theorem. □

Let F be an arbitrary field. Then the homology group $H_k(K, L; F)$ with coefficients in F is a vector space over F. By applying Theorem 5.11, one can easily deduce the following corollary.

Corollary 5.12

$$\chi(K, L) = \sum_{k=0}^{n} (-1)^k \dim H_k(K, L; F).$$

Next we shall take G-actions into consideration. As is well known, the barycentric subdivision does not affect the homology groups. Hence we can assume without loss of generality that K is a regular G-complex.

Let K be a regular G-complex and L a G-invariant subcomplex of K.

Remark Although the results in this section hold for every pair (K, L), we will deal only with K for brevity hereafter.

It follows from the definition of 'regular' that K is a regular H-complex for every subgroup H of G. Accordingly, in view of Lemma 5.4, K^H coincides with ${}^H K$, which is a subcomplex of K. Since the G-action is simplicial, gK^H is also a subcomplex of K for every $g \in G$. Therefore

$$GK^H = \bigcup_{g \in G} gK^H$$

is also a subcomplex of K. Moreover, it is easy to see that GK^H becomes a G-invariant regular G-complex, and hence its Euler characteristic $\chi(GK^H)$ is defined.

Concerning the G-action on K, we make use of the notation $K(H)$ which denotes the union of the orbits of type G/H as in Section 1.8.

Let us consider a set $S = \{(H_i) | i = 1, \ldots, m\}$ of conjugacy classes of subgroups of G which contains the set of isotropy types occurring in K. For example, it could be exactly the set of conjugacy classes of isotropy groups occurring in K, or it could be the set of conjugacy classes of all subgroups of G.

We now arrange $\{(H_i) | i = 1, \ldots, m\}$ as

$$(H_i) \leq (H_j) \Rightarrow i \leq j,$$

which is converse to the ordering mentioned previously.

Lemma 5.13 *For an arbitrary integer i with* $1 \leq i \leq m$, *we have*

$$G^{H_i}K = GK^{H_i} = \bigcup_{(H_j) \geq (H_j)} K(H_j).$$

Proof By combining Proposition 1.77 and Lemma 5.4, we have the theorem. □

Remark $K(H_i)$ is not a subcomplex of K in general.

Definition 5.2 Given $S = \{(H_i) | i = 1, \ldots, m\}$ as above, we put $I = \{1, \ldots, m\}$ and for $i \in I$, we put

$$J[i] = \{j \in I \mid (H_j) \leq (H_i)\},$$
$$J(i) = \{j \in I \mid (H_j) < (H_i)\}.$$

Remark $J(i+1) \neq J[i]$ in general.

Corollary 5.14 *For any* $j \in J[i]$, *we have*

$$K(H_i) \subset GK^{H_j}.$$

Proof If we exchange i for j and vice versa in Lemma 5.13, then we have

$$K(H_i) \subset GK^{H_j} \qquad \text{for} \qquad j \in J[i].$$ □

Lemma 5.15 *For any* $j \in I - J[i]$, *we have*

$$K(H_i) \cap GK^{H_j} = \varnothing.$$

Proof Suppose that there exists an element s of $K(H_i) \cap GK^{H_j}$. Then there exist $g \in G$ and $s' \in K^{H_j}$ with $s = gs'$. Hence we have

$$(H_i) = (G_s) = (G_{s'}) \geq (H_j),$$

which implies $j \in J[i]$, contradicting the assumption. □

Corollary 5.16

(1) *For* $j \in J[i]$, *we have* $K(H_i) = (GK^{H_j})(H_i)$.

(2) *For* $j \in I - J[i]$, *we have* $(GK^{H_j})(H_i) = \varnothing$.

Proof (1) Observe that GK^{H_j} is a G-invariant subcomplex of K. On the other hand, it follows from Corollary 5.14 that $K(H_i) \subset GK^{H_j}$ for $j \in J[i]$. In consequence, we have

$$K(H_i) \subset (GK^{H_j})(H_i) \subset K(H_i),$$

that is, the equality $K(H_i) = (GK^{H_j})(H_i)$ holds.

(2) In view of Lemma 5.15, we have

$$K(H_i) \cap GK^{H_j} = \varnothing \qquad \text{for} \qquad j \in I - J[i],$$

which means that $(GK^{H_j})(H_i) = \varnothing$. □

Definition 5.3 Let G be a finite group and K be a regular G-complex. Let $S = \{(H_i) \mid i = 1, \ldots, m\}$ be a set of conjugacy classes of subgroups of G which contains the set of isotropy types occurring in K. Denote by $|G|$ the order of G and by h_i the order of H_i. Put $h'_i = |G|/h_i$. Then we define inductively three classes of integers $n_1, \ldots, n_m, n'_1, \ldots, n'_m$, and $r_1, \ldots, r_m$ by

$$n_i = h_i - \sum_{j \in J(i)} n_j, \qquad n'_i = h'_i - \sum_{j \in J(i)} n'_j, \qquad r_i = 1 - \sum_{j \in J(i)} r_j.$$

If (H_i) is minimal, that is, $J(i) = \varnothing$, then the above definition means that $n_i = h_i$, $n'_i = h'_i$, and $r_i = 1$. If $j \in J(i)$, *then* $(H_j) < (H_i)$ by definition, so the inequality $j < i$ holds. Consequently it follows from the inductive assumption that the n_j are already defined for $j \in J(i)$ and n_i can be defined by the above formula. Similarly n'_i and r_i are defined.

Notice that for these integers, the following formulae hold:

$$h_i = \sum_{j \in J[i]} n_j, \qquad h'_i = \sum_{j \in J[i]} n'_j, \qquad 1 = \sum_{j \in J[i]} r_j.$$

Theorem 5.17

$$|G|\,\chi(K/G) = \sum_{i=1}^{m} n_i \chi(GK^{H_i}),$$

$$\chi(K) = \sum_{i=1}^{m} n_i' \chi(GK^{H_i}/G),$$

$$\chi(K) = \sum_{i=1}^{m} r_i \chi(GK^{H_i}),$$

$$\chi(K/G) = \sum_{i=1}^{m} r_i \chi(GK^{H_i}/G).$$

Proof Since the proofs of the four formulae are similar, we give a proof only of the first formula.

First we give a proof for the reader who is familiar with the notion of the Möbius inversion formula. Later we shall give an alternative proof of it without using the Möbius inversion formula. However the ideas of the two proofs are related implicitly.

We start by recalling the Möbius function in general. Let P be a partially ordered finite set with order $\leq$. Then there exists the Möbius function $\mu(\,,)$ of P which is defined inductively for $a \leq b$ by

$$\mu(a, a) = 1,$$

$$\mu(a, b) = -\sum_{a \leq z < b} \mu(a, z) = -\sum_{a < z \leq b} \mu(z, b) \quad \text{if} \quad a < b.$$

See Aigner[2] for a discussion of this function. In our case, we take the above S as P and consider the Möbius function of S.

We now generalize the notions of $a_k(\quad)$ and the Euler characteristic as follows. For a subset A of K (which consists of simplexes of K, but is not necessarily a subcomplex of K), we put

$$a_k(A) = \text{the number of } k\text{-simplexes in } A,$$

$$\chi(A) = \sum_{k=0}^{n} (-1)^k a_k(A).$$

On account of Theorem 5.11, this notation is consistent with that in the case where A is a subcomplex of K. A merit of this generalization is that we have the following additive property of $\chi(A)$ without assuming the A to be subcomplexes. If K is written as the disjoint union

$$K = \coprod_i A_i$$

of subsets of K, then we obviously have

$$a_k(K) = \sum_i a_k(A_i) \qquad \text{(additivity)},$$

and hence we have

$$\chi(K) = \sum_i \chi(A_i) \qquad \text{(additivity)}.$$

Our starting point is the following formula:

$$\chi(GK^{H_j}) = \sum_{(H_j) \le (H_i)} \chi(K(H_i)). \tag{5.1}$$

This follows from the disjoint union decomposition

$$GK^{H_j} = \coprod_{(H_j) \le (H_i)} K(H_i)$$

and the additivity of the generalized Euler characteristic $\chi(\ )$ above.

We now apply the Möbius inversion formula to (5.1) above and deduce

$$\sum_{(H_j) \le (H_i) \in S} \mu((H_j), (H_i))\chi(GK^{H_i}) = \chi(K(H_j)). \tag{5.2}$$

Each orbit in $K(H_j)$ has type G/H_j, so that we have

$$\chi(K(H_j)/G) = (h_j/|G|)\chi(K(H_j)). \tag{5.3}$$

Since $K/G = \coprod_{(H_j) \in S} K(H_j)/G$ is a disjoint union decomposition, it follows from the additivity of $\chi(\ )$ that

$$\chi(K/G) = \sum_{(H_j) \in S} \chi(K(H_j)/G). \tag{5.4}$$

Putting all this together, we have

$$\begin{aligned}
\chi(K/G) &= \sum_{(H_j) \in S} \chi(K(H_j)/G) \qquad \text{by (5.4)} \\
&= \sum_{(H_j) \in S} (h_j/|G|)\chi(K(H_j)) \qquad \text{by (5.3)} \\
&= \sum_{(H_j) \in S} (h_j/|G|) \sum_{(H_j) \le (H_i) \in S} \mu((H_j), (H_i))\chi(GK^{H_i}) \qquad \text{by (5.2)} \\
&= \sum_{(H_i) \in S} \left\{ \sum_{S \ni (H_j) \le (H_i)} (h_j/|G|)\mu((H_j), (H_i)) \right\} \chi(GK^{H_i}).
\end{aligned}$$

Thus we have

$$|G|\chi(K/G) = \sum_{(H_i) \in S} \left\{ \sum_{S \ni (H_j) \le (H_i)} h_j \mu((H_j), (H_i)) \right\} \chi(GK^{H_i}). \tag{5.5}$$

We now put

$$d_i = \sum_{S \ni (H_j) \le (H_j)} h_j \mu((H_j), (H_i)),$$

so that we have

$$|G|\chi(K/G) = \sum_{(H_i) \in S} d_i \chi(GK^{H_i}). \tag{5.6}$$

In the following, we shall investigate d_i. For fixed i, consider the following summation of d_k:

$$\begin{aligned}
\sum_{k\in J[i]} d_k &= \sum_{k\in J[i]} \left\{ \sum_{S\ni(H_j)\le(H_k)} h_j\mu((H_j),(H_k)) \right\} \\
&= \sum_{\substack{S\ni(H_j),(H_k)\\(H_j)\le(H_k)\le(H_i)}} h_j\mu((H_j),(H_k)) \\
&= h_i + \sum_{S\ni(H_j)<(H_i)} h_j \left\{ \sum_{\substack{S\ni(H_k)\\(H_j)\le(H_k)\le(H_i)}} \mu((H_j),(H_k)) \right\} \\
&= h_i.
\end{aligned}$$

On the other hand, $\sum_{k\in J[i]} n_k = h_i$ as observed previously. Hence it follows from the uniqueness that we have

$$d_i = n_i \qquad \text{for any } 1 \le i \le m.$$

Thus we have the first formula of the theorem.

Next we shall indicate an alternative proof of the first formula of the theorem without using the Möbius inversion formula. We begin by showing the following claim.

Claim For fixed k, the following equality holds:

$$\sum_{i\in I} h_i a_k(K(H_i)) = \sum_{i\in I} n_i a_k(GK^{H_i}).$$

First we fix i arbitrarily. It follows from the definition of n_i that

$$\begin{aligned}
h_i a_k(K(H_i)) &= \sum_{j\in J[i]} n_j a_k(K(H_i)) \\
&= \sum_{j\in J[i]} n_j a_k((GK^{H_j})(H_i)) \qquad \text{by Corollary 5.16 (1)} \\
&= \sum_{j\in I} n_j a_k((GK^{H_j})(H_i)) \qquad \text{by Corollary 5.16 (2).}
\end{aligned}$$

Next, by making use of the decomposition

$$GK^{H_j} = \coprod_{i\in I} (GK^{H_j})(H_i)$$

and the additivity of $a_k(\ \ )$, we compute

$$\begin{aligned}
\sum_{i\in I} h_i a_k(K(H_i)) &= \sum_{i\in I}\sum_{j\in I} n_j a_k((GK^{H_j})(H_i)) \\
&= \sum_{j\in I} n_j \sum_{i\in I} a_k((GK^{H_j})(H_i)) \\
&= \sum_{j\in I} n_j a_k(GK^{H_j}) = \sum_{i\in I} n_i a_k(GK^{H_i}),
\end{aligned}$$

which completes the proof of the claim.

Observe that K can be written as the disjoint union

$$K = \coprod_{i \in I} K(H_i)$$

of G-invariant subsets $K(H_i)$ of K but note that some of them may be empty. Accordingly we have the disjoint union decomposition

$$K/G = \coprod_{i \in I} K(H_i)/G.$$

It follows from the additivity of $a_k(\quad)$ that

$$a_k(K/G) = \sum_{i \in I} a_k(K(H_i)/G).$$

Hence by the Euler–Poincaré formula, we have

$$\begin{aligned}\chi(K/G) &= \sum_{k=0}^{n} (-1)^k a_k(K/G) \\ &= \sum_{k=0}^{n} (-1)^k \sum_{i=1}^{m} a_k(K(H_i)/G).\end{aligned}$$

On account of Lemma 5.10, the projection $\pi\colon K \to K/G$ preserves the dimension of simplexes. Consequently we have

$$a_k(K(H_i)/G) = (h_i/|G|)a_k(K(H_i)).$$

Putting all this together, we compute:

$$\begin{aligned}|G|\chi(K/G) &= |G| \sum_{k=0}^{n} (-1)^k a_k(K/G) \\ &= |G| \sum_{k=0}^{n} (-1)^k \sum_{i=1}^{m} a_k(K(H_i)/G) \\ &= |G| \sum_{k=0}^{n} (-1)^k \sum_{i=1}^{m} (h_i/|G|)a_k(K(H_i)) \\ &= \sum_{k=0}^{n} (-1)^k \sum_{i=1}^{m} h_i a_k(K(H_i)) \\ &= \sum_{k=0}^{n} (-1)^k \sum_{i=1}^{m} n_i a_k(GK^{H_i}) \qquad \text{by Claim} \\ &= \sum_{i=1}^{m} n_i \sum_{k=0}^{n} (-1)^k a_k(GK^{H_i}) \\ &= \sum_{i=1}^{m} n_i \chi(GK^{H_i}).\end{aligned}$$

This completes the alternative proof of the first formula. □

In order to understand the meaning of the theorem, let us apply the theorem to special types of transformation groups. In the following, we suppose that a regular G-complex K is given.

Corollary 5.18 *Suppose that* $\{(H_i)\}$ *of Theorem* 5.17 *satisfies* $(H_1) < (H_2) < \cdots < (H_m)$. *Then we have*

$$|G|\chi(K/G) = h_1\chi(GK^{H_1}) + (h_2 - h_1)\chi(GK^{H_2}) + \cdots + (h_m - h_{m-1})\chi(GK^{H_m}).$$

Proof It follows from the assumption that for $i \geq 2$ we have

$$J[i-1] = \{1, 2, \ldots, i-1\} = J(i).$$

Accordingly we have

$$n_i = h_i - \sum_{j \in J(i)} n_j = h_i - \sum_{j \in J[i-1]} n_j = h_i - h_{i-1}. \qquad \square$$

Corollary 5.19 *If G is commutative, then we have*

$$|G|\chi(K/G) = \sum_i n_i\chi(K^{H_i}).$$

Proof If G is commutative, then the equality $GK^{H_i} = K^{H_i}$ holds. Hence the theorem yields the corollary.

Corollary 5.20 *If G is a cyclic group $\mathbb{Z}_{p^r}$ of order p^r where p is a prime number, then we have*

$$p^r\chi(K/\mathbb{Z}_{p^r}) = \chi(K) + (p-1)\chi(K^{\mathbb{Z}_p}) + (p^2 - p)\chi(K^{\mathbb{Z}_{p^2}}) + \cdots + (p^r - p^{r-1})\chi(K^{\mathbb{Z}_{p^r}}).$$

Therefore we obtain the congruence:

$$\chi(K) + (p-1)\chi(K^{\mathbb{Z}_p}) + \cdots + (p^r - p^{r-1})\chi(K^{\mathbb{Z}_{p^r}}) \equiv 0 \bmod p^r.$$

Proof The corollary is proved by combining Corollaries 5.18 and 5.19. $\square$

Corollary 5.21 *If the action is semi-free, then we have*

$$|G|\chi(K/G) = \chi(K) + (|G| - 1)\chi(K^G).$$

Therefore we obtain the congruence:

$$\chi(K^G) \equiv \chi(K) \bmod |G|.$$

Corollary 5.22 *If the action is free, then we have*

$$|G|\chi(K/G) = \chi(K).$$

Remark Note that $S = \{(H_i)\}$ can be any set of conjugacy classes of subgroups of G which contains the set of isotropy types occurring in K. However the formulae above depend on the choice of $S = \{(H_i)\}$ and the coefficients n_i, n'_i, r_i depend on this choice as well.

In the following we shall give examples which illustrate such phenomena with respect to the first formula in the theorem. Similar phenomena occur for the rest of the formulae in the theorem.

Example 1 Let $Z_{p^{i_1}} < \cdots < Z_{p^{i_s}}$ be the set of isotropy types which occur actually in K in Corollary 5.20. Then we have

$$p^r\chi(K/Z_{p^r}) = p^{i_1}\chi(K^{Z_{p^{i_1}}}) + (p^{i_2} - p^{i_1})\chi(K^{Z_{p^{i_2}}}) + \cdots + (p^{i_s} - p^{i_{s-1}})\chi(K^{Z_{p^{i_s}}}).$$

This formula coincides with that of Corollary 5.20 by collecting the first i_1 terms, the next $i_2 - i_1$ terms, and so on, and the last $r - i_s$ terms do not occur.

Example 2 Let H_2 and H_3 be two subgroups of a finite group G such that there is no ordering relation between (H_2) and (H_3). Put $H_1 = \{e\}$ and $H_4 = G$. For instance, $G = \mathbb{Z}_{15}$, $H_2 = \mathbb{Z}_3$, and $H_3 = \mathbb{Z}_5$. Suppose that the set of isotropy types of a regular G-complex K is exactly $\{(H_1), (H_2), (H_4)\}$. Then it follows from Lemma 5.13 that

$$GK^{H_3} = \bigcup_{(H_j) \geq (H_3)} K(H_j) = K(H_4) = K^G.$$

If we apply the theorem to K with respect to the set $S = \{(H_1), (H_2), (H_3), (H_4)\}$, we have

$$h_4\chi(K/G) = \chi(K) + (h_2 - 1)\chi(GK^{H_2}) + (h_3 - 1)\chi(GK^{H_3}) + (h_4 - h_2 - h_3 + 1)\chi(K^G). \tag{5.7}$$

On the other hand, if we apply the theorem to K with respect to the set $S = \{(H_1), (H_2), (H_4)\}$, then we have

$$h_4\chi(K/G) = \chi(K) + (h_2 - 1)\chi(GK^{H_2}) + (h_4 - h_2)\chi(K^G). \tag{5.8}$$

By substituting $\chi(K^G)$ for $\chi(GK^{H_3})$ in (5.7), we get the formula (5.8).

Example 3 Let G and H_i be as in Example 2. Suppose that the set of isotropy types of a regular G-complex K is exactly $S = \{(H_2), (H_3), (H_4)\}$. If we apply the theorem to K with respect to the set S above, we have

$$h_4\chi(K/G) = h_2\chi(GK^{H_2}) + h_3\chi(GK^{H_3}) + (h_4 - h_2 - h_3)\chi(K^G). \tag{5.9}$$

In view of Lemma 5.13, we easily find that

$$K = GK^{H_2} \cup GK^{H_3},$$
$$K^G = GK^{H_2} \cap GK^{H_3}.$$

Hence we have

$$\chi(K) = \chi(GK^{H_2}) + \chi(GK^{H_3}) - \chi(K^G) \tag{5.10}$$

by the basic property of the Euler characteristic. By substituting the right-hand side of (5.10) for $\chi(K)$ of (5.7) of Example 2, we obtain the formula (5.9).

Next we shall prove further formulae for Euler characteristics. We begin by preparing the following lemma. For a finite set A, we denote its cardinal number by $|A|$.

Lemma 5.23 *Let G be a finite group and H a subgroup of G. Then we have*

$$\sum_{g \in G} |(G/H)^g| = |G|.$$

Proof Fix an arbitrary coset $aH \in G/H$. We first compute the number of $g \in G$ such that the coset aH is fixed by the map $g\colon G/H \to G/H$, that is,

$$|\{g \in G | gaH = aH\}| = |\{g \in G | g \in aHa^{-1}\}|$$
$$= |aHa^{-1}| = |H|.$$

Accordingly we have

$$\sum_{g \in G} |(G/H)^g| = \sum_{aH \in G/H} |\{g \in G | gaH = aH\}| = \sum_{aH \in G/H} |H|$$
$$= |G/H| \cdot |H| = |G|. \qquad \square$$

Let K be a regular G-complex. Then we put

$$S(K) = \bigcup_{g \in G - \{e\}} K^g$$

which is called the *singular set* of K.

Theorem 5.24 *Let K be a regular G-complex. Then we have:*

(1) $\sum_{g \in G} \chi(K^g) = |G| \chi(K/G)$;

(2) $\chi(K) - \chi(S(K)) = |G| \{\chi(K/G) - \chi(S(K)/G)\}$;

(3) $\chi(K) \equiv \chi(S(K)) \equiv - \sum_{g \in G - \{e\}} \chi(K^g) \bmod |G|$.

Proof We now regard each simplex s of K as an element of the power set $\mathscr{P}(X)$ as mentioned in Section 5.1 where X is the set of vertices of K. Then the orbit $G(s)$ of s can be identified with G/G_s by Theorem 1.5 where G_s denotes the isotropy group at s. Accordingly it follows from Lemma 5.23 that if s is a k-simplex, then

$$\sum_{g\in G} a_k(G(s)^g) = \sum_{g\in G} |(G/G_s)^g| = |G|$$

where $a_k(G(s)^g)$ denotes the number of k-simplexes in $G(s)^g$. In the following we temporarily denote by $(K/G)^k$ the set of representatives in K of k-simplexes in K/G. Then the equality $|(K/G)^k| = a_k(K/G)$ clearly holds. Let n be the dimension of K.

By making use of Lemma 5.10 and the above observation, we compute:

$$\begin{aligned}
\sum_{g\in G} \chi(K^g) &= \sum_{g\in G}\sum_{k=0}^{n} (-1)^k a_k(K^g)\\
&= \sum_{g\in G}\sum_{k=0}^{n} (-1)^k \sum_{s\in(K/G)^k} a_k(G(s)^g)\\
&= \sum_{k=0}^{n} (-1)^k \sum_{s\in(K/G)^k}\sum_{g\in G} a_k(G(s)^g)\\
&= \sum_{k=0}^{n} (-1)^k \sum_{s\in(K/G)^k} |G|\\
&= |G| \sum_{k=0}^{n} (-1)^k a_k(K/G)\\
&= |G|\chi(K/G),
\end{aligned}$$

which completes the proof of (1).

Since G acts freely on $K - S(K)$, it follows that

$$\begin{aligned}
\chi(K) - \chi(S(K)) &= \sum_{k=0}^{n} (-1)^k a_k(K) - \sum_{k=0}^{n} (-1)^k a_k(S(K))\\
&= \sum_{k=0}^{n} (-1)^k a_k(K - S(K))\\
&= \sum_{k=0}^{n} (-1)^k |G| a_k(K/G - S(K)/G)\\
&= |G| \left\{ \sum_{k=0}^{n} (-1)^k a_k(K/G) - \sum_{k=0}^{n} (-1)^k a_k(S(K)/G) \right\}\\
&= |G|\{\chi(K/G) - \chi(S(K)/G)\},
\end{aligned}$$

which completes the proof of (2).

Finally, the two congruences in (3) follow from (1) and (2). □

5.3 The transfer

The concept of the transfer was originally considered for covering spaces. It has since been widely generalized as an important notion and applied to, for instance, the solution of the Adams conjecture and Riemann–Roch type theorems. In this section, we shall introduce the transfer in a form close to the original.

Let G be a finite group and $\mathbb{Z}$ be the ring of integers. Then we denote by $\mathbb{Z}G$ the integral group ring of G. That is, $\mathbb{Z}G$ is the set of all formal sums $\sum_i n_i g_i$, $n_i \in \mathbb{Z}$, $g_i \in G$, with addition and multiplication given by

$$\sum_i n_i g_i + \sum_i m_i g_i = \sum_i (n_i + m_i) g_i,$$

$$\left(\sum_i n_i g_i\right) \cdot \left(\sum_j m_j g_j\right) = \sum_{i,j} (n_i m_j)(g_i g_j).$$

Let K be a regular G-complex and $C(K)$ be the oriented chain complex of K. The G-action on K induces a G-action on $C(K)$, so that $C(K)$ becomes a module over $\mathbb{Z}G$.

We now define a special element α of $\mathbb{Z}G$ by

$$\alpha = \sum_{g \in G} g$$

and put

$$\alpha c = \sum_{g \in G} gc \qquad \text{for } c \in C(K).$$

We also use α to denote the chain map

$$\alpha\colon C(K) \to C(K)$$

defined by the correspondence $c \mapsto \alpha c$. Its image is a subcomplex of $C(K)$ and is denoted by $\alpha C(K)$.

If L is a G-invariant subcomplex of K, then G acts on $C(K, L) = C(K)/C(L)$, so that a chain map $\alpha\colon C(K, L) \to C(K, L)$ and its image $\alpha C(K, L)$ are defined similarly.

Let $\pi\colon K \to K/G$ be the canonical simplicial map obtained in Section 5.1 and use π to also denote the induced chain map

$$\pi\colon C(K, L) \to C(K/G, L/G).$$

Lemma 5.25

$$\operatorname{Ker}\{\alpha\colon C(K, L) \to C(K, L)\} = \operatorname{Ker}\{\pi\colon C(K, L) \to C(K/G, L/G)\}.$$

Proof For any given simplex s_1 of K, let $\{s_1, \ldots, s_k\}$ be the orbit of s_1. By regularity, we can orient each s_i and $[s_1]$ such that the map $\pi\colon s_i \to [s_1]$ preserves these orientations. In order to prove the lemma, it suffices to consider chains of the form:

$$c = \sum_{i=1}^{k} n_i s_i \qquad n_i \in \mathbb{Z}.$$

Since $\pi c = \left(\sum_{i=1}^{k} n_i\right)[s_1]$, we have

$$\pi c = 0 \Leftrightarrow \sum_{i=1}^{k} n_i = 0.$$

On the other hand, it is easy to see that

$$\alpha s_i = \frac{|G|}{k} \sum_{j=1}^{k} s_j$$

for any s_i with $1 \leq i \leq k$. Consequently,

$$\alpha c = \left(\sum_{i=1}^{k} n_i\right) \frac{|G|}{k} \sum_{j=1}^{k} s_j.$$

Thus we have

$$\alpha c = 0 \Leftrightarrow \sum_{i=1}^{k} n_i = 0.$$ □

Corollary 5.26 *The correspondence $\alpha c \mapsto \pi c$ gives a chain isomorphism*

$$h\colon \alpha C(K, L) \to C(K/G, L/G).$$

Proof Evidently π is surjective. Hence it follows from Lemma 5.25 that we have the following isomorphisms of groups:

$$\alpha C(K, L) \cong C(K, L)/\operatorname{Ker} \alpha = C(K, L)/\operatorname{Ker} \pi \cong C(K/G, L/G).$$

Since these isomorphisms are all chain maps, they are chain isomorphisms as well. It follows from the homomorphism theorem of groups that the composition $h\colon \alpha C(K, L) \to C(K/G, L/G)$ of the above isomorphisms is given by taking αc to πc. □

Definition 5.4 First we define a map

$$\tau\colon C(K/G, L/G) \to C(K, L)$$

by the composition

$$C(K/G, L/G) \xrightarrow{\cong} \alpha C(K, L) \subset C(K, L)$$

of the inverse of the above isomorphism with the inclusion, that is, we put $\tau(\pi c) = \alpha c$. Then the homomorphism

$$\tau_*\colon H_*(K/G, L/G) \to H_*(K, L)$$

of homology groups induced by τ is called the *transfer homomorphism*.

On the other hand, π induces the usual homomorphism

$$\pi_*: H_*(K, L) \to H_*(K/G, L/G).$$

Lemma 5.27

$$\pi_* \cdot \tau_* = |G|: H_*(K/G, L/G) \to H_*(K/G, L/G),$$
$$\tau_* \cdot \pi_* = \alpha_* = \sum_{g \in G} g_*: H_*(K, L) \to H_*(K, L).$$

Proof Given $\bar{c} \in C(K/G, L/G)$, let c be an element of $C(K, L)$ with $\bar{c} = \pi c$. Since $\tau \cdot \pi c = \alpha c$ by definition and since $\pi \cdot g = \pi$, we have

$$\pi \cdot \tau(\bar{c}) = \pi \cdot \tau \cdot \pi(c) = \pi(\alpha c) = |G| \pi c = |G| \bar{c},$$

that is, $\pi \cdot \tau = |G|$. Consequently $\pi_* \cdot \tau_* = (\pi \cdot \tau)_* = |G|$. Also, the equality $\tau \cdot \pi = \alpha$ implies that

$$\tau_* \cdot \pi_* = (\tau \cdot \pi)_* = \alpha_* = \sum_{g \in G} g_*. \qquad \square$$

Remark Since Lemma 5.27 was proved at the chain level, the lemma holds for homology groups with coefficients in arbitrary abelian groups.

The following theorem gives the homological relation between a space with a finite group action and its orbit space. In particular, this will be employed for the computation of the homology groups of covering spaces.

Theorem 5.28 *Let Λ be a field of characteristic* 0 *or prime to* $|G|$. *Then the following two maps are isomorphisms*:

$$\pi_*: H_*(K, L; \Lambda)^G \to H_*(K/G, L/G; \Lambda)$$
$$\tau_*: H_*(K/G, L/G; \Lambda) \to H_*(K, L; \Lambda)^G.$$

Here $H_*(K, L; \Lambda)^G$ denotes the fixed point set of the induced action of G on $H_*(K, L; \Lambda)$. Moreover, we have

$$\operatorname{Ker}\{\pi_*: H_*(K, L; \Lambda) \to H_*(K/G, L/G; \Lambda)\}$$
$$= \operatorname{Ker}\{\alpha_*: H_*(K, L; \Lambda) \to H_*(K, L; \Lambda)\}.$$

Proof Obviously, $\operatorname{Im}\tau$ is included in the fixed point set $C(K, L; \Lambda)^G$ of the induced action of G on $C(K, L; \Lambda)$, so that we have

$$\operatorname{Im}\tau_* \subset H_*(K, L; \Lambda)^G.$$

We now restrict the domain of π_* to $H_*(K, L; \Lambda)^G$ and the range of τ_* to $H_*(K, L; \Lambda)^G$ as well, so that we have

$$\tau_* \cdot \pi_* = |G|: H_*(K, L; \Lambda)^G \to H_*(K, L; \Lambda)^G,$$

since $\tau_* \cdot \pi_* = \sum_{g \in G} g_*$ by Lemma 5.27. On the other hand, it follows also from Lemma 5.27 that $\pi_* \cdot \tau_* = |G|$. Consequently, if Λ is a field of characteristic 0 or prime to $|G|$, then both π_* and τ_* are isomorphisms. In particular, $\tau_*: H_*(K/G, L/G; \Lambda) \to H_*(K, L: \Lambda)$ is injective. Therefore the equality $\tau_* \cdot \pi_* = \alpha_*$ implies that $\operatorname{Ker} \pi_* = \operatorname{Ker} \alpha_*$. □

For later use, we generalize the transfer homomorphism somewhat. Let G be a finite group, H a subgroup of G, and K a regular G-complex. Consider the following commutative diagram

$$\begin{array}{ccccc} & & K & & \\ & \swarrow^{\pi_H} & & \searrow^{\pi} & \\ K/H & & \xrightarrow{\pi_{G/H}} & & K/G \end{array}$$

where the maps are all projections. As before π_H and $\pi_{G/H}$ stand also for the induced chain maps. Also we define a special element α_H of the group ring $\mathbb{Z}H$ by

$$\alpha_H = \sum_{h \in H} h \in \mathbb{Z}H.$$

Let $\{g_i\}$ be a set of representatives of the right cosets of H in G. Then we define a chain map

$$\tau_{G/H}: C(K/G, L/G) \to C(K/H, L/H)$$

by the following commutative diagram

$$\begin{array}{ccc} C(K/G, L/G) & \xrightarrow{\tau_{G/H}} & C(K/H, L/H) \\ \cong \downarrow & & \uparrow \cong \\ \alpha C(K, L) & \xrightarrow{\subset} & \alpha_H C(K, L) \end{array}$$

For $c \in C(K, L)$, the correspondences are given by

$$\begin{array}{ccc} \pi(c) & \longmapsto & \pi_H(\Sigma g_i c) \\ \downarrow & & \uparrow \\ \alpha c & \overset{=}{\longmapsto} & \alpha_H(\Sigma g_i c) \end{array}$$

Then the homomorphism

$$(\tau_{G/H})_*: H_*(K/G, L/G) \to H_*(K/H, L/H)$$

of homology groups induced by $\tau_{G/H}$ is also called the *transfer homomorphism*. Observe that

$$\begin{aligned}\pi_{G/H} \cdot \tau_{G/H} \cdot \pi(c) &= \pi_{G/H} \cdot \pi_H(\textstyle\sum g_i c) \\ &= \pi(\textstyle\sum g_i c) = (|G|/|H|)\pi(c).\end{aligned}$$

In consequence, we have the following lemma as in Lemma 5.27.

Lemma 5.29

$$(\pi_{G/H})_* \cdot (\tau_{G/H})_* = |G|/|H|: H_*(K/G, L/G) \to H_*(K/G, L/G).$$

The above results can be dualized so that similar results hold in cohomology.

Theorem 5.30 $\tau^* \cdot \pi^* = |G|$, $\pi^* \cdot \tau^* = \alpha^* = \sum_{g \in G} g^*$. *If* Λ *is a field of characteristic* 0 *or prime to* $|G|$, *then the following two maps are isomorphisms*:

$$\pi^*: H^*(K/G, L/G; \Lambda) \to H^*(K, L; \Lambda)^G$$

$$\tau^*: H^*(K, L; \Lambda)^G \to H^*(K/G, L/G; \Lambda).$$

Moreover, we have

$$\begin{aligned}&\operatorname{Ker}\{\tau^*: H^*(K, L; \Lambda) \to H^*(K/G, L/G; \Lambda)\} \\ &= \operatorname{Ker}\{\alpha^*: H^*(K, L; \Lambda) \to H^*(K, L; \Lambda)\}.\end{aligned}$$

Example 1 When G is a finite group, any G-manifold is G-triangulable.[51] We consider the free $\mathbb{Z}_2$-action on S^n ($n \geq 1$) of Example 1 in Section 4.4 and apply Theorem 5.28 to the projection $\pi: S^n \to \mathbb{R}P^n$. Let Λ be a field of characteristic 0 (for instance the rational numbers $\mathbb{Q}$) or prime to 2. It follows from the above theorem that we have the following isomorphism

$$H_*(S^n; \Lambda)^{\mathbb{Z}_2} \cong H_*(\mathbb{R}P^n; \Lambda).$$

Then we investigate the $\mathbb{Z}_2$-action on $H_*(S^n; \Lambda)$. By induction, one verifies easily that

$$H_n(S^n; \Lambda)^{\mathbb{Z}_2} = \begin{cases} H_n(S^n; \Lambda) \cong \Lambda & \text{for } n \text{ odd}, \\ \{0\} & \text{for } n \text{ even}. \end{cases}$$

Hence we have that, if n is odd, then

$$H_i(\mathbb{R}P^n; \Lambda) = \begin{cases} \Lambda & \text{for } i = 0 \text{ and } i = n, \\ \{0\} & \text{otherwise}, \end{cases}$$

and, if n is even, then

$$H_i(\mathbb{R}P^n; \Lambda) = \begin{cases} \Lambda & \text{for } i = 0, \\ \{0\} & \text{otherwise.} \end{cases}$$

Similar results hold in cohomology.

Example 2 Let us consider the free $\mathbb{Z}_k$-action on S^{2n+1} of Example 2 in Section 4.4. Let Λ be a field of characteristic 0 or prime to k. According to Theorem 5.28, we have

$$H_*(S^{2n+1}; \Lambda)^{\mathbb{Z}_k} \cong H_*(L(k; l_1, \ldots, l_n); \Lambda).$$

Since the $\mathbb{Z}_k$-action on S^{2n+1} can be extended to an S^1-action, the map defined by each $g \in \mathbb{Z}_k$ is homotopic to the identity. Hence we have

$$H_*(L(k; l_1, \ldots, l_n); \Lambda) \cong H_*(S^{2n+1}; \Lambda)^{\mathbb{Z}_k} = H_*(S^{2n+1}; \Lambda).$$

A similar result holds in cohomology.

Remark We shall deal with the case $G = \Lambda = \mathbb{Z}_p$ (p prime) in Sections 5.4, 5.5, and 5.8.

5.4 $\mathbb{Z}_p$-actions and the Smith homology group

In this section, we shall consider a cyclic group G of prime order p and homology with coefficients in $\mathbb{Z}_p$. Here we regard G as a multiplicative group, while we regard $\mathbb{Z}_p$ as a field.

Fix a generator g of G and define two special elements α and β of the group ring $\mathbb{Z}_pG$ of G over $\mathbb{Z}_p$ by

$$\alpha = 1 + g + g^2 + \cdots + g^{p-1},$$
$$\beta = 1 - g.$$

Since $g^p = 1$, we have

$$\alpha\beta = \beta\alpha = 0.$$

By induction on i, one verifies easily that

$$(-1)^i \binom{p-1}{i} \equiv 1 \bmod p,$$

which means that

$$\alpha = \beta^{p-1}.$$

If $\gamma = \beta^i$, we put $\bar{\gamma} = \beta^{p-i}$. In particular, $\beta = \bar{\alpha}$ and $\alpha = \bar{\beta}$.

Let K be a regular G-complex and L a G-invariant subcomplex of K. Corresponding to $\gamma = \beta^i$ $(1 \le i \le p-1)$, we consider the chain subcomplex

$$\gamma C(K, L; \mathbb{Z}_p)$$

of $C(K, L; \mathbb{Z}_p)$. Hereafter, for brevity, we omit the coefficients $\mathbb{Z}_p$ from the notation except when we want to emphasize.

Lemma 5.31 *For each $\gamma = \beta^i$ $(1 \le i \le p-1)$,*

$$0 \to \bar{\gamma}C(K, L) \oplus C(K^G, L^G) \xrightarrow{f} C(K, L) \xrightarrow{\gamma} \gamma C(K, L) \to 0$$

is an exact sequence of chain complexes, where f is the sum of the inclusions and γ denotes also the map given by $c \mapsto \gamma c$.

Proof By the natural inclusion maps, we regard both $\bar{\gamma}C(K, L)$ and $C(K^G, L^G)$ as subcomplexes of $C(K, L)$.

For convenience we shall introduce some temporary notation. Given a subset A of $K - L$, we denote by $C(A)$ the subgroup of $C(K, L)$ generated by the simplexes belonging to A. Note that $C(A)$ is not a chain complex in general.

Evidently γ is surjective.

Now we show that f is injective. We first remark that $K - K^G$ is G-invariant. Suppose that $f(\bar{\gamma}c \oplus c') = 0$ for $c \in C(K, L)$ and $c' \in C(K^G, L^G)$. We write c as $c = c_1 + c_2$ where $c_1 \in C(K - L - K^G)$ and $c_2 \in C(K^G - L^G)$. Then $\beta c_2 = 0$, and hence $\bar{\gamma}c_2 = 0$. It follows that

$$0 = f(\bar{\gamma}c \oplus c') = \bar{\gamma}c + c' = \bar{\gamma}c_1 + \bar{\gamma}c_2 + c' = \bar{\gamma}c_1 + c'.$$

Since $\bar{\gamma}c_1 \in C(K - L - K^G)$ and $c' \in C(K^G - L^G)$ by the above remark, this means that $\bar{\gamma}c_1 = c' = 0$. That is, f is injective.

Next we shall show that $\gamma f = 0$. First notice that $\gamma\bar{\gamma} = \beta^p = \beta\alpha = 0$. For any $c \in C(K, L)$ and $c' \in C(K^G, L^G)$, we have

$$\gamma f(\bar{\gamma}c \oplus c') = \gamma(\bar{\gamma}c + c') = \gamma\bar{\gamma}c + \gamma c' = \gamma c' = 0.$$

That is, $\gamma f = 0$ holds. In other words, $\operatorname{Im} f$ is included in $\operatorname{Ker} \gamma$.

Finally, we show that $\operatorname{Ker} \gamma = \operatorname{Im} f$. Let s be an arbitrary simplex of K. Then we consider the orbit $G(s)$ of s. Since $\gamma C(G(s)) \subset C(G(s))$, it suffices to prove that $\operatorname{Ker} \gamma \subset \operatorname{Im} f$ on $C(G(s))$. If s is in K^G, then $G(s) = s$ so that $\gamma(ns) = 0$ for any $n \in \mathbb{Z}_p$, that is $\gamma\,(C(G(s))) = 0$. On the other hand, for any $n \in \mathbb{Z}_p$, we have $f(0 \oplus ns) = ns$, that is, $\operatorname{Im} f \supset C(G(s))$. Thus $\operatorname{Ker} \gamma \subset \operatorname{Im} f$ on $C(G(s))$.

If s is in $K - K^G$, then any element c of $C(G(s))$ has the form

$$c = \sum n_i g^i s, \qquad n_i \in \mathbb{Z}_p$$

and the correspondence $\sum n_i g^i s \mapsto \sum n_i g^i$ defines a Λ-module isomorphism of $C(G(s))$ onto $\Lambda = \mathbb{Z}_p G$. Therefore we need only prove that

$$\operatorname{Im}\{f: \bar{\gamma}\Lambda \to \Lambda\} = \operatorname{Ker}\{\gamma: \Lambda \to \gamma\Lambda\}.$$

Note that $\bar{\gamma}\Lambda$, Λ, and $\gamma\Lambda$ are vector spaces over $\mathbb{Z}_p$ and that f is injective and γ

is surjective. Since $\operatorname{Im} f \subset \operatorname{Ker} \gamma$ holds in general as shown above, it suffices to show that $\dim \bar{\gamma}\Lambda + \dim \gamma\Lambda = \dim \Lambda = p$ as vector spaces over $\mathbb{Z}_p$. For any $\lambda = \sum n_i g^i \in \Lambda$, we have

$$\beta\lambda = (n_0 - n_{p-1}) + (n_1 - n_0)g + (n_2 - n_1)g^2 + \cdots + (n_{p-1} - n_{p-2})g^{p-1}.$$

Accordingly, we have

$$\lambda \in \operatorname{Ker} \beta \Leftrightarrow n_0 = n_1 = \cdots = n_{p-1},$$

that is, $\operatorname{Ker} \beta$ is the subspace $\mathbb{Z}_p\alpha$ of dimension 1 generated by α. Hence $\dim \beta\Lambda = \dim \Lambda - 1 = p - 1$. Since $\alpha = \beta^{p-1} = \beta^i \beta^{p-i-1}$, we have

$$\operatorname{Ker} \beta = \mathbb{Z}_p\alpha = \mathbb{Z}_p\beta^i\beta^{p-i-1} \subset \beta^i\Lambda.$$

It turns out that

$$\dim \beta^{i+1}\Lambda = \dim \beta(\beta^i\Lambda) = \dim \beta^i\Lambda - 1.$$

Since $\dim \Lambda = p$, it can be shown by a simple induction that $\dim \beta^i\Lambda = p - i$. Thus we have the required equality

$$\dim \bar{\gamma}\Lambda + \dim \gamma\Lambda = p - (p - i) + p - i = p. \qquad \square$$

Definition 5.5 For $\gamma = \beta^i$, $1 \le i \le p - 1$, we put

$$H_*^\gamma(K, L; \mathbb{Z}_p) = H_*(\gamma C(K, L; \mathbb{Z}_p)),$$

which is called the *Smith homology group*. For brevity, $H_*^\gamma(K, L; \mathbb{Z}_p)$ is often written as $H_*^\gamma(K, L)$.

It follows from the general theory of homology groups that Lemma 5.31 yields the following theorem.

Theorem 5.32 *For $\gamma = \beta^i$, $1 \le i \le p - 1$, there is an exact sequence*

$$\longrightarrow H_n^{\bar{\gamma}}(K, L) \oplus H_n(K^G, L^G) \xrightarrow{f_*} H_n(K, L) \xrightarrow{\gamma_*} H_n^\gamma(K, L) \xrightarrow{\partial}$$

of homology groups with coefficients in $\mathbb{Z}_p$. $\square$

The exact sequence in Theorem 5.32 is called the *Smith sequence*. We now interpret $H_*^\gamma(K, L)$ in the case where $\gamma = \beta^{p-1} = \alpha$.

Theorem 5.33 $H_*^\alpha(K, L) \cong H_*(K/G, K^G \cup L/G)$.

Proof First we show that the map

$$\alpha\colon C(K, L) \to C(K, L)$$

and the composition of the following maps

$$C(K, L) \xrightarrow{j} C(K, K^G \cup L) \xrightarrow{\pi} C(K/G, K^G \cup L/G)$$

have the same kernel. Let s be a simplex of $K - L$. As before we need only consider elements of $C(G(s))$. Since $\alpha(\sum n_i g^i s) = (\sum n_i)\alpha(s)$, we have

$$\alpha(\sum n_i g^i s) = 0 \Leftrightarrow \sum n_i = 0 \qquad \text{or} \qquad s \in K^G,$$

that is, both kernels coincide. Since both j and π are surjective, $\alpha C(K, L)$ and $C(K/G, K^G \cup L/G)$ are isomorphic as chain complexes. Accordingly their homology groups are also isomorphic. □

Remark The results in this section can be dualized to cohomology. Specifically, the Smith cohomology group

$$H^*_\gamma(K, L; \mathbb{Z}_p) = H^*(\gamma C(K, L; \mathbb{Z}_p); \mathbb{Z}_p)$$

is defined and similar theorems hold.

5.5 The Smith theory

In this section we shall present the celebrated Smith theory following Bredon.[15] Let G be a cyclic group of prime order p. We deal with homology with coefficients in $\mathbb{Z}_p$. First we study dimensions of homology groups as vector spaces over $\mathbb{Z}_p$.

Theorem 5.34 *Let K be a regular G-complex and L a G-invariant subcomplex of K. For any non-negative integer n and any $\gamma = \beta^i$, $1 \leq i \leq p - 1$, we have*

$$\dim H^\gamma_n(K, L) + \sum_{i \geq n} \dim H_i(K^G, L^G) \leq \sum_{i \geq n} \dim H_i(K, L).$$

Proof It follows from the part

$$H^{\bar{\gamma}}_{n+1}(K, L) \to H^\gamma_n(K, L) \oplus H_n(K^G, L^G) \to H_n(K, L)$$

of the exact sequence in Theorem 5.32 that

$$\dim H^\gamma_n(K, L) + \dim H_n(K^G, L^G) \leq \dim H^{\bar{\gamma}}_{n+1}(K, L) + \dim H_n(K, L).$$

Moreover γ and $\bar{\gamma}$ can be interchanged in the above inequality. For brevity we introduce the following temporary notation:

$$a_n = \dim H_n(K^G, L^G), \qquad b_n = \dim H_n(K, L)$$

$$c_n = \dim H^\gamma_n(K, L), \qquad \bar{c}_n = \dim H^{\bar{\gamma}}_n(K, L).$$

By alternating γ and $\bar{\gamma}$, we have

$$c_n + a_n \leq \bar{c}_{n+1} + b_n$$
$$\bar{c}_{n+1} + a_{n+1} \leq c_{n+2} + b_{n+1}$$
$$\vdots$$

Since K is finite, terms of dimension greater than $\dim K$ vanish. Therefore by adding these inequalities and cancelling, we have

$$c_n + \sum_{i \geq n} a_i \leq \sum_{i \geq n} b_i. \qquad \square$$

A complex K is called a mod p *homology n-sphere* if $H_i(K; \mathbb{Z}_p) \cong H_i(S^n; \mathbb{Z}_p)$ for any i. A pair (K, L) consisting of a complex K and its subcomplex L is called a mod p *homology n-disc* if $H_i(K, L; \mathbb{Z}_p) = 0$ for $i \neq n$ and $H_n(K, L; \mathbb{Z}_p) \cong \mathbb{Z}_p$. A finite group is called a *p-group* if its order is a power of p.

Theorem 5.35 (The Smith theorem) *Let G be a p-group and K a regular G-complex which is a* mod p *homology n-sphere. Then K^G is a* mod p *homology r-sphere for some $-1 \leq r \leq n$ (where $r = -1$ means that K^G is empty). If p is odd, then $n - r$ is even.*

Proof As is well known, for any p-group G with $|G| > p$, there exists a normal subgroup N of G with $\{e\} \neq N \neq G$. Then K^N becomes a regular G/N-complex and the equality

$$K^G = (K^N)^{G/N}$$

holds. Therefore if we prove the theorem only in the case where $|G| = p$, the theorem follows by induction in general. Hence we assume that $|G| = p$ hereafter. In this case, we can apply Theorem 5.34 to deduce

$$\dim H_*(K^G) \leq \dim H_*(K) = 2$$

and

$$H_i(K^G) = 0 \qquad \text{for } i > n.$$

Suppose that $\dim H_*(K^G) = 1$. Then $\dim H_0(K^G) = 1$ and the rest of the homology groups vanish, so that $\chi(K^G) = 1$. Moreover since $|G| = p$, any G-action is semi-free. It follows from Corollary 5.21 that

$$1 = \chi(K^G) \equiv \chi(K) = 0 \text{ or } 2 \bmod p,$$

which is a contradiction. Note that $\dim H_*(K^G) = 0$ implies $K^G = \emptyset$.

Further, $\dim H_*(K^G) = 2$ implies that K^G is a mod p homology r-sphere for some $0 \leq r \leq n$.

Suppose that p is odd, and hence p is greater than 2. Since $\chi(K^G) = 0$ or 2 and $\chi(K) = 0$ or 2 and since $\chi(K^G) \equiv \chi(K) \pmod p$, we can conclude that $\chi(K^G) = \chi(K)$. Hence $n - r$ must be even. □

Theorem 5.36 *Let G be a p-group, K a regular G-complex and L a G-invariant subcomplex of K. If (K, L) is a* mod *p homology n-disc, then (K^G, L^G) is a* mod *p homology r-disc for some $0 \leq r \leq n$. If p is odd, then $n - r$ is even.*

Proof If we just notice that $\dim H_*(K^G, L^G) \neq 0$ which follows from the relations of Euler characteristics, the proof is very similar to that of the previous theorem. □

A complex K is said to be *acyclic* if $H_0(K; \mathbb{Z}) \cong \mathbb{Z}$ and $H_i(K; \mathbb{Z}) = 0$ for $i > 0$. A complex K is said to be mod p *acyclic* if $H_0(K; \mathbb{Z}_p) \cong \mathbb{Z}_p$ and $H_i(K; \mathbb{Z}_p) = 0$ for $i > 0$.

Corollary 5.37 *Let G be a p-group. If a regular G-complex K is* mod *p acyclic, then K^G is also* mod *p acyclic.*

Proof Put $L = \phi$. Then (K, L) is a mod p homology 0-disc if and only if K is mod p acyclic. Hence the corollary is immediate from Theorem 5.36. □

Theorem 5.38 *Let G be an arbitrary finite group and K a regular G-complex. If K is acyclic, then so is K/G.*

Proof First we shall show that, for each prime p, K/G is mod p acyclic. Let P be a p-Sylow subgroup of G. Then we show for the present that K/P is mod p acyclic. As in the proof of Theorem 5.35, if $|P| > p$, then we take a normal subgroup N of P with $\{e\} \neq N \neq P$. Since we have

$$K/P = (K/N)/(P/N),$$

an inductive argument reduces this case to the case where $|P| = p$. In this case, Theorem 5.33 with $L = \phi$ shows that

$$H_i(K/P, K^P; \mathbb{Z}_p) \cong H_i^{\alpha}(K; \mathbb{Z}_p)$$

and the right-hand side of the above isomorphism vanishes for all i by Theorem 5.34 and Corollary 5.37. Since K^P is mod p acyclic by Corollary 5.37 again, it follows from the homology exact sequence of the pair $(K/P, K^P)$ that K/P is also mod p acyclic.

Next we deal with K/G. In view of Lemma 5.29, the composition

$$H_i(K/G; \mathbb{Z}_p) \xrightarrow{(\tau_{G/P})_*} H_i(K/P; \mathbb{Z}_p) \xrightarrow{(\pi_{G/P})_*} H_i(K/G; \mathbb{Z}_p)$$

is multiplication by $|G|/|P|$. Since $|G|/|P|$ is prime to p, the map of scalar multiplication by $|G|/|P|$ in a vector space over $\mathbb{Z}_p$ is an isomorphism. But $H_i(K/P;\ \mathbb{Z}_p)=0$ for $i>0$. This implies that $H_i(K/G;\ \mathbb{Z}_p)=0$ for $i>0$. Furthermore, since $\pi_*: H_0(K;\mathbb{Z}_p)\to H_0(K/G;\mathbb{Z}_p)$ is surjective and $H_0(K;\ \mathbb{Z}_p)\cong\mathbb{Z}_p$, we have $H_0(K/G;\ \mathbb{Z}_p)\cong\mathbb{Z}_p$. Thus we have shown that K/G is mod p acyclic for all primes p.

The exact sequence

$$0\to\mathbb{Z}\xrightarrow{p}\mathbb{Z}\to\mathbb{Z}_p\to 0$$

induces the exact sequence

$$\to H_i(K/G;\ \mathbb{Z})\xrightarrow{p} H_i(K/G;\ \mathbb{Z})\to H_i(K/G;\ \mathbb{Z}_p)\to.$$

Since $H_i(K/G;\ \mathbb{Z}_p)=0$ for $i>0$ as shown above, this shows that any element of $H_i(K/G;\ \mathbb{Z})$ is divisible by p^r for any r. It turns out that any element of $H_i(K/G;\ \mathbb{Z})$ is divisible by any positive integer for $i>0$, because p is an arbitrary prime number. By Lemma 5.27, the composition

$$H_i(K/G;\ \mathbb{Z})\xrightarrow{\tau_*} H_i(K;\ \mathbb{Z})\xrightarrow{\pi_*} H_i(K/G;\ \mathbb{Z})$$

is multiplication by $|G|$, so that it is surjective for $i>0$ by the above observation. But $H_i(K;\ \mathbb{Z})=0$ for $i>0$ by assumption. This implies that $H_i(K/G;\ \mathbb{Z})=0$ for $i>0$. On the other hand, it can be shown, as in the case of $\mathbb{Z}_p$ coefficients, that $H_0(K/G;\ \mathbb{Z})\cong\mathbb{Z}$. □

5.6 Applications to G-manifolds

We have thus far been dealing with G-complexes in this chapter. As is well known, any compact G-manifold can be G-triangulable,[51] so that all the results in this chapter hold for compact G-manifolds. Here we would like to emphasize the following remark. If we consider G-manifolds directly, some of the arguments in the previous sections do not work in general. However these difficulties are overcome by considering G-triangulations. For instance, let H be a subgroup of a finite group G and M a compact G-manifold. Then $M(H)$ is not compact in general. Hence there is no finite triangulation of $M(H)$ in general. Accordingly, the arguments of Section 5.2 are invalid. Let K be a G-complex giving a G-triangulation of M and K'' be the second barycentric subdivision of K. Although $K''(H)$ is not a subcomplex of K in general, $K''(H)$ consists of finite simplexes and possesses desirable properties as shown in the previous sections. When G is a compact Lie group, we derive information by restricting G to various finite subgroups. For instance we have the following theorem.

Theorem 5.39 *Let T^n be the torus group and M a compact T^n-manifold. Then we have*

$$\chi(M^{T^n}) = \chi(M).$$

Proof As shown in Section 4.7, the number of the orbit types occurring in M is finite. Hence the number of the isotropy groups occurring in M is also finite, since T^n is abelian. Accordingly it is easily verified that there exists a subgroup G of T^n satisfying:

(1) G is isomorphic to S^1 as Lie groups;

(2) $M^{T^n} = M^G$.

Hence it suffices to prove the theorem in the case where $T^n = S^1$. Furthermore, it follows from the same argument as above that $M^{S^1} = M^{\mathbb{Z}_m}$ for a sufficiently large positive integer m where $\mathbb{Z}_m$ is regarded canonically as a subgroup of S^1. On the other hand, $\chi(M^{\mathbb{Z}_p}) \equiv \chi(M) \bmod p$ for prime p by Corollary 5.20. Thus we see that the number $\chi(M^{S^1}) - \chi(M)$ is divisible by sufficiently large arbitrary primes p. But such an integer must be zero. □

A complex K is called a *homology n-sphere* if $H_i(K; \mathbb{Z}) \cong H_i(S^n; \mathbb{Z})$ for all i.

Proposition 5.40 *Let K be a simplicial complex. Then K is a* mod *p homology sphere for all primes p if and only if K is a homology sphere.*

Proof If K is a homology sphere, then it follows from the universal coefficient theorem that K is a mod p homology sphere for all primes p. We now show the converse. Suppose that K is a mod p homology sphere for all primes p. In view of Corollary 5.12, we have $\chi(K) = 0$ or 2. Since $H_0(K; \mathbb{Z}) \supset \mathbb{Z}$, there exists at least one more $\mathbb{Z}$ in $H_*(K; \mathbb{Z})$, that is, $H_*(K; \mathbb{Z}) \supset \mathbb{Z} \oplus \mathbb{Z}$. By applying the universal coefficient theorem, we have $H_*(K; \mathbb{Z}_p) \supset \mathbb{Z}_p \oplus \mathbb{Z}_p$ for any prime p. Since K is a mod p homology sphere by assumption, there is no additional $\mathbb{Z}_p$ in $H_*(K; \mathbb{Z}_p)$. Hence the free part of $H_*(K; \mathbb{Z})$ consists exactly of $\mathbb{Z} \oplus \mathbb{Z}$. If $H_*(K; \mathbb{Z})$ contains a non-trivial torsion, then it follows also from the universal coefficient theorem that there exists a prime p with $H_*(K; \mathbb{Z}_p) \neq \mathbb{Z}_p \oplus \mathbb{Z}_p$. But this contradicts the assumption. Thus we have $H_*(K; \mathbb{Z}) \cong \mathbb{Z} \oplus \mathbb{Z}$. □

The Smith theorem also develops into the following form.

Theorem 5.41 *Let M be a compact T^n-manifold. If M is a homology sphere, then so is the fixed point set M^{T^n} and the difference of the dimensions of M and M^{T^n}, respectively, is even.*

Proof When $M^{T^n} = \varnothing$, we regard it as a (-1)-dimensional homology sphere. In this case, it follows from Theorem 5.39 that $\chi(M) = \chi(M^{T^n}) = 0$.

Hence $\dim M$ is odd. Accordingly $\dim M - (-1)$ is even and the theorem holds in this case.

Next we consider the case where $M^{T^n} \neq \emptyset$. As remarked in the proof of Theorem 5.39, it suffices to prove the theorem in the case where $n = 1$. Also, for any given prime p, there exists a positive integer r such that $M^{\mathbb{Z}_{p^r}} = M^{S^1}$. By the universal coefficient theorem, M is a mod p homology sphere. Hence it follows from Theorem 5.35 that $M^{\mathbb{Z}_{p^r}}$ is a mod p homology sphere. Thus we have shown that, for any prime p, M^{S^1} is a mod p homology sphere. Since M^{S^1} is a compact manifold, it is triangulable. Therefore M^{S^1} is a homology sphere by Proposition 5.40. Moreover it is immediate from Theorem 5.35 that the difference of the dimensions of M and M^{S^1}, respectively, is even. □

Remark Concerning a $\mathbb{Z}_p$-action on a homology sphere M, the fixed point set $M^{\mathbb{Z}_p}$ is not necessarily a homology sphere.[127]

Proposition 5.42 *A compact Lie group which acts smoothly and freely on an even dimensional sphere is either $\mathbb{Z}_2$ or the trivial group.*

Proof Suppose that a finite group G acts freely on S^{2n}. Then we apply Corollary 5.22 to deduce that

$$|G|\chi(S^{2n}/G) = \chi(S^{2n}) = 2.$$

In consequence, $|G|$ must be either 1 or 2, that is, G is either the trivial group or $\mathbb{Z}_2$.

Suppose next that a compact Lie group G of positive dimension acts freely on S^{2n}. Since G contains S^1 as a Lie subgroup in this case, any cyclic subgroup acts freely on S^{2n}. But this contradicts the above observation. □

Proposition 5.43 *Let G be a finite group and M a G-manifold. If the action is free and $\chi(M) = 0$, then $\chi(M/G) = 0$.* (See, for example, lens spaces of Example 2 in Section 4.4.)

Proof Since $\chi(M) = 0$ by assumption, the proposition is immediate from Corollary 5.22. □

Remark The proposition does not hold for a compact Lie group of positive dimension. For instance, consider Example 3 in Section 4.4, that is, a natural free S^1-action on S^{2n+1}. Then $\chi(S^{2n+1}) = 0$, but $\chi(S^{2n+1}/S^1) = \chi(\mathbb{C}P^n) = n + 1$.

Let G be a finite group and M be a compact G-manifold. Then we put

$$S(M) = \bigcup_{g \in G - \{e\}} M^g$$

which is called the *singular set* of M.

Theorem 5.44

(1) $\sum_{g \in G} \chi(M^g) = |G| \chi(M/G)$.

(2) $\chi(M) - \chi(S(M)) = |G| \{\chi(M/G) - \chi(S(M)/G)\}$.

(3) $\chi(M) \equiv \chi(S(M)) \equiv - \sum_{g \in G - \{e\}} \chi(M^g) \mod |G|$.

Proof Recall that M is G-triangulable, that is, there exists a G-complex K such that M is G-homeomorphic to $|K|$. We may assume that $M = |K|$ and that K is a regular G-complex. Hence Theorem 5.24 yields Theorem 5.44. For details, we refer the reader to the proof of Corollary 5.60. □

The rest of the section is devoted to the investigation of the relations among the Euler characteristics in the case where compact Lie groups act on manifolds. We shall make use of the decomposition theorem of G-manifolds (Theorem 4.20) freely. In particular, homology groups and Euler characteristics which we shall deal with are defined by making use of the decomposition theorem. For further information, the reader is referred to Kawakubo.[67]

We begin with the following lemma.

Lemma 5.45 *Let X and F be compact smooth manifolds and let $\pi\colon E \to X$ be a smooth fibre bundle with fibre F. Then we have*

$$\chi(E) = \chi(X) \cdot \chi(F).$$

Proof First recall the product formula and the sum formula for Euler characteristics:

$$\chi(X_1 \times X_2) = \chi(X_1) \cdot \chi(X_2), \tag{5.11}$$

$$\chi(A \cup B) = \chi(A) + \chi(B) - \chi(A \cap B). \tag{5.12}$$

Here X_1, X_2, and $A \cup B$ are polyhedra and A, B, and $A \cap B$ are subpolyhedra of $A \cup B$.

As is well known, smooth manifolds are triangulable and homology does not depend on the choice of triangulations. Without loss of generality, we may assume that $X = |K|$ for some simplicial complex K and that for any simplex s of K, $|s|$ is contained in an open set over which the bundle π is trivial. Let n be the dimension of the manifold X and let $\{s_1, \ldots, s_k\}$ be the set of n-simplices in K. Denote by K_i the subcomplex of K consisting of $s_1, \ldots, s_i$ and their faces. Then we prove the lemma by induction on i.

If $i = 1$, then it follows from (5.11) that

$$\chi(\pi^{-1}(|K_1|)) = \chi(|K_1|)\cdot\chi(F).$$

Suppose that the equality

$$\chi(\pi^{-1}(|K_i|)) = \chi(|K_i|)\cdot\chi(F)$$

holds. Then by making use of (5.11) and (5.12) above, we compute:

$$\begin{aligned}\chi(\pi^{-1}(|K_{i+1}|)) &= \chi(\pi^{-1}(|K_i|)\cup\pi^{-1}(|s_{i+1}|))\\ &= \chi(\pi^{-1}(|K_i|)) + \chi(\pi^{-1}(|s_{i+1}|)) - \chi(\pi^{-1}(|K_i|)\cap\pi^{-1}(|s_{i+1}|))\\ &= \chi(|K_i|)\cdot\chi(F) + \chi(|s_{i+1}|)\cdot\chi(F) - \chi(|K_i|\cap|s_{i+1}|)\cdot\chi(F)\\ &= \{\chi(|K_i|) + \chi(|s_{i+1}|) - \chi(|K_i|\cap|s_{i+1}|)\}\cdot\chi(F)\\ &= \chi(|K_i|\cup|s_{i+1}|)\cdot\chi(F)\\ &= \chi(|K_{i+1}|)\cdot\chi(F).\end{aligned}$$

Thus the proof follows by induction. □

Let G be a compact Lie group and let M be a compact G-manifold. In view of Theorem 4.23, the set of isotropy types occurring in M is finite. Let $S = \{(H_i)|\ i = 1, \ldots, m\}$ be a finite set of conjugacy classes of closed subgroups of G including the set of isotropy types occurring in M. For each $(H_i)\in S$, choose a representative H_i of (H_i) and put

$$M^{>H_i} = \bigcup_{S\ni(H_j)>(H_i)} M^{H_j}.$$

Then we make use of the notation

$$GM^{H_i} = \{gx\,|\,g\in G,\ x\in M^{H_i}\},$$
$$GM^{>H_i} = \{gx\,|\,g\in G,\ x\in M^{>H_i}\}$$

introduced in Section 1.4.

A subset T of S is said to be *full* if each element (H_i) of S satisfying $(H_i) \geq (H_j)$ for some $(H_j)\in T$ belongs to T.

Lemma 5.46 *If T is a full subset of S, then we have*

$$\chi\Big(\bigcup_{(H_i)\in T} GM^{H_i}\Big) = \sum_{(H_i)\in T} \chi(GM^{H_i}, GM^{>H_i}),$$

$$\chi\Big(\bigcup_{(H_i)\in T} GM^{H_i}/G\Big) = \sum_{(H_i)\in T} \chi(GM^{H_i}/G, GM^{>H_i}/G).$$

Proof We now arrange $T = \{(H_i) \mid i = 1, \ldots, t\}$ as:

$$(H_i) \geq (H_j) \Rightarrow i \leq j.$$

Then we shall prove that

$$\chi\left(\bigcup_{i=1}^{r} GM^{H_i}\right) = \sum_{i=1}^{r} \chi(GM^{H_i}, GM^{>H_i}) \tag{5.13}$$

by induction on r ($1 \leq r \leq t$). Since $GM^{>H_1} = \varnothing$, the above equality holds for $r = 1$. Suppose that the above equality (5.13) holds. Since T is full in S by assumption, one verifies easily that

$$\left(\bigcup_{i=1}^{r} GM^{H_i}\right) \cap GM^{H_{r+1}} = GM^{>H_{r+1}}.$$

Then it follows from formula (5.12) in the proof of Lemma 5.45 and the inductive assumption (5.13) that

$$\begin{aligned}\chi\left(\bigcup_{i=1}^{r+1} GM^{H_i}\right) &= \chi\left(\bigcup_{i=1}^{r} GM^{H_i}\right) + \chi(GM^{H_{r+1}}) - \chi\left(\left(\bigcup_{i=1}^{r} GM^{H_i}\right) \cap GM^{H_{r+1}}\right) \\ &= \sum_{i=1}^{r} \chi(GM^{H_i}, GM^{>H_i}) + \chi(GM^{H_{r+1}}, GM^{>H_{r+1}}) \\ &= \sum_{i=1}^{r+1} \chi(GM^{H_i}, GM^{>H_i}).\end{aligned}$$

Thus the proof of the first formula of the lemma follows by induction. The second formula can be shown similarly. □

Corollary 5.47 *For each fixed* $(H_j) \in S$, *we have*

$$\chi(GM^{H_j}) = \sum_{(H_j) \leq (H_i) \in S} \chi(GM^{H_i}, GM^{>H_i}),$$

$$\chi(GM^{H_j}/G) = \sum_{(H_j) \leq (H_i) \in S} \chi(GM^{H_i}/G, GM^{>H_i}/G).$$

Proof Put

$$T = \{(H_i) \in S \mid (H_j) \leq (H_i)\}.$$

Then T is a full subset of S. Observe that

$$\bigcup_{(H_i) \in T} GM^{H_i} = GM^{H_j}.$$

Hence the first formula of the corollary is immediate from Lemma 5.46. The second formula can be proved in a similar way. □

Corollary 5.48

$$\chi(M) = \sum_{(H_i)\in S} \chi(GM^{H_i}, GM^{>H_i})$$

$$\chi(M/G) = \sum_{(H_i)\in S} \chi(GM^{H_i}/G, GM^{>H_i}/G).$$

Proof Evidently we have

$$M = \bigcup_{(H_i)\in S} GM^{H_i} \qquad \text{and} \qquad M/G = \bigcup_{(H_i)\in S} GM^{H_i}/G.$$

Hence Corollary 5.48 is an immediate consequence of Lemma 5.46. □

We now arrange $S = \{(H_i) \mid i = 1, \ldots, m\}$ as:

$$(H_i) \le (H_j) \Rightarrow 1 \le j$$

and define $J[i]$ and $J(i)$ as in Section 5.2. Notice that the arrangement above is different from that of the proof of Lemma 5.46. Then we put

$$h_i' = \chi(G/H_i).$$

When $h_i' \neq 0$, we put $h_i = h_i'^{-1}$. Also, we define inductively three classes of integers $n_1, \ldots, n_m, n_1', \ldots, n_m'$, and $r_1, \ldots, r_m$ as in Section 5.2. Of course, the n_i are defined only when $h_i' \neq 0$.

We are now in a position to state the main theorem.

Theorem 5.49 *Under the above conditions, we have*

$$\chi(M) = \sum_i n_i' \chi(GM^{H_i}/G),$$

$$\chi(M) = \sum_i r_i \chi(GM^{H_i}),$$

$$\chi(M/G) = \sum_i r_i \chi(GM^{H_i}/G),$$

$$\chi(M/G) = \sum_i n_i \chi(GM^{H_i}).$$

Here the last formula is valid only when $\chi(G/H_i) \neq 0$ for all i.

Proof Recall the decomposition of a compact G-manifold according to Theorem 4.20:

$$M = D(v_m) \cup D(v_{m-1}) \cup \cdots \cup D(v_1).$$

However, since $\{(H_i)\}$ are now arranged in reverse order, the indices of the decomposition above are also reversed. Put

$$M(i) = D(v_m) \cup \cdots \cup D(v_i).$$

Notice that $GM^{H_j} \cap D(v_j)$ is a compact G-manifold which consists of orbits of type G/H_j only. Hence it follows from Theorem 4.18 that the projection

$$\pi\colon GM^{H_j} \cap D(v_j) \to (GM^{H_j} \cap D(v_j))/G$$

is a smooth fibre bundle with fibre G/H_j. Furthermore $GM^{H_j} \cap M(j+1) \cap D(v_j)$ is a compact G-invariant submanifold of $GM^{H_j} \cap D(v_j)$. Accordingly Lemma 5.45 together with the excision isomorphism theorem yields the following computation:

$$\begin{aligned}
\chi(GM^{H_j}, GM^{>H_j}) &= \chi(GM^{H_j} \cap D(v_j), GM^{H_j} \cap M(j+1) \cap D(v_j)) \\
&= \chi(GM^{H_j} \cap D(v_j)) - \chi(GM^{H_j} \cap M(j+1) \cap D(v_j)) \\
&= \chi((GM^{H_j} \cap D(v_j))/G) \cdot \chi(G/H_j) \\
&\quad - \chi((GM^{H_j} \cap M(j+1) \cap D(v_j))/G) \cdot \chi(G/H_j) \\
&= \chi((GM^{H_j} \cap D(v_j))/G, (GM^{H_j} \cap M(j+1) \\
&\qquad \cap D(v_j))/G) \cdot \chi(G/H_j) \\
&= \chi(GM^{H_j}/G, GM^{>H_j}/G) \cdot \chi(G/H_j). \qquad (5.14)
\end{aligned}$$

Recall the following formulae of Corollaries 5.47 and 5.48:

$$\chi(GM^{H_j}/G) = \sum_{(H_j) \le (H_i) \in S} \chi(GM^{H_i}/G, GM^{>H_i}/G), \qquad (5.15)$$

$$\chi(M) = \sum_{(H_j) \in S} \chi(GM^{H_j}, GM^{>H_j}). \qquad (5.16)$$

We now apply the Möbius inversion formula to (5.15) above and deduce

$$\sum_{(H_j) \le (H_i) \in S} \mu((H_j), (H_i)) \chi(GM^{H_i}/G) = \chi(GM^{H_j}/G, GM^{>H_j}/G). \qquad (5.17)$$

By combining (5.14), (5.16), and (5.17), we have

$$\begin{aligned}
\chi(M) &= \sum_{(H_j) \in S} \chi(GM^{H_j}, GM^{>H_j}) \\
&= \sum_{(H_j) \in S} \chi(GM^{H_j}/G, GM^{>H_j}/G) \cdot \chi(G/H_j) \\
&= \sum_{(H_j) \in S} \Big\{ \sum_{(H_j) \le (H_i) \in S} \mu((H_j), (H_i)) \chi(GM^{H_i}/G) \Big\} \cdot \chi(G/H_j) \\
&= \sum_{(H_i) \in S} \Big\{ \sum_{S \ni (H_j) \le (H_i)} \mu((H_j), (H_i)) \chi(G/H_j) \Big\} \chi(GM^{H_i}/G).
\end{aligned}$$

In the manner of the proof of Theorem 5.17, it is verified that

$$\sum_{S \ni (H_j) \le (H_i)} \mu((H_j), (H_i)) \chi(G/H_j) = n_i'.$$

This completes the proof of the first formula of the theorem. The proofs for the remainder of the formulae are similar. □

For a subgroup H of G, we denote by WH the quotient group $N(H)/H$ where $N(H)$ denotes the normalizer of H in G.

Corollary 5.50 *Let G be a compact Lie group and M be a compact G-manifold. Let $S = \{(H_i)\}$ be the set of isotropy types occurring in M. If* $\dim WH_i > 0$ *for all i, then we have $\chi(M) = 0$.*

Proof Suppose that $\dim WH_i > 0$ for all H_i in S. Then there exists a Lie subgroup K_i of WH_i for each i such that K_i is isomorphic to S^1 as Lie groups. According to Theorem 1.73 WH_i acts freely on G/H_i. It turns out that S^1 acts freely on G/H_i. Hence we have $\chi(G/H_i) = 0$ by Theorem 5.39, that is $h_i' = 0$ for all i. It follows from the definition of n_i' that n_i' vanishes also for all i. We now apply Theorem 5.49 to deduce that $\chi(M) = 0$. □

Corollary 5.51 *Let G be a compact Lie group and M be a compact G-manifold. Let $S = \{(H_i)\}$ be the set of isotropy types occurring in M. If $\chi(M) \neq 0$, then there exists (H_i) in S such that WH_i is a finite group.*

Proof Corollary 5.51 is an immediate consequence of Corollary 5.50. □

5.7 The Lefschetz fixed point theorem

In this section, we shall introduce the Lefschetz fixed point theorem. Until now, we have been mainly concerned with compact group actions. In contrast with this, we deal with an arbitrary continuous map from a polyhedron to itself in this section. The Lefschetz fixed point theorem applies to such a map and is remarkably powerful.

Let X be a set and $f: X \to X$ be a map. Then we put

$$X^f = \{x \in X \mid f(x) = x\}$$

which is called the *fixed point set of f*. When $X^f \neq \varnothing$, we say that *f has a fixed point.*

Let C be a finitely generated graded abelian group and let $\varphi: C \to C$ be an endomorphism of C of degree 0. Denote by $\varphi_n: C_n \to C_n$ the endomorphism defined by φ in degree n. Then we put

$$\Lambda(\varphi) = \sum_n (-1)^n \operatorname{Tr} \varphi_n$$

which is called the *Lefschetz number* of φ. Here $\operatorname{Tr} \varphi_n$ denotes the trace of φ_n.

We begin by preparing the following theorem.

Theorem 5.52 (The Hopf trace formula) *Let C be a finitely generated chain complex and let $\varphi: C \to C$ be a chain map. Denote by $\varphi_*: H_*(C) \to H_*(C)$ the endomorphism of the homology group of C induced by φ. Then*

$$\Lambda(\varphi) = \Lambda(\varphi_*).$$

Proof The proof is similar to that of the corresponding statement about the Euler characteristic of Theorem 5.11. Details are left to the reader. □

Let X be a polyhedron and let $f: X \to X$ be a continuous map. Then the *Lefschetz number of* f, denoted by $\Lambda(f)$, is defined to be the Lefschetz number of the homomorphism $f_*: H_*(X) \to H_*(X)$ induced by f.

Theorem 5.53 (The Lefschetz fixed point theorem) *Let X be a polyhedron and let $f: X \to X$ be a continuous map. If $\Lambda(f) \neq 0$, then f has a fixed point.*

Proof We assume that f has no fixed point and show that $\Lambda(f) = 0$. By assumption there exists a simplicial complex K such that X is homeomorphic to $|K|$. Without loss of generality, we may assume $X = |K|$. Since $|K|$ is realized as a subset of some Euclidean space $\mathbb{R}^N$, X is equipped with a metric $d(\,,)$. If f has no fixed point, then there exists a positive number ε such that $d(x, f(x)) \geq \varepsilon$ for all $x \in X$, since X is compact. Then there is a positive integer q with mesh $K^{(q)} < \varepsilon/3$ where $K^{(q)}$ denotes the qth barycentric subdivision of K and

$$\text{mesh } K^{(q)} = \max\{\text{diam } |s| \mid s \in K^{(q)}\}.$$

As is well known, there is a barycentric subdivision $K^{(r)}$ $(r \geq q)$ of K for which there exists a simplicial map $\varphi: K^{(r)} \to K^{(q)}$ which is a simplicial approximation to $f: X \to X$.

Denote by $|\varphi|: |K| = |K^{(r)}| \to |K| = |K^{(q)}|$ the continuous map induced by $\varphi: K^{(r)} \to K^{(q)}$. It follows from the definition of a simplicial approximation that

$$d(|\varphi|(x), f(x)) < \varepsilon/3 \qquad \text{for all } x \in X.$$

Since for all $x \in X$,

$$d(x, |\varphi|(x)) > d(x, f(x)) - d(|\varphi|(x), f(x)) > 2\varepsilon/3,$$

we have

$$|s| \cap |\varphi|(|s|) = \varnothing \tag{5.18}$$

for all $s \in K^{(q)}$.

Let $S: C(K^{(q)}) \to C(K^{(r)})$ be a subdivision chain map. We now consider the composition

$$(C(\varphi) \cdot S)_n: C_n(K^{(q)}) \to C_n(K^{(q)})$$

where $C(\varphi): C(K^{(r)}) \to C(K^{(q)})$ is the chain map induced by φ. It follows from

(5.18) above that, for an oriented n-simplex σ on an n-simplex s of $K^{(q)}$, $(C(\varphi)\cdot S)(\sigma)$ is an n-chain having coefficient 0 on σ. Since this is so for every σ, we have

$$\mathrm{Tr}(C(\varphi)\cdot S)_n = 0 \qquad \text{for all } n,$$

which implies $\Lambda(C(\varphi)\cdot S) = 0$.

On the other hand, since $f \simeq |\varphi|$, we have $f_* = |\varphi|_*$. Moreover it is well known that the homomorphism

$$(C(\varphi)\cdot S)_* : H_*(K^{(q)}) \to H_*(K^{(q)})$$

coincides with the homomorphism

$$|\varphi|_* : H_*(|K|) \to H_*(|K|)$$

via the canonical isomorphism $H_*(K^{(q)}) \cong H_*(|K|)$.

Finally, recall the formula

$$\Lambda(C(\varphi)\cdot S) = \Lambda((C(\varphi)\cdot S)_*)$$

of Theorem 5.52.

Putting all this together, we have

$$\Lambda(f) = \Lambda(f_*) = \Lambda(|\varphi|_*) = \Lambda((C(\varphi)\cdot S)_*) = \Lambda(C(\varphi)\cdot S) = 0. \qquad \square$$

Remark Since any compact manifold with or without boundary is a polyhedron, the theorem is applicable to any continuous map from a compact manifold to itself. On the other hand the result does not hold for non-compact manifolds.

Corollary 5.54 *Let X be a contractible polyhedron. Then every continuous map $f: X \to X$ has a fixed point.*

Proof If X is contractible, $H_n(X) = 0$ for $n > 0$ and $H_0(X) \cong \mathbb{Z}$. Since $f_* = \mathrm{id}: H_0(X) \to H_0(X)$ for any continuous map $f: X \to X$, we have $\Lambda(f) = 1$. Hence f has a fixed point by the theorem. $\square$

Remark When X is the unit disc D^n, Corollary 5.54 is called the *Brouwer fixed point theorem.* Hence Corollary 5.54 is a generalization of the Brouwer fixed point theorem.

Given a continuous map $f: S^n \to S^n$, we denote by $\deg f$ the *degree* of f. That is, $\deg f$ is the integer defined by

$$f_*(z) = (\deg f)z \qquad \text{for} \qquad z \in H_n(S^n).$$

Then the Lefschetz number $\Lambda(f)$ is obviously given by

$$\Lambda(f) = 1 - 0 + \cdots + (-1)^n \deg f = 1 + (-1)^n \deg f.$$

Then we have

Corollary 5.55 *If* $\deg f \neq (-1)^{n+1}$, *then f has a fixed point.*

Corollary 5.56 *When the wind blows on the surface of the earth* (S^2), *there is at least one place where the wind is not blowing.*

Proof Suppose that the wind is blowing everywhere. Then we get a nowhere zero vector field on S^2. Since S^2 is compact, it follows from Corollary 3.20 that the vector field generates a 1-parameter group φ of diffeomorphisms of S^2. Then $\varphi_t: S^2 \to S^2$ has no fixed point, if t is a sufficiently small positive number. Since φ_t is homotopic to the identity, we have $\deg \varphi_t = 1$. But this contradicts Corollary 5.55. □

Since the antipodal map $A: S^n \to S^n$ has no fixed point, it follows from Theorem 5.53 that the Lefschetz number $\Lambda(A)$ vanishes. Thus we have

Corollary 5.57 *The antipodal map* $A: S^n \to S^n$ *has degree* $(-1)^{n+1}$.

The following proposition is a topological version of Proposition 5.42.

Proposition 5.58 *If n is even, then* $\mathbb{Z}_2$ *is the only nontrivial group that acts freely on* S^n.

Proof Suppose that G acts freely on S^n. Let g_1 and g_2 be arbitrary elements of G which are different from the unit. It follows from Corollary 5.55 that $\deg g_i = (-1)^{n+1} = -1$ for $i = 1, 2$. Hence we have

$$\deg g_2 g_1 = (\deg g_2)(\deg g_1) = 1.$$

It follows again from Corollary 5.55 that the map $g_2 g_1$ has a fixed point. As a consequence, $g_2 g_1$ must be the unit e by assumption. In particular, $g_i^2 = e$. Therefore we have $g_1 = g_2$, which completes the proof of Proposition 5.58. □

Remark If two continuous maps $f_1, f_2: X \to X$ are homotopic, then $\Lambda(f_1) = \Lambda(f_2)$. In particular, if $f: X \to X$ is homotopic to the identity, then we have

$$\Lambda(f) = \chi(X).$$

Next we apply the Lefschetz fixed point theorem to G-complexes.

Theorem 5.59 *Let G be a finite group and K be a regular G-complex. Then we have*:

(1) $\Lambda(g) = \chi(K^g)$ *for any* $g \in G$;

(2) $\sum_{g \in G} \Lambda(g) = |G| \chi(K/G)$.

Proof For any given $g \in G$, the simplicial isomorphism $g\colon K \to K$ induces an automorphism $g_{\#}\colon C(K) \to C(K)$ of the chain complex of K and an automorphism $g_{*}\colon H_{*}(K) \to H_{*}(K)$ of the homology group of K. According to Theorem 5.52, we have

$$\Lambda(g_{\#}) = \Lambda(g_{*}).$$

Since the map $g\colon K \to K$ is a simplicial isomorphism and K is a regular G-complex, the trace of the automorphism

$$g_{\#n}\colon C_n(K) \to C_n(K)$$

is equal to the number of n-simplexes in K^g. That is, if we employ the notation of Section 5.2, we have

$$\operatorname{Tr} g_{\#n} = a_n(K^g).$$

Putting all this together, we have

$$\Lambda(g) = \Lambda(g_{*}) = \Lambda(g_{\#}) = \sum(-1)^n \operatorname{Tr} g_{\#n} = \sum(-1)^n a_n(K^g) = \chi(K^g),$$

which proves (1).

By combining (1) and Theorem 5.24, we have

$$\sum_{g \in G} \Lambda(g) = \sum_{g \in G} \chi(K^g) = |G|\chi(K/G),$$

which completes the proof of (2).

In the following we shall give an alternative proof of (2) without using (1). First recall that, for any representation ψ of a finite group G on a finite dimensional vector space V over $\mathbb{R}$, we have

$$\sum_{g \in G} \operatorname{Tr} \psi(g) = |G| \dim V^G,$$

which follows from the Schur orthogonality relations. Applying this to each representation $H_n(K; \mathbb{R})$ of G, we deduce

$$\sum_{g \in G} \operatorname{Tr} g_{*n} = |G| \dim H_n(K; \mathbb{R})^G$$

for each n. On the other hand, it follows from Theorem 5.28 that $H_n(K; \mathbb{R})^G$ is isomorphic to $H_n(K/G; \mathbb{R})$. Thus we have

$$\begin{aligned}
\sum_{g \in G} \Lambda(g) &= \sum_{g \in G} \sum_{n} (-1)^n \operatorname{Tr} g_{*n} \\
&= \sum_{n} (-1)^n \sum_{g \in G} \operatorname{Tr} g_{*n} \\
&= \sum_{n} (-1)^n |G| \dim H_n(K; \mathbb{R})^G \\
&= \sum_{n} (-1)^n |G| \dim H_n(K/G; \mathbb{R}) \\
&= |G|\chi(K/G).
\end{aligned}$$

□

Corollary 5.60 *Let G be a finite group and M be a compact G-manifold. Then we have:*

(1) $\Lambda(g) = \chi(M^g)$ *for any* $g \in G$;

(2) $\sum_{g \in G} \Lambda(g) = |G| \, \chi(M/G)$.

Proof As explained in Section 5.6, any compact G-manifold M can be G-triangulable. That is, there exists a G-complex K such that its polyhedron $|K|$ is G-homeomorphic to M. Accordingly we may assume without loss of generality that $M = |K|$. Further, by virtue of Theorem 5.8, we may assume that K is a regular G-complex. For any given $g \in G$, the automorphism

$$g_*: H_*(M) \to H_*(M)$$

is identified with

$$g_*: H_*(K) \to H_*(K)$$

by definition. It turns out that

$$\Lambda(g: K \to K) = \Lambda(g: M \to M).$$

We denote by $\langle g \rangle$ the cyclic group generated by g. Note that $M^g = M^{\langle g \rangle}$ and $K^g = K^{\langle g \rangle}$. By combining Lemma 5.4 and Corollary 5.5, we have that $K^{\langle g \rangle}$ is a subcomplex of K and

$$M^{\langle g \rangle} = |K|^{\langle g \rangle} = |{}^{\langle g \rangle}K| = |K^{\langle g \rangle}|.$$

As is well known, the homology group does not depend on the choice of triangulations, so that

$$\chi(M^g) = \chi(M^{\langle g \rangle}) = \chi(|K^{\langle g \rangle}|) = \chi(K^{\langle g \rangle}) = \chi(K^g).$$

Further, it follows from Lemma 5.10 that K/G is a simplicial complex and $|K/G| = |K|/G$ (see the exercise in Section 5.1). Hence we have

$$\chi(M/G) = \chi(|K|/G) = \chi(|K/G|) = \chi(K/G).$$

Therefore the corollary follows from Theorem 5.59. □

5.8 Free and semi-free transformation groups on spheres

In this section, we shall investigate fundamental properties of finite groups which act freely or semi-freely on some sphere S^n along the lines of the Smith theory. We shall deal with homology with coefficients in $\mathbb{Z}_p$ where p is prime.

We begin by preparing the following lemma.

Lemma 5.61 *Let G be a cyclic group of prime order p. Then the following diagram is commutative:*

$$\begin{array}{ccc} H_*(K/G, K^G) \cong H_*^{\alpha}(K) & \xrightarrow{\quad f_* \quad} & H_*(K) \\ {\scriptstyle j_*}\Big\uparrow & \nearrow{\scriptstyle \tau_*} & \\ H_*(K/G) & & \end{array}$$

where $\cong$ stands for the isomorphism in Theorem 5.33, f_* is the restriction of the homomorphism f_* in Theorem 5.32, j_* is the homomorphism induced by the inclusion, and τ_* is the transfer homomorphism of homology with coefficients in $\mathbb{Z}_p$.

Proof Consider the following commutative diagram:

$$\begin{array}{ccccc} C(K;\mathbb{Z}_p) & \xrightarrow{\quad \alpha \quad} & \alpha C(K;\mathbb{Z}_p) & \overset{f}{\subset} & C(K;\mathbb{Z}_p) \\ {\scriptstyle \pi}\Big\downarrow & & \Big\uparrow{\scriptstyle A} & & \\ C(K/G;\mathbb{Z}_p) & \xrightarrow{\quad j \quad} & C(K/G, K^G;\mathbb{Z}_p) & & \end{array}$$

where A denotes the chain isomorphism given in the proof of Theorem 5.33. For any $c \in C(K;\mathbb{Z}_p)$, the correspondence $\pi c \mapsto f \cdot \alpha c$ induces the transfer homomorphism

$$\tau_*\colon H_*(K/G) \to H_*(K)$$

with $\mathbb{Z}_p$ coefficients by definition. On the other hand, the correspondences

$$\pi c \mapsto j \cdot \pi c \mapsto \alpha c \mapsto f \cdot \alpha c$$

induce the homomorphisms

$$H_*(K/G) \xrightarrow{j_*} H_*(K/G, K^G) \xrightarrow{A_*} H_*^{\alpha}(K) \xrightarrow{f_*} H_*(K)$$

in the lemma. Hence their composition coincides with the transfer homomorphism τ_*. □

Proposition 5.62 *Let $\alpha C(K)$ denote the subcomplex of $C(K)$ defined in Section 5.3. Then we have the following commutative diagram:*

$$\begin{array}{ccc} H_*((\alpha C(K))\otimes\mathbb{Z}_p) & \xrightarrow{\quad i_* \quad} & H_*^{\alpha}(K) \\ {\scriptstyle B_*}\Big\uparrow{\scriptstyle \cong} & & {\scriptstyle A_*}\Big\uparrow{\scriptstyle \cong} \\ H_*(K/G;\mathbb{Z}_p) & \xrightarrow{\quad j_* \quad} & H_*(K/G, K^G;\mathbb{Z}_p) \end{array}$$

where the maps i_* and B_* will be given in the proof.

Proof One verifies easily that the following diagram is commutative:

$$\begin{array}{ccccc} C(K)\otimes \mathbb{Z}_p & \xrightarrow{\alpha\otimes \mathrm{id}} & (\alpha C(K))\otimes \mathbb{Z}_p & \xrightarrow{\;i\;} & \alpha C(K;\ \mathbb{Z}_p) \\ \Vert & & \uparrow{\scriptstyle B} & & \uparrow{\scriptstyle A} \\ C(K;\ \mathbb{Z}_p) & \xrightarrow{\;\pi\;} & C(K/G;\ \mathbb{Z}_p) & \xrightarrow{\;j\;} & C(K/G,\ K^G;\ \mathbb{Z}_p) \end{array}$$

Here i is given by $i(\alpha c \otimes 1) = \alpha c \otimes 1$ where the right-hand side is regarded as an element of $C(K)\otimes \mathbb{Z}_p = C(K; \mathbb{Z}_p)$ and B is given by

$$B = h^{-1} \otimes \mathrm{id}\colon C(K/G; \mathbb{Z}_p) = C(K/G)\otimes \mathbb{Z}_p \to (\alpha C(K))\otimes \mathbb{Z}_p$$

where $h\colon \alpha C(K) \to C(K/G)$ denotes the chain isomorphism in Corollary 5.26. Hence B is a chain isomorphism. As shown in the proof of Theorem 5.33, the map A is also a chain isomorphism. Hence the diagram above induces the diagram in the proposition. □

Remark For an arbitrary simplex s in K, $\alpha(s)\otimes 1 \neq 0$ in $(\alpha C(K))\otimes \mathbb{Z}_p$, but $\alpha(s\otimes 1) = i(\alpha(s)\otimes 1) = 0$ in $\alpha C(K; \mathbb{Z}_p)$ if $s \in K^G$, that is, i is not injective in general. Further, as a consequence of the proposition, we see that i_* is not necessarily an isomorphism.

Theorem 5.63 *Let G be a cyclic group of prime order p and let K be a regular G-complex which is a* mod *p homology n-sphere. If the G-action on K is free, then we have*

$$H_i(K/G) \cong \begin{cases} \mathbb{Z}_p & \text{for } 0 \leq i \leq n \\ 0 & \text{otherwise.} \end{cases}$$

Proof Since $H_i(K) \cong 0$ for $1 \leq i \leq n-1$, it follows from the Smith sequence

$$\cdots \to H_i(K) \to H_i^{\gamma}(K) \to H_{i-1}^{\bar{\gamma}}(K) \to H_{i-1}(K) \to \cdots$$

that we have an isomorphism

$$H_i^{\gamma}(K) \cong H_{i-1}^{\bar{\gamma}}(K) \qquad \text{for } 2 \leq i \leq n-1.$$

Recall the following commutative diagram

$$\begin{array}{ccc} H_0(K/G,\ K^G) \cong H_0^{\alpha}(K) & \xrightarrow{\;f_*\;} & H_0(K) \\ \uparrow{\scriptstyle j_*} & \nearrow{\scriptstyle \tau_*} & \\ H_0(K/G) & & \end{array}$$

according to Lemma 5.61. Since $K^G = \emptyset$, j_* is an isomorphism. Evidently τ_* is the zero homomorphism in dimension 0 by definition, so that we have $f_* = 0$. Hence it follows again from the Smith sequence that

$$H_0^\beta(K) \cong H_0(K) \cong \mathbb{Z}_p.$$

Since K/G is connected, $H_0^\alpha(K) \cong H_0(K/G) \cong \mathbb{Z}_p$. We now apply the Smith sequence with $\gamma = \alpha$ or β and deduce by induction that

$$H_i^\alpha(K) \cong H_i^\beta(K) \cong \mathbb{Z}_p$$

for $1 \leq i \leq n-1$. On the other hand, it follows from Theorem 5.34 that $H_i^\gamma(K) \cong 0$ for $i \geq n+1$. Thus we get an exact sequence

$$0 \to H_n^{\bar{\gamma}}(K) \to H_n(K) \to H_n^\gamma(K) \to H_{n-1}^{\bar{\gamma}}(K) \to 0.$$

By putting $\gamma = \alpha$ or β, we have the following two exact sequences:

$$0 \to H_n^\alpha(K) \to \mathbb{Z}_p \to H_n^\beta(K) \to \mathbb{Z}_p \to 0$$
$$0 \to H_n^\beta(K) \to \mathbb{Z}_p \to H_n^\alpha(K) \to \mathbb{Z}_p \to 0.$$

Obviously the only possibility for the groups $H_n^\alpha(K)$ and $H_n^\beta(K)$ to satisfy these exact sequences is that

$$H_n^\alpha(K) \cong H_n^\beta(K) \cong \mathbb{Z}_p.$$

Since $K^G = \emptyset$, it follows from Theorem 5.33 that

$$H_i(K/G) \cong H_i^\alpha(K)$$

which completes the proof. □

Remark Compare Example 2 in Section 5.3 with Theorem 5.63.

Next we shall deal with the case where there exist fixed points.

Theorem 5.64 *Let G be a cyclic group of prime order p and let K be a regular G-complex. Suppose that K is a* mod *p homology n-sphere and that the fixed point set K^G is a* mod *p homology r-sphere with $0 \leq r \leq n$. Then we have*

$$H_i^\gamma(K) \cong \begin{cases} \mathbb{Z}_p & \text{for } r+1 \leq i \leq n, \\ 0 & \text{otherwise,} \end{cases}$$

$$H_i(K/G) \cong \begin{cases} \mathbb{Z}_p & \text{for } i = 0,\ r+2 \leq i \leq n \text{ and } r = i = n, \\ 0 & \text{otherwise.} \end{cases}$$

Proof We shall give a proof for the case where $0 < r < n$. The proof of the remaining case is similar and is left to the reader as an exercise.

From the Smith sequence, we have

$$H^{\gamma}_{i+1}(K) \cong H^{\bar{\gamma}}_i(K) \oplus H_i(K^G) \quad \text{for} \quad 1 \leq i \leq n-2. \tag{5.19}$$

The low dimensional part

$$\begin{array}{ccccccccc} 0 \to H^{\gamma}_1(K) \to & H^{\bar{\gamma}}_0(K) \oplus H_0(K^G) & \to & H_0(K) & \to H^{\gamma}_0(K) \to 0 \\ & \wr\| & & \wr\| & \\ & \mathbb{Z}_p & \xrightarrow{\cong} & \mathbb{Z}_p & \end{array}$$

of the Smith sequence shows that $H^{\gamma}_0(K) \cong 0$, and hence $H^{\bar{\gamma}}_0(K) \cong 0$ and $H^{\gamma}_1(K) \cong 0$. It follows from the isomorphism (5.19) above that

$$H^{\gamma}_i(K) \cong \begin{cases} 0 & \text{for } 0 \leq i \leq r \\ \mathbb{Z}_p & \text{for } r+1 \leq i \leq n-1. \end{cases}$$

On the other hand, Theorem 5.34 implies that $H^{\gamma}_i(K) = 0$ for $i \geq n+1$. In consequence it is shown as in the manner of the observation of the n-dimensional part in Theorem 5.63 that $H^{\gamma}_n(K)$ is isomorphic to $\mathbb{Z}_p$.

The result for $H_i(K/G)$ is easily deduced from the following commutative diagram

$$\begin{array}{ccccccccc} \longrightarrow H_i(K^G) & \longrightarrow & H_i(K/G) & \longrightarrow & H_i(K/G, K^G) & \longrightarrow & H_{i-1}(K^G) \longrightarrow \\ & \searrow & \nearrow{\scriptstyle \pi_*} & & \wr\| & & \\ & H_i(K) & & & H^{\alpha}_i(K) & & \end{array}$$

together with the above computation of $H^{\alpha}_i(K)$. □

We are now in a position to state the main theorem of this section.

Theorem 5.65 *Let G be the group $\mathbb{Z}_p \oplus \mathbb{Z}_p$* ($p$ prime) *and let K be a regular G-complex such that K is a* mod p *homology n-sphere and that K^G is a* mod p *homology r-sphere with $-1 \leq r < n$. Then the action is not semi-free.*

Proof Suppose that the G-action is semi-free. We put $G_i = \mathbb{Z}_p$ for $i = 1, 2$. Then $G \cong G_1 \times G_2$. Since the action is semi-free, we have $K^G = K^{G_1}$ which is a mod p homology r-sphere with $-1 \leq r < n$ by assumption. By suspending twice, we may assume that $r \geq 1$. First we give a proof in the case where $n > r+1$. Note that K/G_1 is a G_2-complex and $(K/G_1)^{G_2} = K^G$. Regarding K/G_1 as a G_2-complex, we get the Smith sequence:

$$\cdots \to H_{i+1}(K/G_1) \to H^{\gamma}_{i+1}(K/G_1) \to H^{\bar{\gamma}}_i(K/G_1) \oplus H_i(K^G) \to H_i(K/G_1) \to \cdots$$

This together with Theorem 5.64 yields the isomorphism

$$H^{\gamma}_{i+1}(K/G_1) \cong H^{\bar{\gamma}}_i(K/G_1) \oplus H_i(K^G) \qquad \text{for} \quad 1 \le i \le r. \tag{5.20}$$

The low dimensional part

$$\begin{array}{c} 0 \to H^{\gamma}_1(K/G_1) \to H^{\bar{\gamma}}_0(K/G_1) \oplus H_0(K^G) \to H_0(K/G_1) \to H^{\gamma}_0(K/G_1) \to 0 \\ \qquad\qquad\qquad\qquad \| \wr \qquad\qquad\qquad \| \wr \\ \qquad\qquad\qquad\qquad \mathbb{Z}_p \xrightarrow{\cong} \mathbb{Z}_p \end{array}$$

of the Smith sequence shows that $H^{\gamma}_0(K/G_1) \cong 0$ and hence $H^{\bar{\gamma}}_1(K/G_1) \cong 0$ and $H^{\gamma}_1(K/G_1) \cong 0$ as well. Accordingly, the isomorphism 5.20 above implies that

$$H^{\gamma}_i(K/G_1) \cong 0 \qquad \text{for} \quad 1 \le i \le r, \qquad \text{and}$$

$$H^{\gamma}_{r+1}(K/G_1) \cong H_r(K^G) \cong \mathbb{Z}_p.$$

Note that the homomorphisms induced by a G_2-map commute with the homomorphisms in the Smith sequences of G_2-actions in general and that the projection

$$\pi \colon K \to K/G_1$$

is a G_2-simplicial map. Thus we have the following commutative diagram:

$$\begin{array}{ccc} H^{\gamma}_{i+1}(K) & \xrightarrow{\partial} & H^{\bar{\gamma}}_i(K) \oplus H_i(K^G) \\ \downarrow{\scriptstyle \pi_*} & & \downarrow{\scriptstyle \pi_* \oplus \mathrm{id}} \\ H^{\gamma}_{i+1}(K/G_1) & \longrightarrow & H^{\bar{\gamma}}_i(K/G_1) \oplus H_i(K^G) \end{array}$$

Since K is a mod p homology n-sphere, ∂ is an isomorphism for $1 \le i \le n - 2$. In view of Theorem 5.64, we have $H^{\bar{\gamma}}_r(K) \cong 0$. Further, $H^{\bar{\gamma}}_r(K/G_1) \cong 0$ as shown above so that $\pi_* \oplus \mathrm{id}$ is an isomorphism for $i = r$. Hence it follows from (5.20) that $\pi_* \colon H^{\gamma}_{r+1}(K) \to H^{\gamma}_{r+1}(K/G_1) \cong \mathbb{Z}_p$ is also an isomorphism. By making use of the above commutative diagram, it can be shown inductively that the π_* are injective up to dimension $n - 1$. For brevity we put $X = K/G_1$ and we have the following commutative diagram:

$$\begin{array}{ccccccccccc} & & \mathbb{Z}_p & \xrightarrow{\cong} & \mathbb{Z}_p & & \mathbb{Z}_p & \xrightarrow{\cong} & \mathbb{Z}_p & & 0 \\ & & \wr\| & & \wr\| & & \wr\| & & \wr\| & & \| \\ 0 & \longrightarrow & H^{\bar{\gamma}}_n(K) & \longrightarrow & H_n(K) & \longrightarrow & H^{\bar{\gamma}}_n(K) & \longrightarrow & H^{\bar{\gamma}}_{n-1}(K) & \longrightarrow & H_{n-1}(K) & \longrightarrow \\ & & \downarrow{\scriptstyle \pi^4_*} & & \downarrow{\scriptstyle \pi^3_*} & & \downarrow{\scriptstyle \pi^2_*} & & \downarrow{\scriptstyle \pi^1_*} & & \downarrow \\ 0 & \longrightarrow & H^{\bar{\gamma}}_n(X) & \longrightarrow & H_n(X) & \longrightarrow & H^{\bar{\gamma}}_n(X) & \longrightarrow & H^{\bar{\gamma}}_{n-1}(X) & \longrightarrow & H_{n-1}(X) & \longrightarrow \end{array}$$

where the data are in accordance with Theorem 5.64. Since π_*^1 is injective, we deduce from the diagram above that π_*^2 is injective. By interchanging γ and $\bar{\gamma}$, we see that π_*^4 is injective as well and hence we can conclude that π_*^3 is injective.

On the other hand, it is shown in the proof of Theorem 5.64 that the f_* in the following commutative diagram:

$$\begin{array}{ccc} H_n(K/G_1, K^{G_1}) \cong H_n^\alpha(K) & \xrightarrow{f_*} & H_n(K) \\ \uparrow{\scriptstyle j_*} & \nearrow{\scriptstyle \tau_*} & \\ H_n(K/G_1) & & \end{array}$$

with respect to the G_1-action on K according to Lemma 5.61 is an isomorphism. In addition to that, j_* is an isomorphism by assumption, so that τ_* is an isomorphism as well.

Putting all this together, we see that the composite map

$$H_n(K/G_1) \xrightarrow{\tau_*} H_n(K) \xrightarrow{\pi_*^3} H_n(K/G_1)$$

is injective. But $\pi_*^3 \cdot \tau_* = p$ by Lemma 5.27, so $\pi_*^3 \cdot \tau_* = 0$, which contradicts the result above. This is caused by the assumption that G acts semi-freely on K.

When $n = r + 1$, it follows from Theorem 5.64 that the isomorphism (5.20) holds for all $i \geq 1$. In consequence, we have

$$H_i^\gamma(K/G_1) = \begin{cases} 0 & \text{for} \quad 0 \leq i \leq r \\ \mathbb{Z}_p & \text{for} \quad i = n. \end{cases}$$

Note that $H_n(K/G_1) = 0$ by Theorem 5.64 and that $H_{n+1}^\gamma(K/G_1) = 0$ by Theorem 5.34. Thus we have the following exact sequence

$$\begin{array}{ccccccc} \to H_{n+1}^\gamma(K/G_1) & \to & H_n^{\bar{\gamma}}(K/G_1) & \to & H_n(K/G_1) & \to \\ \| & & \wr\| & & \| & \\ 0 & & \mathbb{Z}_p & & 0 & \end{array}$$

which is a contradiction. This is also caused by the assumption that G acts semi-freely on K. □

Corollary 5.66 *There is no non-trivial, smooth, semi-free $\mathbb{Z}_p \oplus \mathbb{Z}_p$ (p prime) (or torus T^s, $s \geq 2$) action on a manifold which is a* mod *p homology sphere.*

Proof This is immediate from the fact that every G-manifold is G-triangulable for G a finite group. □

Corollary 5.67 *If the group $\mathbb{Z}_p \oplus \mathbb{Z}_p$ (p prime) (or T^s, $s \geq 2$) acts smoothly and semi-freely on a connected manifold, then the action is either trivial or free.*

Proof If the action is non-trivial and has a fixed point, then each fibre of the sphere bundle associated with the normal bundle of the fixed point set admits a free action. This contradicts Theorem 5.65. □

Exercises for Chapter 5

1. Show that a non-trivial compact Lie group does not act freely on a real projective space of even dimension.

2. If S^1 acts smoothly on a compact manifold M without fixed point, show that $\chi(M) = 0$.

3. Let G be a finite group and K be a regular G-complex. If the induced G-action on $H_*(K; \Lambda)$ is trivial, show that $\chi(K) = \chi(K/G)$ where Λ is a field of characteristic 0 or prime to $|G|$.

4. Let G be a non-trivial finite group and K be a regular G-complex such that the action is free. If the induced G-action on $H_*(K;\Lambda)$ is trivial, show that $\chi(K) - 0$ where Λ is a field of characteristic 0 or prime to $|G|$.

5. Show that there is no S^1-action on $\mathbb{C}P^n$ without fixed point.

6. Consider the $\mathbb{Z}_4$-action on $\mathbb{C}P^2$ stated in Exercise 10 in Chapter 4 and compute $\chi(\mathbb{C}P^2/\mathbb{Z}_4)$.

6 Localization and Riemann–Roch type theorems

In this final chapter, we shall deal with recent research by the author assuming some prerequisite reading. Although the topics treated in this chapter seem special, their concepts are quite fundamental and are of wide application.

Let G be a compact Lie group and h_G be an equivariant multiplicative cohomology theory. Given a G-map $f: M \to N$ between G-manifolds, we shall define a homomorphism

$$f_! : h_G(M) \to h_G(N)$$

which is called an equivariant Gysin homomorphism. Then, by making use of this we shall present several results, including localization theorems, equivariant Riemann–Roch type theorems, the G-signature theorem, the index theorem, and the vanishing theorem of the $\hat{A}$-genus.

Notice that since h_G is a cohomology theory, f induces a homomorphism

$$f^*: h_G(N) \to h_G(M)$$

of the opposite direction in general. Under the condition that M and N are G-manifolds, the homomorphism $f_!$ is defined and is made use of as a powerful tool.

This kind of phenomenon occurs in other fields of mathematics as well. For instance, the notion of induction in representation theory and the notion of the transfer as mentioned in Section 5.3 are considered under certain restrictions, such as finiteness of the index of a subgroup in representation theory, and a finite covering in the transfer.

6.1 Equivariant Gysin homomorphism

Let G be a compact Lie group and $h_G(\ )$ be an equivariant multiplicative cohomology theory. Let $f: M \to N$ be a G-map between G-manifolds. In this section, we shall define a homomorphism

$$f_!: h_G(M) \to h_G(N)$$

and investigate its fundamental properties.

Let X be a compact G-space and ξ be a G-vector bundle over X. We denote by $D(\xi)$ (resp. $S(\xi)$) the disc bundle (resp. sphere bundle) associated with ξ and by $p: D(\xi) \to X$ the projection. An element $t(\xi)$ of $h_G(D(\xi), S(\xi))$ is called a *Thom class* (or *h_G-orientation (class)*) if for every compact G-invariant subspace Y of X, the correspondence $x \mapsto (t(\xi)\,|\,Y)\cdot x$ gives an isomorphism

$$h_G(Y) \to h_G(D(\xi\,|\,Y), S(\xi\,|\,Y)).$$

Then this isomorphism is called a *Thom isomorphism*.

When a G-vector bundle ξ has a Thom class, we say that ξ is *h_G-orientable* and ξ together with a Thom class $t(\xi)$ is called an *h_G-oriented bundle*. Let $s: X \to (D(\xi), S(\xi))$ denote the zero cross-section. Then $s^*t(\xi)$ is called the *equivariant Euler class* of an h_G-oriented bundle ξ, which is denoted by $\chi_G(\xi)$.

In the case of the tangent bundle $T(M)$ of a G-manifold M, a Thom class

$$t(M) \in h_G(D(T(M)), S(T(M)))$$

is called an *h_G-orientation* of M, and M together with $t(M)$ is called an *h_G-oriented G-manifold*.

Let M and N be G-manifolds with h_G-orientations

$$t(M) \in h_G(D(T(M)), S(T(M))),$$

$$t(N) \in h_G(D(T(N)), S(T(N))),$$

respectively.

As mentioned in Theorem 4.12, there exist a G-representation space V and a G-embedding $e: M \to V$. Then the map

$$f \times e: M \to N \times V$$

defined by $(f \times e)(x) = (f(x), e(x))$ for $x \in M$ is also a G-embedding. We now identify M with its image $(f \times e)(M)$. Choose a G-invariant Riemannian metric on $N \times V$ and let ν be a G-invariant open tubular neighbourhood of M in $N \times V$. Then ν may be identified with the normal G-vector bundle of M in $N \times V$ by Theorem 4.8. Choose a G-invariant metric on V and let $D(V)$ (resp. $S(V)$) be the unit disc (resp. unit sphere) in V. Without loss of generality, we may assume that

$$D(\nu) \subset N \times \operatorname{Int} D(V)$$

where $\operatorname{Int} D(V)$ stands for $D(V) - S(V)$ (see Fig. 6.1).

Regarding V as a G-vector bundle over one point, we assume that V is h_G-oriented. Consider the following four homomorphisms:

$$\varphi_1: h_G(M) \to h_G(D(\nu), S(\nu)),$$

$$\varphi_2: h_G(N \times D(V), N \times D(V) - \operatorname{Int} D(\nu)) \to h_G(D(\nu), S(\nu)),$$

$$\varphi_3: h_G(N \times D(V), N \times D(V) - \operatorname{Int} D(\nu)) \to h_G(N \times D(V), N \times S(V)),$$

$$\varphi_4: h_G(N) \to h_G(N \times D(V), N \times S(V)).$$

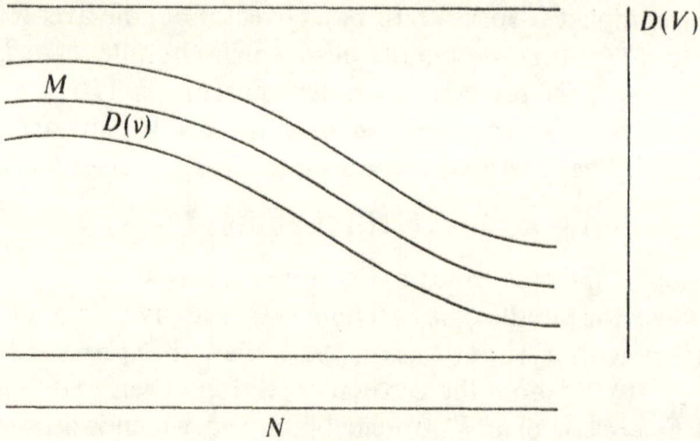

Fig. 6.1

Here $\operatorname{Int} D(v)$ stands for $D(v) - S(v)$. It is easy to see that there is a canonical h_G-orientation class

$$t(v) \in h_G(D(v), S(v))$$

of v such that

$$t(M) \times t(v) = (f \times e)^*(t(N) \times t(V)).$$

Then the homomorphism φ_1 is defined to be the Thom isomorphism by means of the Thom class $t(v)$.

It follows from the excision isomorphism theorem that the natural inclusion

$$i: (D(v), S(v)) \to (N \times D(V), N \times D(V) - \operatorname{Int} D(v))$$

induces an isomorphism

$$i^*: h_G(N \times D(V), N \times D(V) - \operatorname{Int} D(v)) \to h_G(D(v), S(v)).$$

Then we put

$$\varphi_2 = i^*.$$

The homomorphism φ_3 is defined to be the homomorphism j^* which is induced by the natural inclusion

$$j: (N \times D(V), N \times S(V)) \to (N \times D(V), N \times D(V) - \operatorname{Int} D(v)).$$

Finally, φ_4 is again defined to be the Thom isomorphism with respect to the Thom class $t(V)$.

We are now equipped to define our *equivariant Gysin homomorphism* $f_!$ by the composition:

$$f_! = \varphi_4^{-1} \cdot \varphi_3 \cdot \varphi_2^{-1} \cdot \varphi_1 : h_G(M) \to h_G(N).$$

In particular, if N is a point $*$, we put

$$\text{Ind} = f_!: h_G(M) \to h_G(*)$$

which is called the *index homomorphism.*

Theorem 6.1 *The equivariant Gysin homomorphism* $f_!$ *is independent of all choices made except for* $t(M)$, $t(N)$, *and has the following properties:*

(1) $f_!$ *depends only on the G-homotopy class of* f;

(2) $f_!$ *is an* $h_G(*)$*-module homomorphism;*

(3) $(f \cdot g)_! = f_! \cdot g_!$;

(4) $f_!(x \cdot f^*(y)) = f_!(x) \cdot y$ *for* $x \in h_G(M)$, $y \in h_G(N)$;

(5) *if* f *is a G-embedding of class* C^∞ *with a normal bundle* ν, *then* ν *is* h_G*-oriented so that we have*

$$f^* \cdot f_!(x) = \chi_G(\nu) \cdot x \qquad \textit{for} \quad x \in h_G(M).$$

Proof To see that $f_!$ is independent of the choice of the embedding $e: M \to V$, let $e': M \to V'$ be another G-embedding where V' is h_G-oriented. Then consider the following two G-embeddings:

$$\bar{e}: M \xrightarrow{e} V \subset V \oplus V' \subset (V \oplus V')^k,$$

$$\bar{e}': M \xrightarrow{e'} V' \subset V \oplus V' \subset (V \oplus V')^k$$

where k is a positive integer and $V \oplus V' \subset (V \oplus V')^k$ denotes inclusion into the first factor. It follows from the transitivity of the Thom isomorphism that e and $\bar{e}$ give rise to the same equivariant Gysin homomorphism. Similarly e' and $\bar{e}'$ give rise to the same equivariant Gysin homomorphism. Denote by $\nu(f \times \bar{e})$ (resp. $\nu(f \times \bar{e}')$) the normal bundle of the G-embedding:

$$f \times \bar{e}: M \to N \times (V \oplus V')^k$$

$$(\text{resp.} \quad f \times \bar{e}': M \to N \times (V \oplus V')^k).$$

Since $f \times \bar{e}$ and $f \times \bar{e}'$ are G-homotopic by a G-homotopy of smooth G-embeddings, $\nu(f \times \bar{e})$ is G-equivalent to $\nu(f \times \bar{e}')$. If, moreover, $k \geq 2 \dim M + 4$, then two such regular G-homotopies are themselves regularly G-homotopic through regular G-homotopies[135] and the two resulting bundle equivalences are G-homotopic through G-bundle equivalences. Consequently, the Thom class $t(\nu(f \times \bar{e}))$ is chosen uniquely as in the non-equivariant case. Thus $\bar{e}$ and $\bar{e}'$ give rise to the same equivariant Gysin homomorphism.

If f and f' are G-homotopic, then $f \times \bar{e}$ and $f' \times \bar{e}$ are G-homotopic by a G-homotopy of differentiable G-embeddings. Hence (1) is proved as above.

Since each φ_i in the definition of $f_!$ is an $h_G(*)$-module homomorphism, (2) follows.

(3) will be shown by using the multiplicative property of Thom classes.

To show (4), we need to introduce some notation. Let $p\colon D(v) \to M$ be the projection of the bundle, let $i'\colon D(v) \to N \times D(V)$ be the natural inclusion, and let $p'\colon N \times D(V) \to N$ be the projection. Clearly $p' \cdot i'$ is G-homotopic to $f \cdot p$. Using the fundamental properties concerning the product of an equivariant multiplicative cohomology theory and the homomorphism induced by a G-map, we shall make the following computation. For $x \in h_G(M)$ and $y \in h_G(N)$, we have

$$\begin{aligned}
f_!(x \cdot f^*(y)) &= \varphi_4^{-1} \cdot \varphi_3 \cdot \varphi_2^{-1} \cdot \varphi_1(x \cdot f^*(y)) \\
&= \varphi_4^{-1} \cdot \varphi_3 \cdot \varphi_2^{-1}(t(v) \cdot p^*(x \cdot f^*(y))) \\
&= \varphi_4^{-1} \cdot \varphi_3 \cdot i^{*-1}(t(v) \cdot p^*(x) \cdot p^* f^*(y)) \\
&= \varphi_4^{-1} \cdot \varphi_3 \cdot i^{*-1}(t(v) \cdot p^*(x) \cdot i'^* p'^*(y)) \quad (\text{since } p' \cdot i' \underset{G}{\simeq} f \cdot p) \\
&= \varphi_4^{-1} \cdot \varphi_3(i^{*-1}(t(v) \cdot p^*(x)) \cdot p'^*(y)) \\
&= \varphi_4^{-1}((j^* \cdot i^{*-1}(t(v) \cdot p^*(x))) \cdot p'^*(y)) \\
&= (\varphi_4^{-1} \cdot j^* \cdot i^{*-1}(t(v) \cdot p^*(x))) \cdot y \\
&= f_!(x) \cdot y.
\end{aligned}$$

Finally we show (5). Concerning the normal bundle v of the G-embedding $f\colon M \to N$, we consider two natural inclusion maps:

$$i\colon (D(v), S(v)) \to (N, N - \operatorname{Int} D(v)),$$
$$j\colon N \to (N, N - \operatorname{Int} D(v)).$$

Denote by φ_1 the Thom isomorphism of v. By using the multiplicative property of the Thom isomorphism, it is easily seen that

$$f_! = j^* \cdot i^{*-1} \cdot \varphi_1.$$

Let $s\colon M \to (D(v), S(v))$ be the zero cross-section. Then we have

$$j \cdot f = i \cdot s.$$

With this preparation, we can make the following computation. For $x \in h_G(M)$, we have

$$\begin{aligned}
f^* \cdot f_!(x) &= f^* \cdot j^* \cdot i^{*-1} \cdot \varphi_1(x) = (j \cdot f)^* \cdot i^{*-1} \cdot \varphi_1(x) \\
&= (i \cdot s)^* \cdot i^{*-1} \cdot \varphi_1(x) = s^* \cdot i^* \cdot i^{*-1} \cdot \varphi_1(x) \\
&= s^* \cdot \varphi_1(x) = s^*(t(v) \cdot p^*(x)) = s^*(t(v)) \cdot x \\
&= \chi_G(v) \cdot x.
\end{aligned}$$

This proves (5). □

For the reader who is not familiar with equivariant cohomology theories, we shall give examples in the following. In this chapter we keep in mind equivariant K-theory $K_G(\ )$ and the following equivariant cohomology theory.

Let $EG \to BG$ be the universal principal G-bundle. Denote singular cohomology theory by $H^*(\ )$. As shown in Section 2.3, G acts freely on the total space EG of the principal bundle from the right. For a G-space X, we consider the twisted product $EG \underset{G}{\times} X$. Then the cohomology group $H^*(EG \underset{G}{\times} X)$ is called the *Borel cohomology*, which is often written as $H_G^*(X)$. It is well known that this cohomology reflects conditions of the G-action on X fairly well.

As a group, we shall mainly deal with the circle group S^1.

It is easy to see that if $\pi: E \to X$ is a G-vector bundle, then

$$\mathrm{id} \underset{G}{\times} \pi: EG \underset{G}{\times} E \to EG \underset{G}{\times} X$$

becomes a vector bundle. In particular, given a G-manifold M, its tangent G-vector bundle $\pi: T(M) \to M$ yields a vector bundle

$$\mathrm{id} \underset{G}{\times} \pi: EG \underset{G}{\times} T(M) \to EG \underset{G}{\times} M.$$

If M is oriented in the ordinary sense and each element g of G preserves the orientation, then there exists a Thom class of the above bundle with respect to $H^*(\ )$. If we consider the cohomology $H^*(EG \underset{G}{\times} M; \mathbb{Z}_2)$ with $\mathbb{Z}_2$ coefficients, then there exists a Thom class for any G-manifold M.

When we do not take G-actions into consideration, our Gysin homomorphism $f_!$ with respect to the ordinary cohomology $H^*(\ )$ coincides with the classical Gysin homomorphism $\bar{f}$ which is defined by using Poincaré duality. We shall show this in the following.

Theorem 6.2 *Let M and N be oriented manifolds. Regarding them as $\{e\}$-manifolds, we have $f_! = \bar{f}_!$.*

Remark Owing to this theorem, our $f_!$ is called an equivariant Gysin homomorphism.

Proof Let $e: M \to \mathbb{R}^m$ be an embedding. Then the map

$$f \times e: M \to N \times \mathbb{R}^m$$

is also an embedding, by which we identify M with its image $(f \times e)(M)$. Choose a Riemannian metric on $N \times \mathbb{R}^m$ and let v be an open tubular neighbourhood of M in $N \times \mathbb{R}^m$. Then v may be identified with the normal vector bundle of M in $N \times \mathbb{R}^m$. Let D^m (resp. S^{m-1}) be the unit disc (resp.

unit sphere) in $\mathbb{R}^m$. Without loss of generality, we may assume that

$$D(v) \subset N \times \operatorname{Int} D^m$$

where $\operatorname{Int} D^m$ stands for $D^m - S^{m-1}$. Let $p: D(v) \to M$ be the projection of the bundle, let $i': D(v) \to N \times D^m$ be the natural inclusion, and let $p': N \times D^m \to N$ be the projection. Then we have the following commutative diagrams:

$$\begin{array}{ccccc}
H^*(M) & \xrightarrow{\varphi_1} & H^*(D(v), S(v)) & \xleftarrow{\varphi_2} & H^*(N \times D^m, N \times D^m - \operatorname{Int} D(v)) \\
\cong \big\downarrow \cap[M] & (1) & \cong \big\downarrow \cap[D(v), S(v)] & (2) & \big\downarrow \cap[N \times D^m, N \times D^m - \operatorname{Int} D(v)] \\
H_*(M) & \xleftarrow[p_*]{} & H_*(D(v)) & \xrightarrow[i'_*]{} & H_*(N \times D^m)
\end{array}$$

$$\begin{array}{ccccc}
H^*(N \times D^m, N \times D^m - \operatorname{Int} D(v)) & \xrightarrow{\varphi_3} & H^*(N \times D^m, N \times S^{m-1}) & \xleftarrow{\varphi_4} & H^*(N) \\
\big\downarrow \cap[N \times D^m, N \times D^m - \operatorname{Int} D(v)] & (3) & \swarrow \cap[N \times D^m, N \times S^{m-1}] & (4) & \big\downarrow \cap[N] \\
H_*(N \times D^m) & & \xrightarrow[p'_*]{} & & H_*(N)
\end{array}$$

Here $[M]$ and $[N]$ denote the orientation classes of M and N, respectively, and $\cap[A, B]$ denotes the homomorphism defined by the cap product with the element $[A, B]$.

The commutativities of (1) and (4) follow from the theorem of Thom concerning the Thom isomorphism.

Let $i: (D(v), S(v)) \to (N \times D^m, N \times D^m - \operatorname{Int} D(v))$ be the natural inclusion. Since

$$i_*[D(v), S(v)] = [N \times D^m, N \times D^m - \operatorname{Int} D(v)],$$

it follows from the fundamental properties of cap products that, for $x \in H^*(N \times D^m, N \times D^m - \operatorname{Int} D(v))$, we have

$$\begin{aligned}
i'_* \cdot (\cap[D(v), S(v)]) \cdot \varphi_2(x) &= i'_*(i^*(x) \cap [D(v), S(v)]) \\
&= x \cap i_*[D(v), S(v)] \\
&= x \cap [N \times D^m, N \times D^m - \operatorname{Int} D(v)].
\end{aligned}$$

That is, (2) is commutative. The commutativity of (3) is shown similarly. Obviously we have

$$f_* = p'_* \cdot i'_* \cdot p_*^{-1}.$$

It turns out that

$$f_! = (\cap[N])^{-1} \cdot f_* \cdot (\cap[M]) = \bar{f}_!,$$

which completes the proof of the theorem. □

Lemma 6.3 *Let a compact Lie group G act trivially on oriented manifolds M and N. For a map $f\colon M \to N$, its equivariant Gysin homomorphism $f_!\colon H^*(BG \times M) \to H^*(BG \times N)$ is given by*

$$f_!(t \times x) = t \times \bar{f}_!(x) \qquad \text{for} \quad t \in H^*(BG), x \in H^*(M)$$

where $t \times x$ denotes the cross product of t and x and $\bar{f}_!$ denotes the classical Gysin homomorphism.

Proof By making use of the fundamental properties of cross products, cap products, and cup products, it is easily verified that

$$f_!(t \times x) = t \times f_!(x).$$

Here $f_!$ of the right-hand side denotes the equivariant Gysin homomorphism in the case where G is trivial. Therefore the lemma follows from the preceding theorem. □

Corollary 6.4 *In addition to the hypotheses of Lemma 6.3, we suppose that N is a point. Then we have*

$$f_!(t \times x) = \bar{f}_!(x)t = t \times x/[M]$$

for $t \in H^*(BG)$, $x \in H^*(M)$ where $/[M]$ denotes the slant product.

Proof Since $\bar{f}_!(x)$ lies in the cohomology group of dimension zero, t and $\bar{f}_!(x)$ commute. Moreover it is easy to see that $\bar{f}_!(x)$ coincides with the Kronecker product $\langle x, [M] \rangle$. Therefore the equation of the corollary follows from the definition of the slant product. □

Corollary 6.5 *In addition to the hypotheses of Corollary 6.4, we assume that $H^*(BG)$ has no torsion. Then the homomorphism*

$$f_!\colon H^*(BG \times M) \to H^*(BG)$$

is given by the slant product $/[M]$.

Proof If $H^*(BG)$ has no torsion, any element of $H^*(BG \times M)$ can be written as the sum of cross products. Hence Corollary 6.5 is immediate from Corollary 6.4. □

6.2 Localization theorems

In this section, we shall show that there is a beautiful correspondence between algebraic localization and geometric localization. Here we mean by geometric localization that we concentrate on the fixed point set. By combining this localization theorem with Riemann–Roch type theorems which will be dealt with later, we shall obtain interesting results such as the G-signature theorem, the vanishing theorem of the $\hat{A}$-genus, and the index formula.

First we begin with algebraic localization. Let $h_G(\ )$ be an equivariant multiplicative cohomology theory. We regard a representation space V of G as a G-vector bundle over a point. Then we consider the set

$$S = \{\chi_G(V)\}$$

of equivariant Euler classes of h_G-oriented representation spaces V of G without trivial direct summand. The multiplicative property of Thom classes implies the multiplicative property

$$\chi_G(V \oplus W) = \chi_G(V) \cdot \chi_G(W).$$

Hence S is a multiplicatively closed subset of $h_G(*)$. Since $h_G(M)$ is a module over $h_G(*)$, the localization $S^{-1}h_G(M)$ of the module is defined.

Let $f: M \to N$ be a G-map between h_G-oriented G-manifolds. According to Theorem 6.1, $f_!: h_G(M) \to h_G(N)$ is an $h_G(*)$ -module homomorphism, so that we have the localized homomorphism

$$S^{-1}f_!: S^{-1}h_G(M) \to S^{-1}h_G(N).$$

On the other hand, there always exists the localized homomorphism

$$S^{-1}f^*: S^{-1}h_G(N) \to S^{-1}h_G(M).$$

Denote by F_μ each component of the fixed point set of a G-manifold M and by $i_\mu: F_\mu \to M$ the inclusion map. Denote by N_μ the normal bundle of F_μ in M.

It follows from the slice theorem and classical representation theory that there exist a representation space V of G without trivial direct summand and a G-map

$$f: M - \bigcup_\mu \operatorname{Int} D(N_\mu) \to V - \operatorname{Int} D(V).$$

Suppose now that V is h_G-oriented.

Theorem 6.6 *The following homomorphism*

$$\sum_\mu S^{-1}i_\mu^*: S^{-1}h_G(M) \to \sum_\mu S^{-1}h_G(F_\mu)$$

is an isomorphism where the summation is taken over all the components F_μ of the fixed point set.

Proof We begin by introducing some notation. Let

$$i: V - \operatorname{Int} D(V) \to V,$$

$$j: V \to (V, V - \operatorname{Int} D(V)),$$

$$k: * \to V,$$

$$s: * \to (V, V - \operatorname{Int} D(V)),$$

be the inclusion maps where k and s are inclusions to the origin of V. Further we consider the following commutative diagram:

$$\begin{array}{ccc} M - \bigcup_{\mu} \operatorname{Int} D(N_\mu) & \xrightarrow{\ f\ } & V - \operatorname{Int} D(V) \\ & \searrow^{\pi_1} & \downarrow{\scriptstyle \pi_2} \\ & & * \end{array} \tag{6.1}$$

where π_i are the constant maps.

Using this notation, we shall make the following calculations. Since $k \cdot \pi_2$ and i are G-homotopic, we have

$$i^* = (k \cdot \pi_2)^* = \pi_2^* \cdot k^*.$$

Since $j \cdot k = s$, we have $s^* = (j \cdot k)^* = k^* \cdot j^*$. Hence we have the following commutative diagram:

$$\begin{array}{ccccc} h_G(V, V - \operatorname{Int} D(V)) & \xrightarrow{\ j^*\ } & h_G(V) & \xrightarrow{\ i^*\ } & h_G(V - \operatorname{Int} D(V)) \\ & {\scriptstyle s^*}\searrow & \cong\downarrow{\scriptstyle k^*} & \nearrow{\scriptstyle \pi_2^*} & \\ & & h_G(*) & & \end{array} \tag{6.2}$$

Since the horizontal sequence in diagram (6.2) is exact, we have

$$\pi_2^* \chi_G(V) = \pi_2^* \cdot s^* t(V) = i^* \cdot j^* t(V) = 0.$$

It follows from diagram 6.1 that

$$\pi_1^* \chi_G(V) = f^* \cdot \pi_2^* \chi_G(V) = 0.$$

Accordingly, for any

$$x \in h_G\left(M - \bigcup_{\mu} \operatorname{Int} D(N_\mu), \bigcup_{\mu} S(N_\mu) \right),$$

the product $x \cdot \pi_1^* \chi_G(V)$ is trivially zero, which implies that

$$S^{-1} h_G\left(M - \bigcup_{\mu} \operatorname{Int} D(N_\mu), \bigcup_{\mu} S(N_\mu) \right) = 0.$$

Hence it follows from the excision isomorphism theorem that

$$S^{-1} h_G\left(M, \bigcup_{\mu} D(N_\mu) \right) = 0.$$

Since localization is a functor preserving the exactness, we have the following exact sequence:

$$\cdots \to S^{-1}h_G\left(M, \bigcup_\mu D(N_\mu)\right) \to S^{-1}h_G(M) \to S^{-1}h_G\left(\bigcup_\mu D(N_\mu)\right)$$
$$\to S^{-1}h_G\left(M, \bigcup_\mu D(N_\mu)\right) \to \cdots.$$

Thus we obtain an isomorphism

$$S^{-1}h_G(M) \cong S^{-1}h_G\left(\bigcup_\mu D(N_\mu)\right).$$

It follows from the additivity and the G-homotopy type invariance of equivariant cohomology that we have an isomorphism

$$S^{-1}h_G\left(\bigcup_\mu D(N_\mu)\right) \cong \sum_\mu S^{-1}h_G(F_\mu).$$

This completes the proof of the theorem. □

Assumption Each N_μ is h_G-oriented so that its equivariant Euler class $\chi_G(N_\mu)$ is a unit in $S^{-1}h_G(F_\mu)$.

We orient F_μ so that the orientation of N_μ followed by that of F_μ yields the restriction of the orientation of M. Since $F_{\mu'} \subset M - \operatorname{Int} D(N_\mu)$ for $\mu' \neq \mu$, the equality $i_{\mu'}^* \cdot i_{\mu!} = 0$ holds. On the other hand, in view of Theorem 6.1, we have

$$i_\mu^* \cdot i_{\mu!}(x) = \chi_G(N_\mu) \cdot x.$$

Therefore for any $x \in S^{-1}h_G(M)$, we have

$$\left(\sum_\mu S^{-1}i_\mu^*\right)\left(\sum_\mu S^{-1}i_{\mu!}\right)\left(\sum_\mu \frac{S^{-1}i_\mu^* x}{\chi_G(N_\mu)}\right) = \sum_\mu S^{-1}i_\mu^* x.$$

By Theorem 6.6, $\sum_\mu S^{-1}i_\mu^*$ is an isomorphism so that we have

$$\left(\sum_\mu S^{-1}i_{\mu!}\right)\left(\sum_\mu \frac{S^{-1}i_\mu^* x}{\chi_G(N_\mu)}\right) = x.$$

It turns out that $\sum_\mu S^{-1}i_{\mu!}$ is also an isomorphism and the equality

$$\sum_\mu \frac{S^{-1}i_\mu^* x}{\chi_G(N_\mu)} = \left(\sum_\mu S^{-1}i_{\mu!}\right)^{-1}(x)$$

holds. Since our equivariant Gysin homomorphism is unique and functorial, the following theorem is verified.

Theorem 6.7 (Localization theorem). *Under the above assumption, the following diagram is commutative*:

$$\begin{array}{ccc} S^{-1}h_G(M) & \xrightarrow{S^{-1}\mathrm{Ind}} & S^{-1}h_G(*) \\ {\displaystyle\sum_\mu \frac{S^{-1}i_\mu^*}{\chi_G(N_\mu)}}\Big\downarrow & \nearrow{\scriptstyle S^{-1}\mathrm{Ind}} & \\ \displaystyle\sum_\mu S^{-1}h_G(F_\mu) & & \end{array}$$

By combining this theorem with the equivariant Riemann–Roch type theorem which will be dealt with later, we shall obtain various kinds of interesting results. In addition to that, this theorem is interesting in its own right. For instance it connects the global index with the local index. Furthermore, the local index can be expressed in terms of the classical Gysin homomorphism or of the slant product as will be mentioned later. As a consequence, we see that the local index is an easy-to-understand homomorphism.

The remainder of this section will be devoted to ensuring that the assumption of the theorem is satisfied for certain cases and to calculating the Chern classes of the normal bundle of the fixed point set.

As an equivariant cohomology theory, we make use of the Borel cohomology $H^*(EG \underset{G}{\times} M)$. If a G-vector bundle $\xi \to X$ is oriented without considering G-actions and if the G-action preserves the orientation, then the bundle $EG \underset{G}{\times} \xi \to EG \underset{G}{\times} X$ is given the orientation induced by that of ξ. That is, there exists a unique Thom class

$$t(\xi) \in H^*\left(EG \underset{G}{\times} (D(\xi), S(\xi)) \right)$$

in the ordinary sense.

As a Lie group G, we deal with the toral group $T^n = S_1^1 \times \cdots \times S_n^1$. Let $\{t_1, \ldots, t_n\}$ be the set of natural generators of $H^1(T^n)$. We make use of the same notation $\{t_1, \ldots, t_n\}$ for the set of corresponding generators of $H^2(BT^n)$ via the transgression. Given an n-tuple $\alpha = (a_1, \ldots, a_n)$ of integers, we define a complex irreducible representation $T^n \to S^1 = U(1)$ by $(s_1, \ldots, s_n) \mapsto s_1^{a_1} \cdots s_n^{a_n}$ for $s_i \in S_i^1$ which is also denoted by α for brevity. Denote by V_α its representation space. We identify α with the element $a_1 t_1 + \cdots + a_n t_n$ of $H^2(BT^n)$ by the following translations

$$\mathbb{Z}^n = \left\{ \begin{array}{l} \text{complex irreducible} \\ \text{representations} \end{array} \right\} \leftrightarrow H^1(T^n) \leftrightarrow H^2(BT^n).$$

We now investigate the multiplicatively closed subset which will be used for the localization.

As mentioned before, any representation of a compact group can be written as the direct sum of irreducible representations and any non-trivial irreducible representation over $\mathbb{R}$ is given by regarding some complex irreducible representation $\alpha: T^n \to S^1$ as a representation over $\mathbb{R}$. Therefore a T^n-representation without trivial direct summand is given by a homomorphism

$$\psi: T^n \to T^k.$$

Let $\{t'_j | j = 1, \ldots, k\}$ be the set of natural generators of $H^1(T^k)$. Then $w_j = \psi^*(t'_j)$ will be called the *weights* of ψ and can be written as

$$w_j = \sum_{i=1}^{n} a_{ij} t_i \qquad a_{ij} \in \mathbb{Z}.$$

According to Borel and Hirzebruch,[13] the total Chern class $c(ET^n \underset{\psi}{\times} \mathbb{C}^k)$ of the complex vector bundle

$$ET^n \underset{\psi}{\times} \mathbb{C}^k \to BT^n$$

is given by

$$c\left(ET^n \underset{\psi}{\times} \mathbb{C}^k \right) = \prod_{j=1}^{k} (1 + w_j) = \prod_{j=1}^{k} \left(1 + \sum_{i=1}^{n} a_{ij} t_i \right).$$

Accordingly, its Euler class is given by

$$\chi\left(ET^n \underset{\psi}{\times} \mathbb{C}^k \right) = \prod_{j=1}^{k} \left(\sum_{i=1}^{n} a_{ij} t_i \right).$$

Since ψ has no trivial direct summand, $w_j \neq 0$ for all j, that is,

$$\prod_{j=1}^{k} \left(\sum_{i=1}^{n} (a_{ij})^2 \right) \neq 0.$$

Conversely, if an $n \times k$ matrix (a_{ij}) satisfies

$$\prod_{j=1}^{k} \left(\sum_{i=1}^{n} (a_{ij})^2 \right) \neq 0 \qquad a_{ij} \in \mathbb{Z},$$

then the homomorphism $\psi: T^n \to T^k$ defined by

$$\psi(s_1, \ldots, s_n) = (s_1^{a_{11}} \cdots s_n^{a_{n1}}, \ldots, s_1^{a_{1k}} \cdots s_n^{a_{nk}})$$

gives a representation without trivial direct summand such that

$$\psi^*(t'_j) = \sum_{i=1}^{n} a_{ij} t_i \qquad \text{for all } j.$$

Thus we have

Lemma 6.8 *The multiplicative subset $S \subset H^*(BT^n)$ is given by*

$$S = \left\{ \prod_{j=1}^{k} \left(\sum_{i=1}^{n} a_{ij} t_i \right) \right\}$$

where a_{ij} ranges over integers satisfying $\prod_{j=1}^{k} \left(\sum_{i=1}^{n} (a_{ij})^2 \right) \neq 0$ and k may vary.

Remark Since S does not contain the zero element by Lemma 6.8 and since $H^*(BT^n)$ is an integral domain, the localization map

$$\lambda: H^*(BT^n) \to S^{-1} H^*(BT^n)$$

is injective.

Next we shall study the fixed point set and its normal bundle. Let M be a T^n-manifold. Denote by F_μ each component of the fixed point set and by N_μ its normal bundle. As is well known, the normal bundle N_μ has a complex vector bundle structure such that T^n acts on N_μ as complex vector bundle automorphisms.[68] It follows that N_μ has the following decomposition[64]

$$N_\mu = \sum_\alpha E_{\mu\alpha} \otimes V_\alpha$$

where α runs through the complex irreducible representations of T^n and $E_{\mu\alpha}$ denotes complex vector bundles.

Put $\gamma_i = (0, \ldots, 0, 1, 0, \ldots, 0)$ where the ith coefficient is 1 and the others are all 0. As before, we regard γ_i as a complex irreducible representation. Then for $\alpha = (a_1, \ldots, a_n)$, we have obviously

$$V_\alpha = \prod_i (V_{\gamma_i})^{a_i}$$

where the product stands for the tensor product of representations. We now consider the bundles:

$$\Gamma_i = ET^n \underset{T^n}{\times} V_{\gamma_i}$$

$$\Gamma_\alpha = ET^n \underset{T^n}{\times} V_\alpha$$

which are associated with $ET^n \to BT^n$. Then we obviously have

$$\Gamma_\alpha = \prod_i \Gamma_i^{a_i}$$

where the products now stand for the tensor products of complex line bundles.

Lemma 6.9 $ET^n \underset{T^n}{\times} N_\mu \cong \sum_\alpha \left(\prod_i \Gamma_i^{a_i}\right) \hat{\otimes} E_{\mu\alpha}$

where $\alpha = (a_1, \ldots, a_n)$ *and* $\hat{\otimes}$ *denotes the external tensor product.*

Proof It is easily verified that the correspondence

$$[x, y \otimes z] \mapsto [x, z] \hat{\otimes} y$$

gives a vector bundle isomorphism

$$ET^n \underset{T^n}{\times} (E_{\mu\alpha} \otimes V_\alpha) \cong \left(ET^n \underset{T^n}{\times} V_\alpha \right) \hat{\otimes} E_{\mu\alpha}.$$

Therefore we have

$$\begin{aligned} ET^n \underset{T^n}{\times} N_\mu &\cong \sum_\alpha ET^n \underset{T^n}{\times} (E_{\mu\alpha} \otimes V_\alpha) \\ &\cong \sum_\alpha \left(ET^n \underset{T^n}{\times} V_\alpha \right) \hat{\otimes} E_{\mu\alpha} \\ &\cong \sum_\alpha \left(\prod_i \Gamma_i^{a_i} \right) \hat{\otimes} E_{\mu\alpha}. \end{aligned}$$ □

As a consequence of Lemma 6.9 the Chern class and the Euler class of the bundle are calculated by the argument of Borel and Hirzebruch[13] as follows.

Lemma 6.10 *The total Chern class and the Euler class of the complex vector bundle* $ET^n \underset{T^n}{\times} N_\mu \to BT^n \times F_\mu$ *are given by*

$$c\left(ET^n \underset{T^n}{\times} N_\mu \right) = \prod_{\alpha, i} (1 + \alpha + (x_{\mu\alpha})_i),$$

$$\chi\left(ET^n \underset{T^n}{\times} N_\mu \right) = \prod_{\alpha, i} (\alpha + (x_{\mu\alpha})_i)$$

where the total Chern class of $E_{\mu\alpha}$ *is expressed formally as* $\prod_i (1 + (x_{\mu\alpha})_i)$.

We are now ready to show that the assumption for the localization theorem holds in this case.

Lemma 6.11 *The Euler class* $\chi_{T^n}(N_\mu)$ *is a unit in* $S^{-1}H^*(BT^n \times F_\mu)$.

Proof It follows from Lemma 6.10 that $\chi_{T^n}(N_\mu)$ has the form $\prod_{\alpha, i}(\alpha + (x_{\mu\alpha})_i)$. Since the representations α are non-trivial, α are in $\pi^*(S)$ where

$\pi: BT^n \times F_\mu \to BT^n$ is the projection. Consider the formal equation

$$1 = \prod_{\alpha, i} \left\{ \left(1 + \frac{(x_{\mu\alpha})_i}{\alpha} \right) \cdot \sum_{j=0}^{\infty} \left(\frac{-(x_{\mu\alpha})_i}{\alpha} \right)^j \right\}.$$

Since F_μ is a manifold of finite dimension, we have the equality

$$1 = \prod_{\alpha, i} \left\{ \left(1 + \frac{(x_{\mu\alpha})_i}{\alpha} \right) \cdot \sum_{j=0}^{m} \left(\frac{-(x_{\mu\alpha})_i}{\alpha} \right)^j \right\}$$

in $S^{-1}H^*(BT^n \times F_\mu)$ where $m = [\dim F_\mu/2]$. It follows that

$$1 = \chi_{T^n}(N_\mu) \prod_{\alpha} \alpha^{-k(\alpha)} \prod_{\alpha, i} \sum_{j=0}^{m} \left(\frac{-(x_{\mu\alpha})_i}{\alpha} \right)^j$$

where $k(\alpha)$ denotes the dimension of the fibre of the complex vector bundle $E_{\mu\alpha}$. This equality means that $\chi_{T^n}(N_\mu)$ is a unit in $S^{-1}H^*(BT^n \times F_\mu)$. □

Next we shall show that the local index homomorphism is given by the slant product.

Lemma 6.12 *Let F be an oriented manifold on which T^n acts trivially. Then the localized index homomorphism*

$$S^{-1}\mathrm{Ind}: S^{-1}H^*(BT^n \times F) \to S^{-1}H^*(BT^n)$$

is given by the localized slant product $/[F]$ where $[F]$ denotes the orientation class of F.

Proof Since $H^*(BT^n)$ has no torsion, it follows from Corollary 6.5 that $\mathrm{Ind}: H^*(BT^n \times F) \to H^*(BT^n)$ is given by the slant product $/[F]$. Since the slant product $/[F]$ is an $H^*(BT^n)$-module homomorphism, it induces naturally the localized homomorphism

$$S^{-1}H^*(BT^n \times F) \to S^{-1}H^*(BT^n)$$

which is also denoted by $/[F]$. Hence $S^{-1}\mathrm{Ind}$ is given by this localized slant product $/[F]$. □

Let M be an oriented T^n-manifold. Denote by F_μ each component of the fixed point set and by N_μ its normal bundle. As remarked before, N_μ has a complex vector bundle structure such that T^n acts on N_μ as complex vector bundle automorphisms. In consequence N_μ is oriented by the complex structure. We orient F_μ so that the orientation of N_μ followed by that of F_μ yields the restriction of the orientation of M. Then we have

Corollary 6.13 *For* $x \in S^{-1}H^*(ET^n \underset{T^n}{\times} M)$, *we have*

$$S^{-1}\operatorname{Ind} x = \sum_\mu \frac{S^{-1}i_\mu^* x}{\chi_{T^n}(N_\mu)} \bigg/ [F_\mu].$$

Proof Corollary 6.13 is immediate from Theorem 6.7 and Lemma 6.12. □

Observe that in Corollary 6.13, we can take x to be an arbitrary element of $S^{-1}H^*(ET^n \underset{T^n}{\times} M)$. In the remaining sections we shall consider x to be given explicitly by various interesting elements. The remainder of this section is also devoted to giving further examples.

Given a formal power series

$$F(t) = 1 + d_1 t + d_2 t^2 + d_3 t^3 + \cdots$$

with rational coefficients, let K be the multiplicative sequence belonging to the power series $F(t)$ in the sense of Hirzebruch[44] and put $\tilde{F}(t) = F(t^2)$. Given a vector bundle $E \to X$, let $P(E)$ denote the total Pontrjagin class of E. In this chapter, the characteristic class $K(P(E))$ will be denoted briefly by the symbol $K(E)$.

By making use of $H^*(ET^n \underset{T^n}{\times} M; \mathbb{Q})$ as an equivariant cohomology theory, we shall have the following corollary.

Corollary 6.14

$$\operatorname{Ind} K\left(ET^n \underset{T^n}{\times} T(M)\right) = \sum_\mu \frac{\prod_j F(z_j) \prod_{\alpha, i} \tilde{F}(\alpha + (x_{\mu\alpha})_i)}{\prod_{\alpha, i}(\alpha + (x_{\mu\alpha})_i)} \bigg/ [F_\mu].$$

Here the total Pontrjagin class of $T(F_\mu)$ *is expressed formally as* $\prod_j (1 + z_j)$.

Proof By the naturality of characteristic classes, we have

$$i_\mu^* K\left(ET^n \underset{T^n}{\times} T(M)\right) = K\left(ET^n \underset{T^n}{\times} i_\mu^* T(M)\right)$$

where i_μ^* in the right-hand side denotes the induced map of fibre bundles. Since $i_\mu^* T(M) \cong T(F_\mu) \oplus N_\mu$ and since K transforms Whitney sums into products, we have

$$\begin{aligned} K\left(ET^n \underset{T^n}{\times} i_\mu^* T(M)\right) &= K\left(ET^n \underset{T^n}{\times} T(F_\mu) \oplus ET^n \underset{T^n}{\times} N_\mu\right) \\ &= K(BT^n \times T(F_\mu)) K\left(ET^n \underset{T^n}{\times} N_\mu\right). \end{aligned}$$

The projection $p\colon BT^n \times F_\mu \to F_\mu$ induces the injective homomorphism $p^*\colon H^*(F_\mu) \to H^*(BT^n \times F_\mu)$, by which we regard $H^*(F_\mu)$ as a subgroup of $H^*(BT^n \times F_\mu)$. Under this convention, we have

$$K(BT^n \times T(F_\mu)) = K(T(F_\mu)) = \prod_j F(z_j).$$

On account of Lemma 6.10, the Chern class of $ET^n \times_{T^n} N_\mu$ is calculated and we have

$$K\left(ET^n \times_{T^n} N_\mu\right) = \prod_{\alpha, i} \tilde{F}(\alpha + (x_{\mu\alpha})_i).$$

Accordingly, Lemma 6.10 and Corollary 6.13 yield the required equality. □

Next we shall give another kind of application. Given a partition ω of a positive integer and a vector bundle $E \to X$, a characteristic class $s_\omega(P(E))$ is defined.[93] Given an oriented manifold M, a characteristic number $s_\omega(M)$ is also defined.[93]

Corollary 6.15 *Let M be an oriented T^n-manifold of dimension $4k$. For any partition ω of k, we have*

$$s_\omega(M) = \sum_\mu \sum_{\omega = \omega_1 \omega_2} \frac{s_{\omega_1}(\Pi(1 + z_j))s_{\omega_2}(\Pi(1 + (\alpha + (x_{\mu\alpha})_i)^2))}{\Pi(\alpha + (x_{\mu\alpha})_i)} \Bigg/ [F_\mu].$$

Proof Let ξ and ξ' be vector bundles over X. Then the corollary is immediate from Corollary 6.13 and the following fundamental property of s_ω:

$$s_\omega(P(\xi \oplus \xi')) = \sum_{\omega = \omega_1 \omega_2} s_{\omega_1}(P(\xi)) \cdot s_{\omega_2}(P(\xi')).$$ □

Remark A similar formula holds for Stiefel–Whitney classes. Since the oriented bordism class $[M]$ is characterized by these numbers $s_\omega(M)$,[93] Corollary 6.15 means that the class $[M]$ is determined explicitly by the data at the fixed point set.

As a special case of Corollary 6.15, we have the following corollary.

Corollary 6.16 *Let M be an oriented connected manifold of dimension $4k$ on which T^n acts non-trivially. Then we have*

$$s_k(M) = \sum_\mu \frac{\sum_{\alpha, i} (\alpha + (x_{\mu\alpha})_i)^{2k}}{\Pi_{\alpha, i} (\alpha + (x_{\mu\alpha})_i)} \Bigg/ [F_\mu].$$

Proof It follows from the property of s_k that

$$i_\mu^* s_k\left(P\left(ET^n \underset{T^n}{\times} T(M)\right)\right) = s_k\left(P\left(ET^n \underset{T^n}{\times} i_\mu^* T(M)\right)\right)$$
$$= s_k(P(BT^n \times T(F_\mu)))$$
$$+ s_k\left(P\left(ET^n \underset{T^n}{\times} N_\mu\right)\right).$$

Since the action is non-trivial, we have $\dim F_\mu < 4k$. It turns out that $s_k(P(BT^n \times T(F_\mu))) = 0$. Since

$$s_k\left(P\left(ET^n \underset{T^n}{\times} N_\mu\right)\right) = \sum_{\alpha, i} (\alpha + (x_{\mu\alpha})_i)^{2k}$$

by definition, the corollary follows. □

Although the following result has already been shown in Section 5.6 without assuming orientedness, we show that it can also be deduced from the localization theorem.

Corollary 6.17 *Let M be an oriented T^n-manifold. Then we have*

$$\chi(M) = \sum_\mu \chi(F_\mu).$$

Proof As is well known, the evaluation of the Euler class $\chi_{\{e\}}(M)$ at the orientation class $[M]$ is equal to the Euler characteristic. Let $j: M \to ET^n \underset{T^n}{\times} M$ be the natural inclusion and $f: M \to *$ be the constant map. Under the natural identifications $H^0(*) \cong H^0(BT^n) \cong \mathbb{Z}$, we have

$$\begin{aligned}\chi(M) &= \chi_{\{e\}}(M)[M] = j^*\chi_{T^n}(M)[M] \\ &= \bar{f}_! j^* \chi_{T^n}(M) = \operatorname{Ind} \chi_{T^n}(M) \\ &= \sum_\mu \frac{i_\mu^* \chi_{T^n}(M)}{\chi_{T^n}(N_\mu)} \Big/ [F_\mu] = \sum_\mu \frac{\chi_{T^n}(F_\mu) \cdot \chi_{T^n}(N_\mu)}{\chi_{T^n}(N_\mu)} \Big/ [F_\mu] \\ &= \sum_\mu \chi_{T^n}(F_\mu)/[F_\mu] = \sum_\mu \chi(F_\mu).\end{aligned}$$ □

Remark If we make use of ordinary cohomology theory instead of equivariant cohomology, then note that $i_\mu^* \chi_{T^n}(M) = 0$ unless the action is trivial.

6.3 The *G*-signature theorem

The purpose of this section is to give a topological proof of the Atiyah–Singer *G*-signature theorem and to apply the theorem to the study of transformation groups. The Atiyah–Singer theorem is concerned with the actions of topologically cyclic groups, that is, groups of the form $T^n \times \mathbb{Z}_k$. Since E. Ossa[105] dealt with the case of $\mathbb{Z}_k$, we shall treat the case $G = T^n$ in this section.

We begin by preparing several lemmas. Put

$$ET^n_m = S^{2m+1} \times \overset{n}{\cdots} \times S^{2m+1}.$$

Then T^n acts freely and canonically on ET^n_m. We denote its orbit space ET^n_m/T^n by BT^n_m and denote the following two inclusion maps by the same j:

$$j\colon ET^n_m = S^{2m+1} \times \cdots \times S^{2m+1} \to S^\infty \times \cdots \times S^\infty = ET^n,$$

$$j\colon BT^n_m = \mathbb{C}P^m \times \cdots \times \mathbb{C}P^m \to \mathbb{C}P^\infty \times \cdots \times \mathbb{C}P^\infty = BT^n.$$

Let M and N be oriented T^n-manifolds and $f\colon M \to N$ be a T^n-map. Since the T^n-action on ET^n_m is free, both $ET^n_m \underset{T^n}{\times} M$ and $ET^n_m \underset{T^n}{\times} N$ are smooth manifolds. Since ET^n_m is endowed with a canonical orientation, $ET^n_m \underset{T^n}{\times} M$ is given an orientation by means of the bundle

$$ET^n_m \times M \to ET^n_m \underset{T^n}{\times} M.$$

Similarly $ET^n_m \underset{T^n}{\times} N$ is given an orientation. Then

$$\mathrm{id} \underset{T^n}{\times} f\colon ET^n_m \underset{T^n}{\times} M \to ET^n_m \underset{T^n}{\times} N$$

is a map between oriented manifolds, so that it induces the classical Gysin homomorphism

$$\overline{\left(\mathrm{id} \underset{T^n}{\times} f\right)_!}\colon H^*\left(ET^n_m \underset{T^n}{\times} M\right) \to H^*\left(ET^n_m \underset{T^n}{\times} N\right).$$

Lemma 6.18 *The following diagram is commutative:*

$$\begin{array}{ccc}
H^*(ET^n \underset{T^n}{\times} M) & \xrightarrow{\quad f_! \quad} & H^*(ET^n \underset{T^n}{\times} N) \\
\Big\downarrow (j \underset{T^n}{\times} \mathrm{id})^* & & \Big\downarrow (j \underset{T^n}{\times} \mathrm{id})^* \\
H^*(ET^n_m \underset{T^n}{\times} M) & \xrightarrow{\overline{(\mathrm{id} \underset{T^n}{\times} f)_!}} & H^*(ET^n_m \underset{T^n}{\times} N)
\end{array}$$

Proof By making use of $ET^n_m \underset{T^n}{\times} X$ instead of $ET^n \underset{T^n}{\times} X$, we define maps ψ_i in a similar way to φ_i which appeared in the definition of our equivariant Gysin homomorphism $f_!$ and put

$$f_{m!} = \psi_4^{-1}\cdot\psi_3\cdot\psi_2^{-1}\cdot\psi_1 : H^*\left(ET^n_m \underset{T^n}{\times} M\right) \to H^*\left(ET^n_m \underset{T^n}{\times} N\right).$$

For brevity, we denote all the inclusions $ET^n_m \underset{T^n}{\times} X \to ET^n \underset{T^n}{\times} X$ by j as well. Since it is easily verified that

$$j^*\cdot\varphi_i = \psi_i\cdot j^*,$$

we have

$$f_{m!}\cdot j^* = j^*\cdot f_!.$$

Therefore to prove Lemma 6.18, it suffices to show that

$$f_{m!} = \overline{\left(\mathrm{id} \underset{T^n}{\times} f\right)_!}.$$

But this is verified as in the proof of Theorem 6.2 by making use of $ET^n_m \underset{T^n}{\times} X$ instead of X. □

Let $p\colon M \to *$ be the constant map. Denote $\mathrm{id} \underset{T^n}{\times} p\colon ET^n_m \underset{T^n}{\times} M \to ET^n_m \underset{T^n}{\times} * = BT^n_m$ by P for simplicity.

Corollary 6.19 *The following diagram is commutative*:

$$\begin{array}{ccc}
H^*(ET^n \underset{T^n}{\times} M) & \xrightarrow{\ \mathrm{Ind}\ } & H^*(BT^n) \\
\Big\downarrow{\scriptstyle j^*} & & \Big\downarrow{\scriptstyle j^*} \\
H^*(ET^n_m \underset{T^n}{\times} M) & \xrightarrow{\ \overline{P}_!\ } & H^*(BT^n_m)
\end{array}$$

Proof If we put $N = *$ (one point), then the corollary follows from Lemma 6.18. □

The Hirzebruch L-sequence is defined to be the multiplicative sequence belonging to the power series:

$$F(t) = \frac{\sqrt{t}}{\tanh\sqrt{t}} = 1 + \frac{1}{3}t - \frac{1}{45}t^2 + \cdots.$$

Given a T^n-manifold M, denote the L-class of the bundle $ET^n \underset{T^n}{\times} T(M) \to ET^n \underset{T^n}{\times} M$ by $L\left(ET^n \underset{T^n}{\times} T(M)\right)$.

Lemma 6.20 $\operatorname{Ind} L\left(ET^n \underset{T^n}{\times} T(M)\right) \in H^0(BT^n; \mathbb{Q})$.

Proof First note that $P\colon ET^n_m \underset{T^n}{\times} M \to BT^n_m$ is a fibre bundle with fibre M and structure group T^n. This follows from Theorem 4.11 and the remark immediately after Theorem 2.10. According to Chern[21] the classical Gysin homomorphism $\bar{P}_!$ is equivalent to integration over the fibre of the fibre bundle $P\colon ET^n_m \underset{T^n}{\times} M \to BT^n_m$. Moreover the tangent bundle along the fibre of the bundle P is nothing but the bundle

$$ET^n_m \underset{T^n}{\times} T(M) \to ET^n_m \underset{T^n}{\times} M.$$

Since BT^n_m is simply connected, it follows from Borel and Hirzebruch[13] that the L-class of the bundle P is strictly multiplicative.

Putting all this together, we have

$$\bar{P}_! L\left(ET^n_m \underset{T^n}{\times} T(M)\right) \in H^0(BT^n_m; \mathbb{Q}).$$

Obviously there is a natural bundle map

$$\begin{array}{ccc} ET^n_m \underset{T^n}{\times} T(M) & \longrightarrow & ET^n \underset{T^n}{\times} T(M) \\ \downarrow & & \downarrow \\ ET^n_m \underset{T^n}{\times} M & \xrightarrow{\;j\;} & ET^n \underset{T^n}{\times} M \end{array}$$

Accordingly, it follows from Corollary 6.19 that

$$\begin{aligned} j^* \operatorname{Ind} L\left(ET^n \underset{T^n}{\times} T(M)\right) &= \bar{P}_! j^* L\left(ET^n \underset{T^n}{\times} T(M)\right) \\ &= \bar{P}_! L\left(ET^n_m \underset{T^n}{\times} T(M)\right) \in H^0(BT^n_m; \mathbb{Q}). \end{aligned}$$

Since this holds for arbitrary m, we may conclude that

$$\operatorname{Ind} L\left(ET^n \underset{T^n}{\times} T(M)\right) \in H^0(BT^n; \mathbb{Q}).$$ □

We now define the *index* $I(M)$ of a compact oriented manifold M as follows. When $\dim M$ is not a multiple of 4, $I(M)$ is defined to be zero. When

$\dim M = 4k$, we choose a basis $a_1, \ldots, a_r$ for $H^{2k}(M; \mathbb{Q})$ so that the symmetric matrix

$$\langle a_i \cup a_j, [M] \rangle$$

is diagonal, where $\cup$ denotes the cup product and $\langle , \rangle$ denotes the Kronecker index. Then $I(M)$ is the number of positive diagonal entries minus the number of negative ones.

Theorem 6.21 $\operatorname{Ind} L(ET^n \times_{T^n} T(M)) = I(M)$. *Here by Lemma* 6.20 *the left-hand side belongs to* $H^0(BT^n; \mathbb{Q})$ *which we identify with the rational numbers* $\mathbb{Q}$.

Proof By taking m to be 0 in Corollary 6.19, we have the following commutative diagram:

$$\begin{array}{ccc} H^*(ET^n \times_{T^n} M; \mathbb{Q}) & \xrightarrow{\operatorname{Ind}} & H^*(BT^n; \mathbb{Q}) \\ \downarrow j^* & & \downarrow j^* \\ H^*(ET_0^n \times_{T^n} M; \mathbb{Q}) & \xrightarrow{\bar{P}_!} & H^*(BT_0^n; \mathbb{Q}) \end{array}$$

Note that $ET_0^n \times_{T^n} M = M$ and $BT_0^n = *$. Hence it is easy to see that the Gysin homomorphism

$$\bar{P}_!: H^{\dim M}\left(ET_0^n \times_{T^n} M; \mathbb{Q}\right) = \mathbb{Q} \to H^0(*; \mathbb{Q}) = \mathbb{Q}$$

is the identity map and it is the 0-map for other dimensions. The bundle $ET_0^n \times_{T^n} T(M) \to ET_0^n \times_{T^n} M$ is obviously equivalent to the bundle $T(M) \to M$ so that

$$\begin{aligned} j^* \operatorname{Ind} L\left(ET^n \times_{T^n} T(M)\right) &= \bar{P}_! j^* L\left(ET^n \times_{T^n} T(M)\right) \\ &= \bar{P}_! L(T(M)) = L(T(M))[M] = I(M). \end{aligned}$$

Here the last equality is the famous Hirzebruch index theorem. In view of Lemma 6.20 $\operatorname{Ind} L(ET^n \times_{T^n} T(M)) \in H^0(BT^n; \mathbb{Q})$. Since $j^*: H^0(BT^n; \mathbb{Q}) = \mathbb{Q} \to H^0(*; \mathbb{Q}) = \mathbb{Q}$ is the identity map, we can conclude that

$$\operatorname{Ind} L\left(ET^n \times_{T^n} T(M)\right) = I(M). \qquad \square$$

We are now ready to state the G-signature theorem.

Theorem 6.22 *Let M be an oriented T^n-manifold. Denote by F_μ each component of the fixed point set and by N_μ the normal bundle of F_μ in M. Then N_μ has a complex vector bundle structure and has a decomposition*

$$N_\mu = \sum_\alpha E_{\mu\alpha} \otimes V_\alpha$$

where α ranges over the complex irreducible representations of T^n, V_α denotes the representation space of α, and $E_{\mu\alpha}$ denotes a complex vector bundle. We regard each irreducible representation α as an element of $H^2(BT^n)$ via the translations in Section 6.2. Finally, let $L(\ )$ denote the Hirzebruch L-sequence. Then one can orient each F_μ so that we have the equality:

$$I(M) = \sum_\mu L(T(F_\mu)) \prod_{\alpha, i} \left(\frac{\exp 2(\alpha + (x_{\mu\alpha})_i) + 1}{\exp 2(\alpha + (x_{\mu\alpha})_i) - 1} \right) \Big/ [F_\mu].$$

Here the summation is taken over all the components of the fixed point set and the total Chern class of the bundle $E_{\mu\alpha}$ is written formally as $\prod_i (1 + (x_{\mu\alpha})_i)$, and $/[F_\mu]$ denotes the slant product.

Proof It follows from Lemma 6.10 that the total Pontrjagin class of the bundle $ET^n \times_{T^n} N_\mu \to BT^n \times F_\mu$ is given by

$$P\left(ET^n \times_{T^n} N_\mu\right) = \prod_{\alpha, i} (1 + (\alpha + (x_{\mu\alpha})_i)^2).$$

Therefore we have the following equalities in the localized module $S^{-1}H^*(BT^n; \mathbb{Q})$:

$$\begin{aligned}
I(M) &= \operatorname{Ind} L\left(ET^n \times_{T^n} T(M)\right) && \text{(by Theorem 6.21)} \\
&= \sum_\mu \frac{i_\mu^* L\left(ET^n \times_{T^n} T(M)\right)}{\chi_{T^n}(N_\mu)} \Big/ [F_\mu] && \text{(by Corollary 6.13)} \\
&= \sum_\mu \frac{L(T(F_\mu)) \cdot L\left(ET^n \times_{T^n} N_\mu\right)}{\chi_{T^n}(N_\mu)} \Big/ [F_\mu] && \text{(multiplicative property of } L) \\
&= \sum_\mu L(T(F_\mu)) \prod_{\alpha, i} \left(\frac{\exp 2(\alpha + (x_{\mu\alpha})_i) + 1}{\exp 2(\alpha + (x_{\mu\alpha})_i) - 1} \right) \Big/ [F_\mu]. && \square
\end{aligned}$$

Remark Although $I(M) = 0$ in the case where $\dim M = 4k + 2$, each summand of the right-hand side of the equality in the theorem is not zero in general. Even in this case, the equality provides us with important information.

Corollary 6.23 *We can orient each component F_μ of the fixed point set of an oriented T^n-manifold M so that we get*

$$I(M) = \sum_\mu I(F_\mu).$$

Proof Note that the number of the irreducible representations α occurring in some N_μ is finite. As observed in Section 6.2, we regard each representation α as an n-tuple $(a_1, \ldots, a_n)$ of integers as well. Then there exists an n-tuple $\beta = (b_1, \ldots, b_n)$ of integers such that

$$(\alpha, \beta) = a_1 b_1 + \cdots + a_n b_n \neq 0$$

for any α occurring in M. By replacing α by $-\alpha$ in the case where $(\alpha, \beta) < 0$, we may assume that $(\alpha, \beta) > 0$ for all α. We make use of cohomology groups with coefficients in $\mathbb{R}$. Since $H^*(BT^n; \mathbb{R})$ is isomorphic to the polynomial ring $\mathbb{R}[t_1, \ldots, t_n]$ of n-variables over $\mathbb{R}$, it is a unique factorization domain. Hence any equalities in the localized ring $S^{-1}H^*(BT^n; \mathbb{R})$ hold even if we regard each t_i as an indeterminate. As a consequence the equality in Theorem 6.22 holds even if we replace t_i by real numbers such that $\alpha \neq 0$. Corresponding to $\beta = (b_1, \ldots, b_n)$ chosen above, we put $t_i = b_i t$. Then we have $\alpha = \sum_i a_i t_i = \sum_i a_i b_i t = (\alpha, \beta)t$. It turns out that

$$\lim_{t\to\infty} \frac{\exp 2(\alpha + (x_{\mu\alpha})_i) + 1}{\exp 2(\alpha + (x_{\mu\alpha})_i) - 1} = \lim_{t\to\infty} \frac{\exp 2((\alpha, \beta)t + (x_{\mu\alpha})_i) + 1}{\exp 2((\alpha, \beta)t + (x_{\mu\alpha})_i) - 1} = 1 \text{ in } H^*(F_\mu; \mathbb{R}).$$

Thus we have

$$I(M) = \sum_\mu L(T(F_\mu))/[F_\mu] = \sum_\mu I(F_\mu). \qquad \square$$

Denote by $\dim F$ the maximum of $\{\dim F_\mu\}$.

Corollary 6.24 *Let M be a semi-free S^1-manifold. If $I(M) \neq 0$, then* $\dim F \geq \dim M/2$.

Proof In this case, we have $\alpha = t$. We now make use of cohomology groups with coefficients in $\mathbb{C}$. If we replace α by $\sqrt{-1}\pi/2$ in the proof of the previous corollary, we have

$$\frac{\exp 2(\sqrt{-1}\pi/2 + (x_\mu)_i) + 1}{\exp 2(\sqrt{-1}\pi/2 + (x_\mu)_i) - 1} = \frac{-2(x_\mu)_i + \cdots}{-2 - 2(x_\mu)_i - \cdots} = (x_\mu)_i + \text{higher}.$$

Hence it follows from Theorem 6.22 that

$$I(M) = \sum_\mu L(T(F_\mu)) \left\{ \prod_i (x_\mu)_i + \text{higher} \right\} \Big/ [F_\mu].$$

The dimension of the term $\prod_i (x_\mu)_i$ is equal to that of the fibre of N_μ, which is equal to $\dim M - \dim F_\mu$. Suppose that $\dim M > 2\dim F_\mu$. Then we have $\dim M - \dim F_\mu > \dim F_\mu$ so that the right-hand side of the above equality vanishes by considering the dimension. Hence we have $I(M) = 0$, which contradicts the assumption. □

6.4 Equivariant Riemann–Roch theorems

The purpose of this section is to enunciate certain equivariant analogues of the differentiable Riemann–Roch theorems according to Atiyah and Hirzebruch.[6] These 'equivariant Riemann–Roch theorems' will play an important role in the proof of the vanishing of various genera in Sections 6.5 and 6.6. These theorems also yield an equivariant non-embedding theorem.[64]

As equivariant cohomology theories, we shall make use of the equivariant K-theory $K_G(M)$ and the Borel cohomology $H^*(EG \underset{G}{\times} M; \mathbb{Q})$. First we define an equivariant Chern character

$$\mathrm{ch}_G\colon K_G(M) \to H^{**}\left(EG \underset{G}{\times} M; \mathbb{Q}\right)$$

as follows. Given an arbitrary element x of $K_G(M)$, there are complex G-vector bundles ξ and η such that $x = [\xi] - [\eta]$. Let m and n be the complex dimensions of the fibres of ξ and η, respectively. When we write the total Chern classes of $EG \underset{G}{\times} \xi$ and $EG \underset{G}{\times} \eta$ formally as

$$c\left(EG \underset{G}{\times} \xi\right) = \prod_{i=1}^{m} (1 + s_i), \qquad c\left(EG \underset{G}{\times} \eta\right) = \prod_{i=1}^{n} (1 + t_i),$$

respectively, we define $\mathrm{ch}_G(x)$ by

$$\mathrm{ch}_G(x) = \sum_{i=1}^{m} \exp s_i - \sum_{i=1}^{n} \exp t_i.$$

Then we have the following lemma whose proof is left to the reader as an exercise.

Lemma 6.25 ch_G *is a ring homomorphism. That is, for any* $x, y \in K_G(M)$, *we have*

$$\mathrm{ch}_G(x + y) = \mathrm{ch}_G(x) + \mathrm{ch}_G(y),$$

$$\mathrm{ch}_G(x \otimes y) = \mathrm{ch}_G(x) \cdot \mathrm{ch}_G(y).$$

Next, for a G-vector bundle ξ, we define an equivariant $\hat{A}$-class $\hat{A}_G(\xi)$ by

$$\hat{A}_G(\xi) = \hat{A}\left(EG \underset{G}{\times} \xi\right).$$

Here $\hat{A}$ denotes the class belonging to the power series

$$F(t) = \frac{x/2}{\sinh x/2} = \frac{x}{\exp(x/2) - \exp(-x/2)}$$

in the sense of Hirzebruch.[44] For a G-manifold M, we denote $\hat{A}_G(T(M))$ by $\hat{A}_G(M)$ for simplicity. Let

$$\pi\colon \mathrm{Spin}\,(n+2) \to SO(n+2)$$

be the standard double covering. Then $\mathrm{Spin}^c(n)$ is defined to be the subgroup $\pi^{-1}(SO(n) \times SO(2))$ of $\mathrm{Spin}\,(n+2)$, which is isomorphic to $\mathrm{Spin}\,(n) \underset{\mathbb{Z}_2}{\times} S^1$ as Lie groups. Let Q be a principal $SO(n)$-bundle over M. A $\mathrm{Spin}^c(n)$-*reduction* of Q is a principal $\mathrm{Spin}^c(n)$-bundle P over M together with an S^1-bundle $p_1\colon P \to Q$ such that the diagram

$$\begin{array}{ccccc} P \times \mathrm{Spin}^c(n) & \longrightarrow & P & & \\ \downarrow{\scriptstyle p_1 \times \lambda} & & \downarrow{\scriptstyle p_1} & \searrow & \\ & & & & M \\ & & & \nearrow & \\ Q \times SO(n) & \longrightarrow & Q & & \end{array}$$

commutes, where the horizontal arrows are the actions of $\mathrm{Spin}^c(n)$ and $SO(n)$ from the right on P and Q, respectively, and λ is the composite homomorphism $\mathrm{Spin}^c(n) \xrightarrow{\pi} SO(n) \times SO(2) \to SO(n)$. It is easy to see that the above definition of a $\mathrm{Spin}^c(n)$-reduction coincides with that of the reduction with respect to λ defined in Section 2.3.

As is well known, there exists a $\mathrm{Spin}^c(n)$-reduction P of Q if and only if there exists an element c_1 of $H^2(M; \mathbb{Z})$ such that

$$W_2(Q) \equiv c_1 \bmod 2$$

where $W_2(Q)$ denotes the second Stiefel–Whitney class of the bundle $Q \to M$.

Moreover the map which assigns to a $\mathrm{Spin}^c(n)$-reduction P of Q the first Chern class c_1 of the principal S^1-bundle $P/\mathrm{Spin}\,(n) \to M$ gives a bijective correspondence between $\mathrm{Spin}^c(n)$-reductions of Q and $\{c_1 \in H^2(M; \mathbb{Z}) \mid W_2(Q) \equiv c_1 \bmod 2\}$.

We now introduce the concept of a lifting of a G-action. Let G and K be topological groups and $\pi\colon P \to X$ be a principal K-bundle. Then a *lifting* of a G-action on X is a G-action on P satisfying

$$\pi(gy) = g\pi(y), \quad (gy)k = g(yk) \qquad \text{for } g \in G,\ y \in P,\ k \in K.$$

Here the K-action on P from the right is that obtained in Section 2.3.

Let M and N be oriented G-manifolds with respect to the equivariant cohomology group $H^*\left(EG \underset{G}{\times} M; \mathbb{Q}\right)$. Let $f\colon M \to N$ be a G-map. Following Atiyah and Hirzebruch,[6] we call f a *c_1-map* if we are given an element $c_1 \in H^2(M; \mathbb{Z})$ such that

$$c_1 \equiv W_2(M) - f^* W_2(N) \bmod 2$$

where $W_2(X)$ denotes the second Stiefel–Whitney class of X. Hereafter we assume that $\dim M \equiv \dim N \bmod 2$. As stated in Theorem 4.12, there exist a G-representation space V and a G-embedding $e\colon M \to V$. We may assume that V is a complex representation space. Denote by ν the normal bundle of the G-embedding $f \times e\colon M \to N \times V$. Denote by Q the principal $SO(2k)$-bundle associated with ν. Note that the G-action on ν induces a G-action on Q.

If f is a c_1-map, then Q has a $\mathrm{Spin}^c(2k)$-reduction P corresponding to c_1. We are now ready to state our theorem.

Theorem 6.26 (Equivariant Riemann–Roch theorem) *Suppose that there exists a G-action on P which is a lifting of the G-action on M as well as a lifting of the G-action on Q. Then for any $\xi \in K_G(M)$, there exists $\eta \in K_G(N)$ such that*

$$f_!(\exp(c_G/2) \cdot \mathrm{ch}_G(\xi) \cdot \hat{A}_G(M)) = \mathrm{ch}_G(\eta) \cdot \hat{A}_G(N).$$

Here c_G denotes the first Chern class of the following complex line bundle:

$$EG \underset{G}{\times} P \underset{\mathrm{Spin}^c(2k)}{\times} \mathbb{C} \to EG \underset{G}{\times} M.$$

Proof The following brief explanation of the theorem and its proof may be of immense help to the reader.

In short, the theorem gives a formula which relates two kinds of equivariant Gysin homomorphisms with respect to the equivariant K-theory $K_G(X)$ and the equivariant cohomology theory $H^*(EG \underset{G}{\times} X; \mathbb{Q})$ via the equivariant Chern character. A key point of the proof is to construct a Thom class in K_G-theory by making use of a $\mathrm{Spin}^c(2k)$-representation and to investigate the pull-back element of its equivariant Chern character through the Thom isomorphism of $H^*(EG \underset{G}{\times} X; \mathbb{Q})$ theory.

In Section 6.2, we defined the notion of weight. In the following we define the notion of weight and character for a homomorphism of Lie groups. Let G and G' be compact Lie groups, $\psi\colon G \to G'$ a homomorphism, and T and T' be maximal tori of G and G', respectively, with $\psi(T) \subset T'$. Let $\{x_i'\}$ be a basis of $H^1(T')$. Then the elements

$$w_i = \psi^*(x_i')$$

induced by the homomorphism $\psi^*\colon H^1(T') \to H^1(T)$ are called *weights* of ψ.

By the correspondence

$$H^1(T) \leftrightarrow H^2(BT)$$

mentioned in Section 6.2, we regard w_i also as elements of $H^2(BT)$. Then we put

$$\mathrm{ch}(\psi) = \sum \exp w_i \qquad (\in H^{**}(BT; \mathbb{Q}))$$

which is called the *character* of ψ. This completes the definition of weight and character for a homomorphism of Lie groups.

The current aim is to construct a Thom class $t_G(v) \in K_G(D(v), S(v))$ by making use of a Spin^c-representation.

Let T be the standard maximal torus of $SO(2k) \times SO(2)$ and $x_1, \ldots, x_k, y$ be the standard basis of $H^1(T; \mathbb{Z})$. Put $T' = \pi^{-1}(T)$ where π: Spin$(2k+2) \to SO(2k+2)$ is the standard double covering. Then T' is a maximal torus of Spin$^c(2k)$ and covers T twofold. Since there is no torsion in $H^1(T)$ and $H^1(T')$, it follows from Theorem 5.30 that π^*: $H^1(T) \to H^1(T')$ is injective. By this injective homomorphism, we regard $H^1(T)$ as a subgroup of $H^1(T')$. Then we can take $\{x_1, \ldots, x_k, (x_1 + \cdots + x_k + y)/2\}$ as a basis of $H^1(T'; \mathbb{Z})$.

Denote by C_{2k} the *Clifford algebra* and by C^c_{2k} its complexification $C_{2k} \otimes \mathbb{C}$. Previously we defined Spin$^c(2k)$ to be the subgroup $\pi^{-1}(SO(2k) \times SO(2))$ of Spin$(2k+2)$. But we now regard Spin$^c(2k)$ as the multiplicative group involved in C^c_{2k} via the following inclusions: Spin$(2k) \to C_{2k}$ and $S^1 \to \mathbb{C}$.

According to Atiyah, Bott and Shapiro,[5] there exists an irreducible graded C^c_{2k}-module $\Delta_k^+ + \Delta_k^-$ of dimension 2^k over $\mathbb{C}$ which has the following properties:

(1) Δ_k^+ and Δ_k^- are invariant if one restricts C^c_{2k} to Spin$^c(2k)$ and are irreducible Spin$^c(2k)$-representations:

(2) the weights of Δ_k^+ (resp. Δ_k^-) consist of

$$\{(-1)^{\delta_1}x_1 + \cdots + (-1)^{\delta_k}x_k + y\}/2$$

where $\delta_i = 0$ or 1 and the number of 1 in $\{\delta_i\}$ is even (resp. odd).

Then we put

$$E = P \underset{\mathrm{Spin}^c(2k)}{\times} (\Delta_k^+ + \Delta_k^-)$$

where P denotes the principal Spin$^c(2k)$-bundle appeared in the statement of Theorem 6.26. The G-action on P induces a G-action on E which makes E *a* G-vector bundle.

Since v has a Spin$^c(2k)$-bundle structure, E has a $\mathbb{Z}_2$-graded $C(v)$-module structure. This needs some explanation. Consider the following vector bundles

$$E^{\pm} = P \underset{\mathrm{Spin}^c(2k)}{\times} \Delta_k^{\pm}.$$

Then we define vector bundle homomorphisms

$$v \underset{\mathbb{R}}{\otimes} E^+ \to E^- \qquad \text{and} \qquad v \underset{\mathbb{R}}{\otimes} E^- \to E^+$$

by $v \otimes e \mapsto v(e)$, which satisfy

$$v(v(e)) = -\|v\|^2 e.$$

Thus each fibre $E_x = E_x^+ \otimes E_x^-$ becomes naturally a $\mathbb{Z}_2$-graded module over the Clifford algebra $C(v_x)$. Then we say that E is a $\mathbb{Z}_2$-graded *$C(v)$-module*. When we denote by $\pi: D(v) \to M$ the bundle projection, we get the following sequence of bundles:

$$0 \to \pi^* E^- \xrightarrow{\sigma(E)} \pi^* E^+ \to 0$$

where $\sigma(E)$ is the map defined by $\sigma(E)_v(e) = -ve$. The restriction of $\sigma(E)$ to the bundle over $S(v)$ gives a bundle isomorphism. Accordingly it follows from Atiyah, Bott and Shapiro[5] that the sequence defines an element $t_G(v)$ of $K_G(D(v), S(v))$ called the *difference bundle*.

In order to calculate the equivariant Chern character of $t_G(v)$, we consider the universal bundle. Let $E(G, \mathrm{Spin}^c(2k)) \to B(G, \mathrm{Spin}^c(2k))$ be the universal principal $\mathrm{Spin}^c(2k)$-bundle with G-action. Obviously we can take $E(G, \mathrm{Spin}^c(2k))/\mathrm{Spin}^c(2k-1)$ as $B(G, \mathrm{Spin}^c(2k-1))$ and $E(G, \mathrm{Spin}^c(2k))/\mathrm{Spin}^c(2k-1)$ is isomorphic to $E(G, \mathrm{Spin}^c(2k)) \underset{\mathrm{Spin}^c(2k)}{\times} S^{2k-1}$ as sphere bundles over $B(G, \mathrm{Spin}^c(2k))$ with G-action. For brevity, we put

$$D = E(G, \mathrm{Spin}^c(2k)) \underset{\mathrm{Spin}^c(2k)}{\times} D^{2k}$$

$$S = E(G, \mathrm{Spin}^c(2k)) \underset{\mathrm{Spin}^c(2k)}{\times} S^{2k-1} = B(G, \mathrm{Spin}^c(2k-1)).$$

Then D includes S as the total space of the associated sphere bundle. As is well known, we have the following homotopy equivalences and the homotopy commutative diagram:

$$\begin{array}{ccc}
EG \underset{G}{\times} B(G, \mathrm{Spin}^c(2k-1)) & \simeq & BG \times B\,\mathrm{Spin}^c(2k-1) \\
\Big\downarrow {\scriptstyle \mathrm{id} \underset{G}{\times} j} & & \Big\downarrow {\scriptstyle \mathrm{id} \times j'} \\
EG \underset{G}{\times} B(G, \mathrm{Spin}^c(2k)) & \simeq & BG \times B\,\mathrm{Spin}^c(2k)
\end{array}$$

where j and j' are the maps induced naturally by the inclusion $\mathrm{Spin}^c(2k-1) \to \mathrm{Spin}^c(2k)$.

Next define $k: B(G, \mathrm{Spin}^c(2k)) \to (D, S)$ to be the composition of the G-homotopy equivalence $B(G, \mathrm{Spin}^c(2k)) \simeq D$ with the inclusion $D \to (D, S)$. Since $H^*(B\,\mathrm{Spin}^c(2k)) \to H^*(B\,\mathrm{Spin}^c(2k-1))$ is surjective, it follows from the commutative diagram above that

$$\left(\mathrm{id} \underset{G}{\times} k\right)^*: H^*\left(EG \underset{G}{\times} (D, S); \mathbb{Q}\right) \to H^*\left(EG \underset{G}{\times} B(G, \mathrm{Spin}^c(2k)); \mathbb{Q}\right)$$

is injective.

For the Chern character of representations, one easily verifies the following formula:

$$\mathrm{ch}\,\Delta_k^+ - \mathrm{ch}\,\Delta_k^- = x_1 \ldots x_k \cdot \exp(y/2) \prod_i \frac{\sinh x_i/2}{x_i/2}.$$

Concerning the disc bundle $\pi: D \to B(G, \mathrm{Spin}^c(2k))$ associated with the universal bundle, we construct a Thom class in a similar way to $t_G(v)$ above, which is denoted by

$$t_G(D) \in K_G(D, S).$$

Then the equality of Chern characters of representations enables us to calculate the equivariant Chern character so that we have

$$\begin{aligned}
\left(\mathrm{id} \underset{G}{\times} k\right)^* \mathrm{ch}_G t_G(D) &= \mathrm{ch}_G k^* t_G(D) \\
&= \mathrm{ch}\left[EG \underset{G}{\times} \left\{E(G, \mathrm{Spin}^c(2k)) \underset{\mathrm{Spin}^c(2k)}{\times} (\Delta_k^+ - \Delta_k^-)\right\}\right] \\
&= \mathrm{ch}\left[\left\{EG \underset{G}{\times} E(G, \mathrm{Spin}^c(2k))\right\} \underset{\mathrm{Spin}^c(2k)}{\times} (\Delta_k^+ - \Delta_k^-)\right] \\
&= \chi(D) \cdot \exp(c'/2) \cdot \hat{A}_G(D)^{-1}
\end{aligned}$$

where $\chi(D)$ denotes the Euler class of the bundle $\{EG \underset{G}{\times} E(G, \mathrm{Spin}^c(2k))\} \underset{\mathrm{Spin}^c(2k)}{\times} \mathbb{R}^{2k}$ and c' denotes the first Chern class of the complex line bundle $\{EG \underset{G}{\times} E(G, \mathrm{Spin}^c(2k))\} \underset{\mathrm{Spin}^c(2k)}{\times} \mathbb{C}$. It follows from the injectivity of $(\mathrm{id} \underset{G}{\times} k)^*$ that

$$\mathrm{ch}_G t_G(D) = \varphi(\exp(c'/2) \cdot \hat{A}_G(D)^{-1})$$

where φ denotes the Thom isomorphism

$$\varphi: H^*\left(EG \underset{G}{\times} B(G, \mathrm{Spin}^c(2k)); \mathbb{Q}\right) \to H^*\left(EG \underset{G}{\times} (D, S); \mathbb{Q}\right).$$

We now apply these results obtained for the universal bundle to v and deduce that

$$\mathrm{ch}_G t_G(v) = \varphi_1(\exp(c_G/2) \cdot \hat{A}_G(v)^{-1}).$$

Recall the maps φ_i in defining our equivariant Gysin homomorphism. We denote those maps in K_G-theory temporarily by $\varphi_i^{K_G}$ for $i \geq 2$. Then the equalities

$$\varphi_2 \operatorname{ch}_G = \operatorname{ch}_G \varphi_2^{K_G}, \qquad \varphi_3 \operatorname{ch}_G = \operatorname{ch}_G \varphi_3^{K_G}$$

hold. Moreover when we write the total Chern class of $EG \underset{G}{\times} (N \times V) \to EG \underset{G}{\times} N$ formally as $\Pi_i(1 + y_i)$, we have

$$\varphi_4^{-1} \operatorname{ch}_G \varphi_4^{K_G}(x) = \prod_i \left(\frac{1 - \exp y_i}{y_i}\right) \operatorname{ch}_G(x) \qquad \text{for} \quad x \in K_G(N),$$

which follows from the property of the Thom class of the bundle above.[64]

All the results above together with (4) of Theorem 6.1 show that for any $\xi \in K_G(M)$, there exists $\eta \in K_G(N)$ such that

$$f_!(\exp(c_G/2) \cdot \operatorname{ch}_G(\xi) \cdot \hat{A}_G(M)) = \exp(c''/2) \cdot \operatorname{ch}_G(\eta) \cdot \hat{A}_G(N)$$

where c'' is the first Chern class of the bundle $EG \underset{G}{\times} (N \times V) \to EG \underset{G}{\times} N$. Actually, by the following observation we see that the term $\exp(c''/2)$ is unnecessary. If $c'' \neq 0$, then we consider the conjugate representation space $\bar{V}$ of V and replace V by $V \oplus \bar{V}$, so that we have $c'' = 0$.

This completes the outline of the proof of the Equivariant Riemann–Roch theorem. □

6.5 The vanishing theorem of the $\hat{A}$-genus

The purpose of this section is to give a purely topological proof of the following theorem attributed to Atiyah and Hirzebruch.[7] This solves the problem of I. M. Singer.[117]

When the structure group of the tangent bundle $T(M^n)$ of a manifold M^n is reducible to $\operatorname{Spin}(n)$, M^n is called a *Spin manifold*. As is well known, M^n is a Spin manifold if and only if the second Stiefel–Whitney class $W_2(M)$ vanishes.

Theorem 6.27 *Let M be a compact connected oriented Spin manifold. If a compact connected Lie group G acts non-trivially on M, then $\hat{A}(M)[M] = 0$.*

Proof In short the theorem will be proved by combining the localization theorem and the equivariant Riemann–Roch theorem. A sort of finiteness property which is derived from the analytic index in the proof of Atiyah and Hirzebruch is derived instead from the equivariant Riemann–Roch theorem in our proof.

Let T be a maximal torus of G. It follows from the maximal torus theorem that

$$G = \bigcup_{g \in G} gTg^{-1}.$$

This means that if T acts trivially on M, then so does G. Hence the restricted T-action cannot be trivial. As a consequence, there is a subgroup H of T which is isomorphic to S^1 such that H acts non-trivially on M. Thus it suffices to prove the theorem in the case where $G = S^1$. Hereafter we assume that $G = S^1$.

Let $f\colon M \to *$ (one point) be the constant map. Since M is a Spin manifold, f is a c_1-map by taking $c_1 = 0$. Let Q and P be those which appeared just before Theorem 6.26. We now check if there is a lifting of the S^1-action on Q to P. Although there is no lifting in general, there is a lifting such that $c_{S^1} = 0$ if we consider a new action by using the double covering[64] $p\colon S^1 \to S^1 (z \mapsto z^2)$. Thus we can apply Theorem 6.26 to this new action and deduce the formula

$$f_!(\hat{A}_{S^1}(M)) = \mathrm{ch}_{S^1}(\eta)$$

for some $\eta \in K_{S^1}(*) = R(S^1)$. The right-hand side can be written as

$$\sum a_j \exp jt \qquad j, a_j \in \mathbb{Z}$$

where t denotes the first Chern class of the canonical complex line bundle over BS^1 and almost all the a_j are zero.

Next we shall make use of the localization theorem. Let $f_n\colon S^1 \to S^1 = U(1)$ be the complex irreducible representation defined by $z \mapsto z^n$ and denote by V_n the representation space of f_n. Denote by F_μ each component of the fixed point set and by N_μ its normal bundle in M. Then we can impose a complex vector bundle structure on N_μ so that we get a decomposition:

$$N_\mu = \sum_n E_{\mu n} \otimes V_n$$

where n ranges over positive integers and the $E_{\mu n}$ denote complex vector bundles. Then the assumptions in the localization theorem are satisfied and we have the following equality in the localized ring $S^{-1}H^{**}(BS^1; \mathbb{Q})$:

$$f_!(\hat{A}_{S^1}(M)) = \sum_\mu \hat{A}(F_\mu) \prod_{n,i} \left(\frac{1}{\exp(nt + x_{ni})/2 - \exp(-nt - x_{ni})/2} \right) \Big/ [F_\mu]$$

where the total Chern class of $E_{\mu n}$ is written formally as $\Pi_i(1 + x_{ni})$. Combined with the equivariant Riemann–Roch theorem, this will give the following equality

$$\sum_j a_j \exp jt = \sum_\mu \hat{A}(F_\mu) \prod_{n,i} \left(\frac{1}{\exp(nt + x_{ni})/2 - \exp(-nt - x_{ni})/2} \right) \Big/ [F_\mu].$$

As before this equality holds even if we regard t as an indeterminate. As t tends to $\pm\infty$, the limit of the local index (right-hand side) is 0 but the global index (left-hand side) with this property must be identically 0. Since the zero dimensional part of the $f_!(\hat{A}_{S^1}(M))$ coincides with the genus $\hat{A}(M)[M]$ in general, we have in particular $\hat{A}(M)[M] = 0$. □

Remark In the proof of Theorem 6.27, Theorem 6.26 plays an essential role. This is because any non-trivial S^1-action on any connected oriented manifold has the same local property as above and there are many examples of compact connected oriented manifolds, for instance $\mathbb{C}P^{2n}$, with a non-trivial S^1-action such that their $\hat{A}$-genera do not vanish.

6.6 Vanishing theorems of $\exp(c/2)\cdot\hat{A}$-genera

In Section 6.5, we showed that the $\hat{A}$-genus of a Spin manifold with a non-trivial S^1-action vanishes. In this section, we consider generalizations of this result as follows. Although the $\hat{A}$-genus does not vanish in general, it often happens that there exists $c \in H^2(M; \mathbb{Z})$ such that

$$\exp(c/2)\cdot\hat{A}(M)[M] = 0.$$

These phenomena are confirmed by various experiments. The vanishing of such generalized genera provides us with substantial information on Pontrjagin classes. As will be shown later, such information sometimes characterize Pontrjagin classes completely.

First we prove the theorem which motivates our study in this direction.

Theorem 6.28 *Let $\mathbb{C}P^n$ be the complex projective space of dimension n and x be a generator of $H^2(\mathbb{C}P^n)$. We choose an orientation class $[\mathbb{C}P^n]$ of $\mathbb{C}P^n$ such that $x^n[\mathbb{C}P^n] = 1$. Given any integer k, we put $b = n + 1 - 2k$. Then we have*

$$\exp(bx/2)\cdot\hat{A}(\mathbb{C}P^n)[\mathbb{C}P^n] = \begin{cases} {}_{n-k}C_n & \text{for } k \le 0,\\ 0 & \text{for } k = 1, 2, \ldots, n\\ (-1)^n{}_{k-1}C_n & \text{for } k \ge n+1. \end{cases}$$

Proof Since the total Pontrjagin class of $\mathbb{C}P^n$ is given by $(1 + x^2)^{n+1}$, we have

$$\hat{A}(\mathbb{C}P^n) = \left(\frac{x}{\exp(x/2) - \exp(-x/2)}\right)^{n+1}.$$

Hence we have

$$\exp(bx/2)\cdot\hat{A}(\mathbb{C}P^n) = \exp(-kx)\left(\frac{x}{1 - \exp(-x)}\right)^{n+1}.$$

To calculate the number, we need only investigate the coefficient of x^n. For this, we first put

$$z = 1 - \exp(-x).$$

Then it is easily verified that if x rotates once in the positive direction around the origin, then so does $z = 1 - \exp(-x)$. Hence by making use of the formula

$$dx = \frac{dz}{1-z}.$$

We compute:

$$\begin{aligned}
&\exp(bx/2)\cdot \hat{A}(\mathbb{C}P^n)[\mathbb{C}P^n] \\
&\quad = \text{the coefficient of } x^n \text{ in } \exp(-kx)\cdot\left(\frac{x}{1-\exp(-x)}\right)^{n+1} \\
&\quad = \frac{1}{2\pi i}\oint \frac{\exp(-kx)}{x^{n+1}}\cdot\left(\frac{x}{1-\exp(-x)}\right)^{n+1} dx \\
&\quad = \frac{1}{2\pi i}\oint \frac{\exp(-kx)}{(1-\exp(-x))^{n+1}} dx \\
&\quad = \frac{1}{2\pi i}\oint \frac{(1-z)^k}{z^{n+1}}\cdot\frac{1}{1-z} dz \\
&\quad = \frac{1}{2\pi i}\oint \frac{(1-z)^{k-1}}{z^{n+1}} dz \\
&\quad = \text{the coefficient of } z^n \text{ in } (1-z)^{k-1} \\
&\quad = (1-k)(2-k)(3-k)\cdots(n-k)/n!.
\end{aligned}$$

The above value is valid for each of the three cases. □

Next we show that the vanishing of the generalized $\hat{A}$-genera characterizes Pontrjagin classes completely in some cases.

Definition 6.1 When there exists $x \in H^2(M)$ such that

$$H^*(M) \cong \mathbb{Z}[x]/x^{n+1},$$

M is called a *cohomology complex projective space* or *cohomology* $\mathbb{C}P^n$.

Now suppose that we choose an orientation class $[M]$ such that $x^n[M] = 1$.

Theorem 6.29 *Let M be a cohomology $\mathbb{C}P^n$* and *x be a generator of $H^2(M)$. For the total Pontrjagin class $P(M)$ of M, we have*

$$P(M) = (1+x^2)^{n+1} \Leftrightarrow \exp(bx/2)\cdot\hat{A}(M)[M] = 0 \quad \textit{for}$$

$$b = n-1, n-3, n-5, \ldots, (3-(-1)^n)/2.$$

Proof $\Rightarrow$ If the cohomology ring of $H^*(M)$ is isomorphic to that of $\mathbb{C}P^n$ and if the total Pontrjagin class of M coincides with that of $\mathbb{C}P^n$ via the isomorphism, then the form of the $\hat{A}$-class of M coincides with that of the $\hat{A}$-class of $\mathbb{C}P^n$ as well. Accordingly, the proof of Theorem 6.27 is also valid in this case as it is.

$\Leftarrow$ Write $\hat{A}(M)$ as $1 + a_1(M)x^2 + a_2(M)x^4 + \cdots + a_{[n/2]}(M)x^{2[n/2]}$ where $a_i(M) \in \mathbb{Q}$.

In case n is even. In this case, we have

$$\exp(bx/2)\cdot\hat{A}(M)[M] = \frac{1}{n!}\left(\frac{b}{2}\right)^n + \frac{1}{(n-2)!}\left(\frac{b}{2}\right)^{n-2} a_1(M) + \cdots$$
$$+ \frac{1}{2!}\left(\frac{b}{2}\right)^2 a_{n/2-1}(M) + a_{n/2}(M).$$

Suppose that these values are 0 for all $b = n-1, n-3, \ldots, 1$. Then we get the following equation of matrices:

$$\begin{bmatrix} (n-1)^n & (n-1)^{n-2} & \cdots & (n-1)^2 & 1 \\ (n-3)^n & (n-3)^{n-2} & \cdots & (n-3)^2 & 1 \\ \vdots & & & & \vdots \\ 1^n & 1^{n-2} & \cdots & 1^2 & 1 \end{bmatrix} \begin{bmatrix} \dfrac{1}{n!\,2^n} \\ \dfrac{a_1(M)}{(n-2)!\,2^{n-2}} \\ \vdots \\ a_{n/2}(M) \end{bmatrix} = \begin{bmatrix} 0 \\ 0 \\ \vdots \\ 0 \end{bmatrix}.$$

This equation can be altered into the following equation of matrices:

$$\begin{bmatrix} (n-1)^{n-2} & \cdots & (n-1)^2 & 1 \\ (n-3)^{n-2} & \cdots & (n-3)^2 & 1 \\ \vdots & & & \vdots \\ 1^{n-2} & \cdots & 1^2 & 1 \end{bmatrix} \begin{bmatrix} \dfrac{a_1(M)}{(n-2)!\,2^{n-2}} \\ \vdots \\ a_{n/2}(M) \end{bmatrix} = - \begin{bmatrix} \dfrac{(n-1)^n}{n!\,2^n} \\ \vdots \\ \dfrac{1^n}{n!\,2^n} \end{bmatrix}.$$

We now denote by A the left matrix of the left-hand side of the above equation. Since the determinant of A is that of the Vandermonde, A is non-singular. Accordingly it follows from the uniqueness of the solution of linear equations that the equation above determines $a_1(M), \ldots, a_{n/2}(M)$ completely.

On the other hand, it follows from Theorem 6.28 that the above equation holds for $M = \mathbb{C}P^n$. Therefore we can conclude that

$$a_i(M) = a_i(\mathbb{C}P^n) \qquad \text{for } i = 1, 2, \ldots, n/2.$$

Since $H^*(M)$ has no torsion by assumption, $\hat{A}(M)$ characterizes the total Pontrjagin class $P(M)$ so that we have

$$P(M) = (1 + x^2)^{n+1}.$$

In case n is odd. In this case, we have

$$\exp(bx/2)\cdot \hat{A}(M)[M] = \frac{1}{n!}\left(\frac{b}{2}\right)^n + \frac{1}{(n-2)!}\left(\frac{b}{2}\right)^{n-2} a_1(M) + \cdots + \frac{1}{1!}\left(\frac{b}{2}\right) a_{(n-1)/2}(M).$$

As in the case where n is even, we substitute $n-1, n-3, \ldots, 2$ for b so that we have

$$\begin{bmatrix} (n-1)^{n-2} & (n-1)^{n-4} & \cdots & (n-1) \\ (n-3)^{n-2} & (n-3)^{n-4} & \cdots & (n-3) \\ \vdots & & & \vdots \\ 2^{n-2} & 2^{n-4} & \cdots & 2 \end{bmatrix} \begin{bmatrix} \dfrac{a_1(M)}{(n-2)!2^{n-2}} \\ \vdots \\ \dfrac{a_{(n-1)/2}(M)}{1!2^1} \end{bmatrix} = - \begin{bmatrix} \dfrac{(n-1)^n}{n!2^n} \\ \vdots \\ \dfrac{2^n}{n!2^n} \end{bmatrix}.$$

The determinant of the left matrix of the left-hand side of the above equation is equal to

$$(n-1)(n-3)\cdots 2 \times \det \begin{bmatrix} (n-1)^{n-3} & (n-1)^{n-5} & \cdots & 1 \\ (n-3)^{n-3} & (n-3)^{n-5} & \cdots & 1 \\ \vdots & & & \vdots \\ 2^{n-3} & 2^{n-5} & \cdots & 1 \end{bmatrix}$$

which is seen to be non-zero likewise. Hence the rest of the proof proceeds as in the proof of the case where n is even. □

Theorem 6.29 implies that the vanishing of the $\exp(c/2)\cdot\hat{A}$-genera contains considerable information. Therefore it is worthwhile developing vanishing theorems of the generalized $\hat{A}$-genera.

Let M be an oriented S^1-manifold. Denote by F_μ each component of the fixed point set and by $i_\mu\colon F_\mu \to M$ its inclusion. For any integer n, let $f_n\colon S^1 \to S^1 = U(1)$ be the complex irreducible representation defined by $z \mapsto z^n$. Denote by V_n the representation space of f_n. Then we get a decomposition of N_μ as before:

$$N_\mu = \sum_{n>0} E_{\mu n} \otimes V_n$$

where the $E_{\mu n}$ denote complex vector bundles as in Section 6.5. Denote by $\dim E_{\mu n}$ the dimension of the fibre of $E_{\mu n}$.

Let $\pi: \xi \to M$ be a complex line bundle. Suppose that there is a lifting of the S^1-action on M to ξ. Choose a point y_μ from each component F_μ. Then $\pi^{-1}(y_\mu)$ is a complex representation space of S^1 of dimension 1. Hence there exists an integer n_μ with

$$\pi^{-1}(y_\mu) = V_{n_\mu}.$$

We are now ready to state our vanishing theorem.

Theorem 6.30 (Vanishing theorem of $\exp(c/2)\cdot\hat{A}$-genera) *Let M be an oriented S^1-manifold. Suppose that an element c of $H^2(M)$ satisfies the following conditions.*

(1) $c \equiv W_2(M) \bmod 2$.

(2) *Let ξ be the complex line bundle over M with $c_1(\xi) = c$. Then there exists a lifting of the S^1-action on M to ξ.*

(3) *Concerning the lifting in* (2), *the inequality*

$$|n_\mu| < \sum_{n>0} n\cdot\dim E_{\mu n}$$

holds for every μ.

Then we have

$$\exp(c/2)\cdot\hat{A}(M)[M] = 0.$$

Proof Let Q denote the principal $SO(m)$-bundle associated with the tangent bundle $T(M)$ of M. As mentioned in Section 6.4, it follows from assumption (1) that there exists a unique $\mathrm{Spin}^c(m)$-reduction P of Q with

$$c_1(P/\mathrm{Spin}(m)) = c.$$

Note that the complex line bundle associated with the principal S^1-bundle $P/\mathrm{Spin}(m)$ is isomorphic to ξ. By assumption (2), there exists a lifting of the S^1-action on M to ξ and hence to $P/\mathrm{Spin}(m)$. It follows from the observation of Section 6 in Kawakubo[64] that there exists a lifting of this action or its double covering action to the principal $\mathrm{Spin}^c(m)$-bundle P. We now suppose that there exists a lifting of this action. Let $f: M \to *$ be the constant map. Then it follows from Corollary 6.13 that

$$f_!(\exp(c_{S^1}/2)\cdot\hat{A}_{S^1}(M))$$
$$= \sum_\mu \hat{A}(F_\mu)\frac{\exp(n_\mu t + i_\mu^* c)/2}{\prod_{n,i}\{\exp(nt + x_{ni})/2 - \exp(-nt - x_{ni})/2\}}\Bigg/[F_\mu]$$

where $c_{S^1} = c_1\left(ES^1 \underset{S^1}{\times} P \underset{\mathrm{Spin}^c(m)}{\times} \mathbb{C}\right) = c_1\left(ES^1 \underset{S^1}{\times} \xi\right)$ and $c(E_{\mu n})$ is written formally as $\Pi(1 + x_{ni})$ and t is regarded as an element of $H^2(BS^1)$. As before this equation holds even if we regard t as an indeterminate. As t tends to $\pm\infty$, the limit of the local index is 0 by assumption (3).

Next suppose that there exists a lifting of the double covering action to P. In this case, each n_μ (resp. n) is replaced by $2n_\mu$ (resp. $2n$) so that the inequality (3) holds again. Hence the limit of the local index vanishes in this case too.

On the other hand, it follows from the equivariant Riemann–Roch theorem that the left-hand side can be written as

$$\sum_j a_j \exp jt \qquad j,\, a_j \in \mathbb{Z}.$$

The function of this form with the above property must be identically 0, that is,

$$f_!(\exp(c_{S^1}/2)\cdot \hat{A}_{S^1}(M)) = 0.$$

Since the zero-dimensional part of the left-hand side coincides with the genus $\exp(c/2)\cdot \hat{A}(M)[M]$, we have in particular

$$\exp(c/2)\cdot \hat{A}(M)[M] = 0. \qquad \square$$

In the proof of Theorem 6.30, the equivariant Riemann–Roch theorem plays an essential role. The reason for this will be clarified by the following example.

Example 1 We consider the semi-free S^1-action on $\mathbb{C}P^2$:

$$\varphi\colon S^1 \times \mathbb{C}P^2 \to \mathbb{C}P^2$$

given by $\varphi(g, [z_0, z_1, z_2]) = [z_0, z_1, gz_2]$. If we take $c \in H^2(\mathbb{C}P^2)$ to be 0, then ξ is the trivial bundle $\mathbb{C}P^2 \times \mathbb{C} \to \mathbb{C}P^2$. Let S^1 act trivially on $\mathbb{C}$. Then condition (2) of Theorem 6.30 is satisfied. Note that condition (1) of Theorem 6.30 is not satisfied, since $W_2(\mathbb{C}P^2) \neq 0$. Obviously condition (3) is satisfied, so that

$$f_!(\hat{A}_{S^1}(\mathbb{C}P^2)) \to 0 \quad \text{as } t \to \pm\infty.$$

However $\hat{A}(\mathbb{C}P^2)[\mathbb{C}P^2]$ does not vanish. In fact, we have

$$\hat{A}(\mathbb{C}P^2)[\mathbb{C}P^2] = -1/8.$$

In the following, we explain the situation. Since we can not apply the equivariant Riemann–Roch theorem in this case, we cannot write $f_!(\hat{A}_{S^1}(\mathbb{C}P^2))$ in the form $\sum_j a_j \exp jt$. In consequence, the argument in the proof of Theorem 6.30 is invalid.

To see this phenomenon precisely, we will calculate it concretely. First note that the fixed point set is $\mathbb{C}P^1 \cup \{*\}$ where $*$ denotes the point $[0, 0, z_2]$.

Then we have

$$f_!(\hat{A}_{S^1}(\mathbb{C}P^2)) = \hat{A}_{S^1}(\mathbb{C}P^1)\frac{1}{\exp(t+x)/2 - \exp(-t-x)/2}\Big/[\mathbb{C}P^1]$$

$$+ \hat{A}_{S^1}(*)\frac{1}{(\exp(t/2) - \exp(-t/2))^2}\Big/[*]$$

$$= \text{the coefficient of } x \text{ in } \frac{1}{\exp(t+x)/2 - \exp(-x-t)/2}$$

$$+ \frac{1}{(\exp(t/2) - \exp(-t/2))^2}$$

$$= \frac{2 - (\exp(t/2) + \exp(-t/2))}{2(\exp(t/2) - \exp(-t/2))^2}.$$

It may be said in this connection that the constant term of the Laurent expansion of the above function is equal to $-1/8$, which is nothing but $\hat{A}(\mathbb{C}P^2)[\mathbb{C}P^2]$.

When a manifold M has the homotopy type of the complex projective space $\mathbb{C}P^n$, we say that M is a *homotopy complex projective space* or *homotopy* $\mathbb{C}P^n$.

A homotopy complex projective space is a cohomology complex projective space.

Petrie conjecture[109] Let M be a homotopy $\mathbb{C}P^n$. If M admits a non-trivial S^1-action, then the total Pontrjagin class $P(M)$ is equal to that of $\mathbb{C}P^n$, that is,

$$P(M) = (1 + x^2)^{n+1}$$

where x is a generator of $H^2(M)$.

Although a partial answer is obtained here, it is an unsolved problem in general.

We shall show that the conjecture is true in a special case by making use of the vanishing theorem of the generalized $\hat{A}$-genera. Before stating the result, let us make some preliminary definitions and observations. Let M be a cohomology $\mathbb{C}P^n$ and x a generator of $H^2(M)$. Denote by $\pi\colon \eta \to M$ the complex line bundle with $c_1(\eta) = x$ and by $S(\eta)$ the total space of the sphere bundle associated with η. Let $\pi'\colon S(\eta) \to M$ denote the restriction of $\pi\colon \eta \to M$ to $S(\eta)$. It follows from the Gysin exact sequence

$$\cdots \to H^i(M) \xrightarrow{\cup x} H^{i+2}(M) \xrightarrow{\pi'^*} H^{i+2}(S(\eta)) \to H^{i+1}(M) \to \cdots$$

with respect to the bundle $\pi': S(\eta) \to M$ that

$$H^*(S(\eta)) \cong H^*(S^{2n+1}).$$

That is, $S(\eta)$ is a cohomology $(2n+1)$-sphere.

Given a non-trivial S^1-action on M, we can lift it to $S(\eta)$ by the obstruction theory of Su,[125] since $H^1(M) = 0$. We now apply the Smith theory to the S^1-action on the cohomology sphere $S(\eta)$ and deduce the following lemmas.

Lemma 6.31[139] *If the S^1-action on M is semi-free, the fixed point set consists of exactly two components F_1 and F_2, and there are integers n_1 and n_2 with $n = n_1 + n_2 + 1$ such that F_1 and F_2 are cohomology $\mathbb{C}P^{n_1}$ and cohomology $\mathbb{C}P^{n_2}$, respectively. Moreover if we denote by x_i the restriction of x to F_i, then x_i is a generator of $H^2(F_i)$ for $i = 1, 2$.*

Lemma 6.32 *With the conditions of the previous lemma, there exists a lifting of the S^1-action on M to $S(\eta)$ such that the action is trivial on $\pi'^{-1}(F_1)$ and is free on $\pi'^{-1}(F_2)$.*

Choose arbitrary $y_i \in F_i$ for $i = 1, 2$. Then $\pi^{-1}(y_i)$ are representation spaces of S^1. If we take the lifting of Lemma 6.32, we have

$$\pi^{-1}(y_1) = V_0, \qquad \pi^{-1}(y_2) = V_1 \text{ or } V_{-1}.$$

For simplicity of exposition, we shall consider the case where $\pi^{-1}(y_2) = V_1$ only. The case where $\pi^{-1}(y_2) = V_{-1}$ is entirely similar. The S^1-vector bundle obtained in this way is denoted by η_1. If we define a new S^1-vector bundle η_2 by

$$\eta_2 = \eta_1 \otimes V_{-1},$$

it satisfies

$$\pi^{-1}(y_1) = V_{-1}, \qquad \pi^{-1}(y_2) = V_0$$

where the projection of η_2 is also denoted by π.

Given non-negative integers i_j, put

$$\eta_j^{i_j} = \eta_j \otimes \overset{i_j}{\cdots} \otimes \eta_j \qquad \text{for } j = 1, 2.$$

Here the right-hand side denotes the tensor products over $\mathbb{C}$ of i_j copies of the S^1-vector bundle η_j for $j = 1, 2$. Note that η_j^0 denotes the trivial bundle $M \times \mathbb{C} \to M$ where S^1 acts trivially on $\mathbb{C}$.

Then the following lemma can be proved in the manner of Kawakubo.[64]

Lemma 6.33 *The restrictions of the equivariant first Chern class to the components of the fixed point set are given by:*

$$c_1\left(ES^1 \underset{S^1}{\times} (\eta_1^{i_1} \otimes \eta_2^{i_2}) | BS^1 \times F_1\right) = -i_2 t + (i_1 + i_2)x_1,$$

$$c_1\left(ES^1 \underset{S^1}{\times} (\eta_1^{i_1} \otimes \eta_2^{i_2}) | BS^1 \times F_2\right) = i_1 t + (i_1 + i_2)x_2.$$

Here t denotes the first Chern class of the canonical complex line bundle over BS^1.

We are now in a position to state the theorem.

Theorem 6.34 *Let M be a cohomology $\mathbb{C}P^n$. If M admits a non-trivial semi-free S^1-action, then we have*

$$\exp(bx/2)\cdot\hat{A}(M)[M] = 0$$

for $b = n-1, n-3, n-5, \ldots, (3-(-1)^n)/2$.

Proof We will make free use of the notation which we introduced before the theorem. It follows from the Wu formula that the second Stiefel–Whitney class $W_2(M)$ of M satisfies

$$W_2(M) \equiv (n+1)x \bmod 2.$$

Hence we have

$$W_2(M) \equiv bx \bmod 2$$

for b in the theorem so that condition (1) of Theorem 6.30 is satisfied. Corresponding to these numbers b, we choose integers i_1 and i_2 such that

$$0 \le i_1 \le n_1, \qquad 0 \le i_2 \le n_2, \qquad i_1 + i_2 = b.$$

In fact, this is possible, since $0 \le b \le n-1 = n_1 + n_2$. We now put

$$\xi = \eta_1^{i_1} \otimes \eta_2^{i_2}.$$

Since $c_1(\xi) = (i_1 + i_2)x = bx$ and since ξ is an S^1-complex line bundle, ξ satisfies condition (2) of Theorem 6.30 with respect to $c = bx$. Note that $E_{\mu n}$ in condition (3) is E_{j1} ($j = 1, 2$) in this case. Since

$$\dim E_{j1} = n - n_j,$$

we have

$$\dim E_{11} = n_2 + 1, \qquad \dim E_{21} = n_1 + 1.$$

Moreover we have the inequalities:

$$|-i_2| \leq n_2 < n_2 + 1 = \dim E_{11},$$
$$i_1 \leq n_1 < n_1 + 1 = \dim E_{21}.$$

It follows from Lemma 6.33 that condition (3) of Theorem 6.30 is satisfied. Thus Theorem 6.30 yields Theorem 6.34. □

By combining Theorems 6.29 and 6.34, we have the following partial answer to the Petrie conjecture.

Corollary 6.35 *Let M be a cohomology $\mathbb{C}P^n$. If M admits a non-trivial semi-free S^1-action, then we have*

$$P(M) = (1 + x^2)^{n+1},$$

where x denotes a generator of $H^2(M)$.

Remark For various results concerning the Petrie conjecture, the reader is referred to Hattori,[41] Masuda,[87] Tsukada and Washiyama,[131] Watabe,[136] and Yoshida[139] among others.

Exercises for Chapter 6

1. Show that there is no free action of a non-trivial group on $\mathbb{C}P^{2n}$.

2. By analogy with the equivariant Pontrjagin classes, define equivariant Stiefel–Whitney classes of G-vector bundles and show that those of the tangent bundles of G-manifolds are G-homotopy type invariants.

Answers to exercises

Chapter 1

1. The isotropy group consists of matrices of the form

$$\begin{bmatrix} 1 & * \cdots * \\ 0 & \\ \vdots & A \\ 0 & \end{bmatrix} \qquad A \in GL(n-1, \mathbb{R})$$

where $*$ denote arbitrary real numbers.

2. Let G' denote the group generated by U. For any $g' \in G'$, $g'U$ is a neighbourhood of g' in G and satisfies $g'U \subset G'$. Hence G' is open in G. Note that

$$G' = G - \bigcup_{g \in G - G'} gG'.$$

Since $\bigcup_{g \in G - G'} gG'$ is open, we see that G' is closed in G. Hence we have $G = G'$ by assumption.

3. Given $g \in G$, define a map f_g: $G \to G$ by $f_g(h) = ghg^{-1}h^{-1}$. Since f_g is continuous and since the centre C of G is given by

$$C = \bigcap_{g \in G} f_g^{-1}(e),$$

C is closed in G.

Next, let Γ be a discrete subgroup of G. Then there exists a neighbourhood U of e such that $U \cap \Gamma = \{e\}$. It follows from the continuity of the map γ in Lemma 1.6 that for any element g of $G - \Gamma$, there exists a neighbourhood V of g satisfying $V \cdot V^{-1} \subset U$. For any $g_1, g_2 \in V \cap \Gamma$, we have

$$g_1 g_2^{-1} \in V \cdot V^{-1} \cap \Gamma \subset U \cap \Gamma = \{e\},$$

that is, we have $g_1 = g_2$. We now put $V' = V - \{g_1\}$. Then V' is a neighbourhood of g. Since $V' \cap \Gamma = \varnothing$, this means that $G - \Gamma$ is open in G and hence Γ is closed in G.

4. We need only use the fact that the left translation L_g is a homeomorphism and f is a homomorphism.

5. For any $h \in H$, define a continuous map f_h: $G \to G$ by $f_h(g) = ghg^{-1}$. Then it is easily verified that $f_h(G) = \{h\}$.

6. Let O_1 and O_2 be open sets of G satisfying $G = O_1 \cup O_2$ and $O_1 \cap O_2 = \varnothing$. Since each coset gH is connected, it is included entirely either in O_1 or in O_2. This means that the images $\pi(O_1)$ and $\pi(O_2)$ are disjoint open sets of G/H where $\pi\colon G \to G/H$ denotes the projection. Since $\pi(O_1) \cup \pi(O_2) = G/H$, it follows from the assumption that either O_1 or O_2 is empty.

7. Apply Lemma 1.50 and Proposition 1.77.

8. Use the fact that $\bar{A}_g$ is a subgroup of S^1.

9. We need only consider new splittings $s_1\colon \mathbb{Z}_2 \to S_n$ and $s_2\colon \mathbb{Z}_2 \to D_n$ defined by

$$s_1(-1) = \left[\begin{array}{cc|c} 0 & 1 & 0 \\ 1 & 0 & \\ \hline \multicolumn{2}{c|}{0} & I \end{array}\right], \qquad s_2(-1) = \begin{bmatrix} 0 & 1 \\ 1 & 0 \end{bmatrix}$$

where I denotes the unit matrix of degree $n-2$.

10. It is easy to see that φ gives a G-action on $\mathbb{H}$. For any $x \in \mathbb{R}$, $y > 0$, we have

$$\varphi\left(\begin{bmatrix} \sqrt{y} & \dfrac{x}{\sqrt{y}} \\ 0 & \dfrac{1}{\sqrt{y}} \end{bmatrix}, i \right) = x + iy,$$

that is, the action is transitive. The isotropy group at i is $SO(2)$.

11. The following example is due to Professor K. Uno. Let G be the general linear group $GL(2, \mathbb{R})$. Put

$$H = \left\{ \begin{bmatrix} 1 & n \\ 0 & 1 \end{bmatrix} \middle| \, n \in \mathbb{Z} \right\}.$$

Then it is easy to see that H is a closed subgroup of G. For a non-zero integer k, put

$$g_k = \begin{bmatrix} k & 0 \\ 0 & 1 \end{bmatrix}.$$

Then g_k belongs to G and satisfies

$$g_k \begin{bmatrix} 1 & n \\ 0 & 1 \end{bmatrix} g_k^{-1} = \begin{bmatrix} 1 & kn \\ 0 & 1 \end{bmatrix}.$$

It follows that

$$g_k H g_k^{-1} = \left\{ \begin{bmatrix} 1 & kn \\ 0 & 1 \end{bmatrix} \middle| n \in \mathbb{Z} \right\} \subset H.$$

If $k \neq \pm 1$, then $g_k H g_k^{-1} \neq H$. Hence $N(H, H) \neq N(H)$.

Chapter 2

1. Since $\mathbb{Z}$ is a subgroup of $\mathbb{R}$, $\mathbb{Z}$ acts on $\mathbb{R}$ by addition. Define two subsets A and B of $\mathbb{R}$ by

$$A = \mathbb{Z} + (1/6, 5/6), \qquad B = \mathbb{Z} + (-1/3, 1/3)$$

and put $U = p(A)$, $V = p(B)$. Then we have

$$\mathbb{R} = A \cup B, \qquad S^1 = U \cup V, \qquad A = p^{-1}(U), \qquad B = p^{-1}(V).$$

Define two maps $\varphi: p^{-1}(U) = \mathbb{Z} + (1/6, 5/6) \to U \times \mathbb{Z}$ and $\psi: p^{-1}(V) = \mathbb{Z} + (-1/3, 1/3) \to V \times \mathbb{Z}$ by

$$\varphi(i + x) = (p(x), i) \qquad \text{for } i \in \mathbb{Z},\ x \in (1/6, 5/6)$$

$$\psi(i + x) = (p(x), i) \qquad \text{for } i \in \mathbb{Z},\ x \in (-1/3, 1/3).$$

Then φ and ψ give coordinate functions. Since the map $\theta: U \cap V \to \mathbb{Z}$ defined by $\theta(p(x)) = \psi_{p(x)} \cdot \varphi_{p(x)}^{-1}$ satisfies

$$\theta(p(x)) = \begin{cases} 0 & p(x) \in p((1/6, 1/3)) \\ 1 & p(x) \in p((2/3, 5/6)), \end{cases}$$

the projection $p: \mathbb{R} \to S^1$ is a principal $\mathbb{Z}$-bundle, and hence a regular covering.

2. This can be shown as in Exercise 1 above.

3. Since the principal $\mathbb{Z}_n$-bundle associated with $S^1 \times_{\mathbb{Z}_n} S^1 \to S^1$ is not trivial, the bundle $S^1 \times_{\mathbb{Z}_n} S^1 \to S^1$ is not trivial as a bundle with structure group $\mathbb{Z}_n$. The latter half of Exercise 3 is immediate from Theorem 2.29.

4. The map $f: P/H \to P \times_G G/H$ defined by $[x] \mapsto [x, eH]$ gives a bundle isomorphism. If H is a closed normal subgroup of G, a principal G/H-bundle structure on $P \times_G G/H$ is transferred to P/H by the bundle isomorphism f.

5. Given $\varepsilon > 0$, let S_ε^1 denote the circle of radius ε with centre at the origin of $\mathbb{R}^2$. The restriction $p|: S_\varepsilon^1 \to S_\varepsilon^1/\mathbb{Z}_2$ of p is a non-trivial principal $\mathbb{Z}_2$-bundle so that $p|$ has no cross-section. Hence p has no cross-section around $p(0)$.

Chapter 3

1. Obviously there is a diffeomorphism $\mathbb{R}P^1 = S^1/\mathbb{Z}_2 \approx S^1$. Put $U_i = \{[z_0, z_1] \in \mathbb{C}P^1 \mid z_i \neq 0\}$ and define maps $f_i\colon U_i \to \mathbb{C}$ by

$$f_0([z_0, z_1]) = z_1/z_0 \qquad \text{for } [z_0, z_1] \in U_0,$$

$$f_1([z_0, z_1]) = z_0/z_1 \qquad \text{for } [z_0, z_1] \in U_1.$$

Referring to Example 2 in Section 3.1, one can construct a diffeomorphism $f\colon \mathbb{C}P^1 \to S^2$ from f_i.

2. Suppose that N is a submanifold of $\mathbb{R}^2$. Then there exist two linearly independent tangent vectors to N at the origin. It follows from the inverse function theorem that a neighbourhood of the origin is included in N, which is a contradiction.

3. The fixed point set is nothing but $\mathbb{R}P^n$.

4. This follows easily from the definition.

5. $\varphi(t, x) = \{(2x + 1)\exp 2t - 1\}/2$.

6, 7. Easy and omitted.

Chapter 4

1. We consider the following special case: $n = 2$ and $a_0 = a_1 < a_2$. Put

$$M_1 = \{[z_0, z_1, z_2] \in \mathbb{C}P^2 \mid z_2 = 0\} \approx \mathbb{C}P^1,$$

$$* = \{[z_0, z_1, z_2] \in \mathbb{C}P^2 \mid z_0 = z_1 = 0\} \text{ (1 point)}.$$

Then we have $\mathbb{C}P^{2S^1} = M_1 \cup *$ and $\mathbb{C}P^2(\mathbb{Z}_{(a_2 - a_1)}) = \mathbb{C}P^2 - (M_1 \cup *)$. The principal orbit is $S^1/\mathbb{Z}_{(a_2 - a_1)}$. The application of the decomposition theorem is left to the reader.

2, 3. These results are easily deduced from the fact that the number of orbit types occurring in M is finite (Theorem 4.23).

4. Since G is commutative, the principal isotropy group is the smallest one among the isotropy groups occurring in M. Since the action is effective, this means that the principal isotropy group must be trivial.

5. Since G is finite, all the slices are open in M. If H is a principal isotropy group, the slice representation of H is trivial by Corollary 4.29. In consequence, we see that M^H is non-empty and open in M. On the other hand, it follows from Theorem 1.46 that M^H is closed in M. Accordingly, we have $M^H = M$. Since the action is effective by assumption, we can conclude that $H = \{e\}$.

6. This is immediate from Exercise 5 above.

7. Let H denote the isotropy group at $(a, b) \in \mathbb{R}^n \oplus \mathbb{R}^n$. Regarding a, b as vectors of the same vector space $\mathbb{R}^n$, we denote by $\mathrm{Sp}\{a, b\}$ the subspace spanned by a and b. Then, corresponding to $\dim \mathrm{Sp}\{a, b\} = 0, 1, 2$, we have $(H) = (O(n)), (O(n-1)), (O(n-2))$, respectively. The general case is left to the reader.

8. This can be shown in a similar way to Exercise 7.

9. The natural $O(n)$-action on $\mathbb{C}^n$ is identified with the diagonal action on $\mathbb{R}^n \oplus \mathbb{R}^n$ so that we can apply Exercise 7. Since the case $a = b = 0$ does not occur, $(O(n-1))$ and $(O(n-2))$ are the required types.

10. $M(\mathbb{Z}_4) = \{[*, 0, 0] \in \mathbb{C}P^2\}$ (one point),

$M(\mathbb{Z}_2) = \{[0, *, *] \in \mathbb{C}P^2\} \approx \mathbb{C}P^1$, $M(\{e\}) = M - M(\mathbb{Z}_4) - M(\mathbb{Z}_2)$.

Chapter 5

1. Since $\mathbb{R}P^{2n}$ is the orbit space of a free $\mathbb{Z}_2$-action on S^{2n}, it follows from Corollary 5.22 that $\chi(\mathbb{R}P^{2n}) = 1$. Hence it follows again from Corollary 5.22 that there is no free action of a non-trivial compact Lie group.

2. In view of Exercise 3 in Chapter 4, M supports a free $\mathbb{Z}_n$-action for n sufficiently large. Hence we have $\chi(M) = 0$ by Corollary 5.22.

3. This is an immediate consequence of Theorem 5.28.

4. This follows from the equalities $\chi(K) = |G|\chi(K/G) = |G|\chi(K)$ which are consequences of Corollary 5.22 and Exercise 3.

5. Since $\chi(\mathbb{C}P^n) = n + 1$, this follows from Exercise 2.

6. By Corollary 5.20 and Exercise 10 in Chapter 4, we have $4\chi(\mathbb{C}P^2/\mathbb{Z}_4) = \chi(\mathbb{C}P^2) + (2-1)\chi(\mathbb{C}P^{2\mathbb{Z}_2}) + (4-2)\chi(\mathbb{C}P^{2\mathbb{Z}_4}) = 8$. It follows that $\chi(\mathbb{C}P^2/\mathbb{Z}_4) = 2$.

Chapter 6

1. Note that a non-trivial compact Lie group includes a cyclic group of prime order. Since $\chi(\mathbb{C}P^{2n}) = 2n + 1$, it follows from Corollary 5.22 that there is no free $\mathbb{Z}_2$-action on $\mathbb{C}P^{2n+1}$. Suppose that $G = \mathbb{Z}_p$, p odd prime, acts freely on $\mathbb{C}P^{2n}$. Then it follows from Conner and Floyd[24] that the oriented bordism class $[\mathbb{C}P^{2n}]$ represented by $\mathbb{C}P^{2n}$ is divisible by p. In consequence, the index $I(\mathbb{C}P^{2n})$ is divisible by p. But $I(\mathbb{C}P^{2n})$ is equal to 1, which is a contradiction.

2. Put $W_G(T(M)) = W(EG \times_G T(M))$ where $W(\ )$ denotes the total Stiefel–Whitney class. By approximating EG by G-manifolds, we can apply the proof of the homotopical invariance of the ordinary Stiefel–Whitney classes so that we get the desired invariance. The reader is referred to Kawakubo[64], for more details.

Bibliography

1. Adams, J. F. (1969). *Lectures on Lie groups.* Benjamin.
2. Aigner, M. (1979). *Combinatorial theory.* Springer-Verlag.
3. Araki, S. and Kawakubo, K. (1988). Equivariant s-cobordism theorems. *Journal of the Mathematical Society of Japan*, **40**, 349–67.
4. Assadi, A. H. (1982). Finite group actions on simply connected manifolds and CW complexes. *Memoirs of the American Mathematical Society* **35**.
5. Atiyah, M. F., Bott, R., and Shapiro, A. (1964). Clifford modules. *Topology*, **3** (Supplement 1), 3–38.
6. Atiyah, M. F. and Hirzebruch, F. (1959). Riemann–Roch theorems for differentiable manifolds. *Bulletin of the American Mathematical Society*, **65**, 276–81.
7. Atiyah, M. F. and Hirzebruch, F. (1970). Spin manifolds and group actions. In *Essays on topology and related topics* (Mémoirs dédiés à Georges de Rham), Springer-Verlag, 17–28.
8. Atiyah, M. F. and Segal, G. B. (1965). Equivariant K-theory. University of Warwick, mimeographed note.
9. Atiyah, M. F. and Segal, G. B. (1967). The index of elliptic operators II. *Annals of Mathematics*, **87**, 531–45.
10. Atiyah, M. F. and Singer, I. M. (1968). The index of elliptic operators I, III. *Annals of Mathematics*, **87**, 484–530; **87**, 546–604.
11. Boothby, W. M. (1986). *An introduction to differentiable manifolds and Riemannian geometry* (2nd edn). Academic Press.
12. Borel, A. (1960). Seminar on transformation groups. *Annals of Mathematics Studies*, **46**, Princeton University Press.
13. Borel, A. and Hirzebruch, F. (1958–60). Characteristic classes and homogeneous spaces. *American Journal of Mathematics*, **80** (1958), 458–538; **81** (1959), 315–82; **82** (1960), 491–504.
14. Bredon, G. E. (1967). *Equivariant cohomology theories.* Lecture Notes in Mathematics, **34**, Springer-Verlag.
15. Bredon, G. E., (1972). *Introduction to compact transformation groups.* Academic Press.
16. Brieskorn, E. (1966). Beispiele zur Differentialtopologie von Singularitäten. *Invent. Math.*, **2**, 1–14.
17. Bröcker, T. and tom Dieck, T. (1970). *Kobordismentheorie.* Lecture Notes in Mathematics, **178**, Springer-Verlag.
18. Browder, W. (1983). Cohomology and group actions. *Invent. Math.*, **71**, 599–608.
19. Brown, K. S. (1974). Euler characteristics of discrete groups and G-spaces. *Invent. Math.*, **27**, 229–64.
20. Capell, S. E. and Shaneson, J. L. (1981). Non-linear similarity. *Annals of Mathematics* **113**, 315–55.
21. Chern, S. S. (1953). On the characteristic classes of complex sphere bundles and algebraic varieties. *American Journal of Mathematics.*, **75**, 565–97.
22. Chevalley, C. (1946). *Theory of Lie groups.* Princeton University Press.

23. Conner, P. E. and Floyd, E. E. (1964). *Differentiable periodic maps*. Springer-Verlag.
24. Conner, P. E. and Floyd, E. E. (1966). Maps of odd period. *Annals of Mathematics*, **84**, 132–56.
25. Conner, P., Raymond, F. and Weinberger, P. J. (1972). Manifolds with no periodic maps. *Proceedings of the Second Conference on Compact Transformation Groups, Amherst, Mass. 1971*, Lecture Notes in Mathematics, **299**, Springer-Verlag, 81–108.
26. tom Dieck, T. (1969). Faserbündel mit Gruppenoperation. *Arch. Math.*, **20** 136–43.
27. tom Dieck, T. (1970). Bordism of *G*-manifolds and integrality theorems. *Topology*, **9**, 345–58.
28. tom Dieck, T. (1971). Characteristic numbers of *G*-manifolds I. *Invent. Math.*, **13**, 213–24.
29. tom Dieck, T. (1972). Kobordismentheorie, klassifizierender Räume und Transformations gruppen. *Mathematische Zeitschrift* **126**, 31–9.
30. tom Dieck, T. (1975). The Burnside ring of a compact Lie group I. *Mathematische Annalen.*, **215**, 235–50.
31. tom Dieck, T. (1978). Homotopy equivalent group representations. *Journal für die reine und angewandte Mathematik*, **298**, 182–95.
32. tom Dieck, T. (1979). *Transformation groups and representation theory*. Lecture Notes in Mathematics, **766**, Springer-Verlag.
33. tom Dieck, T. (1987). *Transformation groups*. Walter de Gruyter.
34. tom Dieck, T. and Petrie, T. (1978). Geometric modules over the Burnside ring. *Invent. Math.*, **47**, 273–87.
35. Dold, A. (1963). Partitions of unity in the theory of fibrations. *Annals of Mathematics*, **78**, 223–55.
36. Dold, A. (1972). *Lectures on algebraic topology*. Springer-Verlag.
37. Dold, A. (1976). The fixed point index of fibre-preserving maps. *Mathematische Zeitschrift*, **148**, 215–44.
38. Dold, A. (1983). Fixed point indices of iterated maps. *Invent. Math.* **74**, 419–35.
39. Dovermann, K. H. and Petrie, T. (1982). *G*-surgery II. *Memoirs of the American Mathematical Society*, **260**.
40. Dovermann, K. H. and Rothenberg, M. (1989). The generalized Whitehead torsion of a *G* fiber homotopy equivalence. *Proceedings of the Conference on Transformation Groups, Osaka 1987*, Lecture Notes in Mathematics **1375**, Springer-Verlag, 60–88.
41. Hattori, A. (1978). Spin^c-structures and S^1-actions. *Invent. Math.* **48**, 7–31.
42. Hattori, A. and Taniguchi, H. (1972). Smooth S^1-action and bordism. *Journal of the Mathematical Society of Japan*, **24**, 701–31.
43. Hattori, A. and Yoshida, T. (1976). Lifting compact group actions in fiber bundles. *Japanese Journal of Mathematics*, **2**, 13–25.
44. Hirzebruch, F. (1966). *Topological method in algebraic geometry*. Springer-Verlag.
45. Hsiang, W. C. (1966). A note on free differentiable actions of S^1 and S^3 on homotopy spheres. *Annals of Mathematics*, **83**, 266–72.
46. Hsiang, W. C. (1989). A personal perspective of differentiable transformation groups. *Proceedings of the Conference on Transformation Groups, Osaka, 1987*, Lecture Notes in Mathematics, **1375**, Springer-Verlag, 1–5.
47. Hsiang, W. C. and Hsiang, W. Y. (1967, 70). Differentiable actions of compact

connected classical groups I, II. *American Journal of Mathematics* **89** (1967), 705–86; *Annals of Mathematics*, **92** (1970), 189–223.

48. Hsiang, W. Y. (1975). *Cohomology theory of topological transformation groups*. Springer-Verlag.
49. Illman, S. (1973). Equivariant algebraic topology. *Ann. Inst. Fourier*, **23**, 87–91.
50. Illman, S. (1974). Whitehead torsion and group actions. *Ann. Acad. Sci. Fenn. Ser.*, **A1588**, 1–44.
51. Illman, S. (1978). Smooth equivariant triangulations of G-manifolds for G a finite group. *Mathematische Annalen*, **233**, 199–220.
52. Jackowski, S. (1977). Equivariant K-theory and cyclic subgroups. *Proceedings of the Conference on Transformation Groups, Newcastle upon Tyne, 1976*, Cambridge University Press, 76–91.
53. James, I. M. and Segal, G. B. (1978). On equivariant homotopy type. *Topology*, **17**, 267–72.
54. Jänich, K. (1968). *Differenzierbare G-Mannigfaltigkeiten*. Lecture Notes in Mathematics **59**, Springer-Verlag.
55. Kawakubo, K. (1969). Smooth structures on $S^p \times S^q$, *Osaka Journal of Mathematics* **6**, 165–96.
56. Kawakubo, K. (1970). On homotopy spheres which admit differentiable actions II. *Osaka Journal of Mathematics*, **7**, 179–92.
57. Kawakubo, K. (1972). Topological S^1 and $\mathbb{Z}_{2k}$ actions on spheres. *Proceedings of the Second Conference on Compact Transformation Groups, Amherst, Mass. 1971*, Lecture Notes in Mathematics, **298**, Springer-Verlag, 14–8.
58. Kawakubo, K. (1971). Invariants for semi-free S^1-action. *Bulletin of the American Mathematical Society*, 77, 987–91.
59. Kawakubo, K. (1971). Knotted fixed point sets of semi-free S^1-actions. *Osaka Journal of Mathematics*, **8**, 385–92.
60. Kawakubo, K. (1975). The index and the generalized Todd genus of $\mathbb{Z}_p$-actions. *American Journal of Mathematics*, **97**, 182–204.
61. Kawakubo, K. (1977). Equivariant Riemann–Roch type theorems and related topics. *Proceedings of the Conference on Transformation Groups, Newcastle, 1976*, London Mathematical Society Lecture Note Series, **no. 26**, Cambridge University Press 284–94.
62. Kawakubo, K. (1980). Equivariant homotopy equivalence of group representations. *Journal of the Mathematical Society of Japan*, **32**, 105–18.
63. Kawakubo, K. (1980). Global and local equivariant characteristic numbers of G-manifolds. *Journal of the Mathematical Society of Japan*, **32**, 301–23.
64. Kawakubo, K. (1980). Equivariant Riemann–Roch theorems, localization and formal group law. *Osaka Journal of Mathematics*, **17**, 531–71.
65. Kawakubo, K. (1980). Weyl group actions and equivariant homotopy equivalence. *Proceedings of the American Mathematical Society*, **80**, 172–6.
66. Kawakubo, K. (1980). Cancellation law for G-homotopy equivalent representations. *Japanese Journal of Mathematics*, **6**, 259–66.
67. Kawakubo, K. (1981). Compact Lie group actions and fiber homotopy type. *Journal of the Mathematical Society of Japan*, **33**, 295–321.
68. Kawakubo, K. (1982). ΛG-structure of G-vector bundles and groups $KO_G(X)$, $KSp_G(X)$ and $J_G(X)$. *Osaka Journal of Mathematics*, **19**, 695–715.
69. Kawakubo, K. (1986). Topological invariance of the equivariant rational Pontrjagin classes. *Proceedings of the Conference on Transformation Groups, Poznań 1985*, Lecture Notes in Mathematics, **1217**, Springer-Verlag, 183–95.

70. Kawakubo, K. (1986). Induction theorems for equivariant K-theory and J-theory. *Journal of the Mathematical Society of Japan*, **38**, 173–98.
71. Kawakubo, K. (1986). Stable equivalence of G-manifolds, Homotopy theory and related topics. *Advanced Studies in Pure Mathematics*, **9**, 26–40.
72. Kawakubo, K. (1988). An s-cobordism theorem for semi-free S^1-manifolds. In *A fete of topology*, Academic Press, 565–83.
73. Kawakubo, K. (1989). G–s-cobordism theorems do not hold in general for many compact Lie groups G. *Proceedings of the Conference on Transformation Groups, Osaka 1987*, Lecture Notes in Mathematics **1375**, Springer-Verlag, 183–90.
74. Kawakubo, K. G–s-cobordant manifolds are not necessarily G-homeomorphic for arbitrary compact Lie groups G, preprint.
75. Kawakubo, K. and Raymond, F. (1972). The index of manifolds with toral actions and geometric interpretations of $\sigma(\infty, (S^1, M^n))$ invariant of Atiyah and Singer. *Invent. Math.* **15**, 53–66.
76. Kawakubo, K. and Uchida, F. (1971). On the index of a semi-free S^1-action. *Journal of the Mathematical Society of Japan*, **23**, 351–5.
77. Komiya, K. (1987). The Lefschetz number for equivariant maps. *Osaka Journal of Mathematics*, **24**, 299–305.
78. Komiya, K. (1988). Fixed point indices of equivariant maps and Möbius inversion. *Invent Math.*, **91**, 129–35.
79. Kreck, M. (1984). *Bordism of diffeomorphisms and related topics*. Lecture Notes in Mathematics, Springer-Verlag.
80. Kulkarni, R. S. (1982). Pseudofree actions and Hurwitz's $84(g-1)$ theorem. *Mathematische Annalen*, **261**, 209–26.
81. Laitinen, E. and Lück, W. (1989). Equivariant Lefschetz classes. *Osaka Journal of Mathematics*, **26**, 491–525.
82. Lee, K. B. and Raymond, F. (1989). Seifert manifolds modelled on principal bundles. *Proceedings of the Conference on Transformation Groups, Osaka 1987*, Lecture Notes in Mathematics, **1375**, Springer-Verlag, 207–15.
83. Löffler, P. (1980). Äquivariante Homotopie, äquivarianter Bordismus und freie differenzierbare Involutionen auf Sphären. *Mathematische Zeitschrift*, **170**, 233–46.
84. Lück, W. (1989). *Transformation groups and algebraic K-theory*. Lecture Notes in Mathematics, **1408**, Springer-Verlag.
85. Madsen, I. and Rothenberg, M. (1985). On the classification of G-spheres I: Equivariant transversality; II: *PL* automorphism groups; III: Top automorphism groups. Aarhus Universitet, Preprint series.
86. Madsen, I., Thomas, C. B. and Wall, C. T. C. (1976). The topological spherical space form problem II. *Topology*, **15**, 375–82.
87. Masuda, M. (1981). On smooth S^1-actions on cohomology complex projective spaces. The case where the fixed point set consists of four connected components. *Journal of the Faculty of Science, University of Tokyo, Section 1A, Mathematics*, **28**, 127–67.
88. Matumoto, T. (1971). On G–CW complexes and a theorem of J. H. C. Whitehead. *Journal of the Faculty of Science, University of Tokyo, Section 1A*, **18**, 363–74.
89. Matumoto, T. and Shiota, M. (1987). Unique triangulation of the orbit space of a differentiable transformation group and its applications. *Advanced Studies in Pure Mathematics*, **9**, 41–55.

90. May, J. P. (1982). Equivariant homotopy and cohomology theory. *Symposium in Honor of J. Adem, Contemporary Mathematics*, **12**, 202–17.
91. Milnor, J. (1957). Groups which act on S^n without fixed points. *American Journal of Mathematics*, **79**, 623–30.
92. Milnor, J. (1958). *Differential topology.* Princeton University, mimeographed note.
93. Milnor, J. and Stasheff, J. (1974). *Lectures on characteristic classes.* Princeton University Press.
94. Montgomery, D. and Yang, C. T. (1957). The existence of a slice. *Annals of Mathematics*, **65**, 108–16.
95. Morimoto, M. (1982). On the groups $J_G(*)$ for $G = SL(2, p)$. *Osaka Journal of Mathematics*, **19**, 57–78.
96. Morimoto, M. (1989). Most of the standard spheres have one fixed point actions of A_5. *Proceedings of the Conference on Transformation Groups, Osaka 1987*, Lecture Notes in Mathematics, **1375**, Springer-Verlag, 240–58.
97. Murakami, S. (1969). Manifolds. Kyorits Shuppan, Tokyo (in Japanese).
98. Nagao, H. (1967). The foundations of groups. Asakura Shoten, Tokyo (in Japanese).
99. Nagasaki, I. (1985). Homotopy representations and spheres of representations. *Osaka Journal of Mathematics*, **22**, 895–905.
100. Nagasaki, I. (1989). Semilinear G-spheres and homotopy representation groups. *Proceedings of the Conference on Transformation Groups, Osaka, 1987*, Lecture Notes in Mathematics, **1375**, Springer-Verlag, 259–68.
101. Nakaoka, M. (1977). Fixed point theorems and related topics. Iwanami Shoten, (in Japanese).
102. Nishida, G. (1978). The transfer homomorphism in equivariant generalized cohomology theories. *Journal of Mathematics of Kyoto University*, **18**, 435–51.
103. Oliver, R. (1976). A proof of the Conner conjecture. *Annals of Mathematics*, **103**, 637–44.
104. Oliver, R. (1977). G-actions on disks and permutation representations II. *Mathematische Zeitschrift*, **157**, 237–63.
105. Ossa, E. (1967). Äquivariante Cobordismustheorie, Diplom-Arbeit, Bonn.
106. Palais, R. (1957). Imbedding of compact differentiable transformation groups in orthogonal representations. *Journal of Mathematics and Mechanics*, **6**, 673–8.
107. Pawałowski, K. (1984). Group actions with inequivalent representations at fixed points. *Mathematische Zeitschrift*, **187**, 29–47.
108. Pawałowski, K. (1989). Normal representations over the connected components of fixed point sets. *Proceedings of the Conference on Transformation Groups, Osaka, 1987*, Lecture Notes in Mathematics, **1375**, Springer-Verlag, 276–90.
109. Petrie, T. (1972). Smooth S^1-actions on homotopy complex projective spaces and related topics. *Bulletin of the American Mathematical Society*, **78**, 105–53.
110. Petrie, T. and Randall, J. D. (1984). *Transformation groups on manifolds.* Marcel Dekker.
111. Pontrjagin, L. S. (1946). *Topological groups.* Princeton University Press.
112. Raymond, F. (1968). A classification of the actions of the circle on 3-manifolds. *Transactions of the American Mathematical Society*, **131**, 51–78.
113. Rothenberg, M. (1978). Torsion invariants and finite transformation groups. *Proceedings of the Conference on Algebraic and Geometric Topology, Stanford 1976, Proceedings of Symposia in Pure Mathematics*, **32 Part 1**, 267–311.

114. Rubinsztein, R. L. (1973). On the equivariant homotopy of spheres. Preprint 58, Polish Academy of Science.
115. Schultz, R. (1981). Differentiable group actions on homotopy spheres II: Ultrasemifree actions. *Transactions of the American Mathematical Society*, **268**, 255–97.
116. Schultz, R. (1977). On the topological classification of linear representations. *Topology*, **16**, 263–9.
117. Singer, I. M. (1973). Recent applications of index theory for elliptic operators. *Proceedings of Symposia in Pure Mathematics, American Mathematical Society*, **23**, 11–31.
118. Smith, P. A. (1938). Transformations of finite period. *Annals of Mathematics*, **39**, 127–64.
119. Spanier, E. H. (1966). *Algebraic topology*. McGraw-Hill.
120. Steenrod, N. E. (1951). *The topology of fibre bundles*. Princeton University Press.
121. Steinberger, M. (1988). The equivariant topological s-cobordism theorem. *Invent. Math.* **91**, 61–104.
122. Steinberger, M. and West, J. (1985). Equivariant h-cobordisms and finiteness obstruction. *Bulletin of the American Mathematical Society*, **12**, 217–20.
123. Stewart, T. E. (1961). Lifting group actions in fibre bundles. *Annals of Mathematics*, **74**, 192–8.
124. Stong, R. E. (1970). Unoriented bordism and actions of finite groups. *Memoirs of the American Mathematical Society*, **103**.
125. Su, J. C. (1963). Transformation groups on cohomology projective spaces. *Transactions of the American Mathematical Society*, **106**, 305–18.
126. Sumners, D. W. (1975). Smooth $\mathbb{Z}_p$-actions on spheres which leave knots pointwise fixed. *Transactions of the American Mathematical Society*, **205**, 193–203.
127. Tamura, I. (1969). Fixed point sets of differentiable periodic transformations on spheres. *Journal of the Faculty of Science, University of Tokyo, Sect. 1A*, **16**, 101–14.
128. Tamura, I. (1978). Differential topology. Iwanami Shoten, Tokyo (in Japanese).
129. Thomas, C. B. (1973). Riemann–Roch formulae for group representations. *Mathematika*, **20**, 253–62.
130. Traczyk, P. (1978). On the G-homotopy equivalence of spheres of representations. *Mathematische Zeitschrift*, **161**, 257–61.
131. Tsukada, E. and Washiyama, R. (1979). Smooth S^1-actions on cohomology complex projective spaces with three components of the fixed point set. *Hiroshima Mathematical Journal*, **9**, 41–6.
132. Uchida, F. (1974). Transformation groups and cobordism theory. Kinokuniya Shoten, Tokyo (in Japanese).
133. Ushitaki, F. (1989). On extensions of non-linear actions on spheres. *Proceedings of the Conference on Transformation Groups, Osaka 1987*, Lecture Notes in Mathematics, **1375**, Springer-Verlag, 339–46.
134. Wall, C. T. C. (1970). *Surgery on compact manifolds*. Academic Press.
135. Wasserman, A. (1969). Equivariant differential topology. *Topology*, **8**, 127–50.
136. Watabe, T. (1974). On the degree of symmetry of complex quadric and homotopy complex projective spaces. *Science Reports of Niigata University*, **11**, 85–94.
137. Weintraub, S. H. (1975). Semi-free $\mathbb{Z}_p$-actions on highly-connected manifolds. *Mathematische Zeitschrift*, **145**, 163–85.

138. Wu, C. m. (1989). The ring structure of $U_*(\mathbb{Z}_p)$. *Proceedings of the Conference on Transformation Groups, Osaka 1987*, Lecture Notes in Mathematics, **1375**, Springer-Verlag, 368–80.
139. Yoshida, T. (1976). On smooth semi-free S^1-actions on cohomology complex projective spaces. *Publications of the Research Institute for Mathematical Sciences Kyoto University*, **11**, 483–96.
140. Yoshida, T. (1977). On S^1-actions on cohomology complex projective spaces. *Sugaku*, **29**, 154–64 (in Japanese).

Index

The manufacturer's authorised representative in the EU for product safety is Oxford University Press España S.A. of el Parque Empresarial San Fernando de Henares, Avenida de Castilla, 2 – 28830 Madrid (www.oup.es/en or product.safety@oup.com). OUP España S.A. also acts as importer into Spain of products made by the manufacturer.

www.ingramcontent.com/pod-product-compliance
Lightning Source LLC
LaVergne TN
LVHW101458200926
848483LV00001B/155
* 9 7 8 0 1 9 8 5 3 2 1 2 5 *